# Raman Amplification in Fiber Optical Communication Systems

# Raman Amplification in Fiber Optical Communication Systems

**Clifford Headley**

**Govind P. Agrawal**

*Editors*

**ELSEVIER**
ACADEMIC
PRESS

Amsterdam • Boston • Heidelberg • London • New York • Oxford
Paris • San Diego • San Francisco • Singapore • Sydney • Tokyo

**Elsevier Academic Press**
200 Wheeler Road, 6th Floor, Burlington, MA 01803, USA
525 B Street, Suite 1900, San Diego, California 92101-4495, USA
84 Theobald's Road, London WC1X 8RR, UK

Permissions may be sought directly from Elsevier's Science & Technology
Rights Department in Oxford, UK: phone: (+44) 1865 843830,
fax: (+44) 1865 853333, e-mail: permissions@elsevier.com.uk. You may also
complete your request on-line via the Elsevier homepage (http://elsevier.com),
by selecting "Customer Support" and then "Obtaining Permissions."

**Library of Congress Cataloging-in-Publication Data**
Application Submitted

**British Library Cataloguing in Publication Data**
A catalogue record for this book is available from the British Library

ISBN: 0-12-044506-9

For all information on all Academic Press publications
visit our Web site at www.academicpress.com

Printed in the United States of America
04  05  06  07  08  09     9  8  7  6  5  4  3  2  1

# Contents

# Contributors

Numbers in parenthesis indicate the pages on which the authors' contributions begin.

**Govind P. Agrawal** (*1, 33*), *Professor of Optics*, University of Rochester, Institute of Optics Rochester, NY 14627

**Yoshihiro Emori** (*169*), The Furukawa Electric Co., Ltd., 6 Yawata-Kaigandori, Ichihara, Chiba, 290-8555 Japan

**Clifford Headley** (*1, 303*), *Technical Manager*, OFS Laboratories, 17 Aberdeen Circle, Flemington, NJ 08822

**Howard Kidorf** (*215*), 82 Tower Hill Drive, Red Bank, NJ 07701

**Shu Namiki** (*169*), The Furukawa Electric Co., Ltd., 6 Yawata-Kaigandori, Ichihara, Chiba, 290-8555 Japan

**Morton Nissov** (*215*), *Director*, Tyco Telecommunications, 6 Mark Place, Ocean, NJ 07712

**Atsushi Oguri** (*169*), The Furukawa Electric Co., Ltd., 6 Yawata-Kaigandori, Ichihara, Chiba, 290-8555 Japan

**Karsten Rottwitt** (*103*), *Associate Professor*, Technical Univiversity of Denmark, Research Center COM, Sandskraenten 5, DK-2850 Narun, Denmark

**Naoki Tsukiji** (*267*), *Technical Manager*, Furukawa Electric Co., Ltd., 2-4-3, Okano, Nishi-ku, Yokohama 220-0073 Japan

**Junji Yoshida** (*267*), *Engineer*, Furukawa Electric Co., Ltd., 2-4-3, Okano, Nishi-ku, Yokohama 220-0073 Japan

# Preface

Although inelastic scattering of light from molecules, a phenomenon now known as Raman scattering, was observed by C. V. Raman in 1928, the nonlinear phenomenon of stimulated Raman scattering was not demonstrated until 1962. Soon after low-loss silica fibers became available around 1970, Roger Stolen and coworkers used stimulated Raman scattering in such fibers not only for amplification of optical signals but also for constructing fiber-based Raman lasers. The potential of Raman amplification for compensating fiber losses in lightwave systems was demonstrated during the 1980s in several experiments performed by Linn Mollenauer and his coworkers. However, these experiments employed bulky color-center lasers for pumping and were not practical for deployment of Raman amplifiers in commercial optical communication systems. The advent of erbium-doped fiber amplifiers that could be pumped using semiconductor lasers made the adoption of such amplifiers much more practical, and Raman amplification was nearly ignored during the 1990s.

Fortunately, research into pump lasers suitable for Raman amplification continued. Several technological advances during the 1990s made it possible to fabricate single-transverse-mode semiconductor lasers capable of emitting power levels in excess of 0.2 W. It was also realized that several pump lasers may be employed at different wavelengths simultaneously and provide Raman amplification over a wide bandwidth encompassing both the C and the L transmission bands. Moreover, it turned out that distributed Raman amplifiers providing gain over tens of kilometers are inherently less noisy than erbium-doped fiber amplifiers that provide gain over tens of meters. As high-power semiconductor lasers became available commercially near the end of the 20th century, Raman amplifiers were used in a number of record-breaking experiments to show that such amplifiers indeed improve the performance of wavelength-division-multiplexed lightwave systems. By 2003, the use of Raman amplification had become quite common for long-haul systems designed to operate over thousands of kilometers.

This book provides a comprehensive description of the Raman amplification technique as it is applied to telecommunication systems. It covers both the fundamental and the applied aspects and is intended to serve both as a useful reference for researchers engaged in the field and as a teaching manual for those who are planning to enter this exciting field. It may also be used as a teaching tool for graduate students with some knowledge of electromagnetic theory. We strove to avoid having a book with a large number of short chapters because we feel that such a book, although it would contain a collection of invited research papers from the experts, would not serve the objectives that we had in mind. For this reason, we wanted each chapter to be 40 to 50 pages long so that it would cover a specific aspect of Raman-amplification technology in detail. We also asked the contributors to minimize cross-referencing so that each chapter can be read by itself. The book contains seven chapters written by scientists who are experts in their own subfield. It is our hope that the reader will find this book a useful addition to the field of lightwave technology.

# Acknowledgments

**Clifford Headley** (*Editor/Author, Chapters 1 and 7*)

The support of my employers at OFS Fitel Laboratories is gratefully acknowledged, in particular the support of my former and current bosses, Benjamin Eggleton and David DiGiovanni. There have been so many collaborations with colleagues throughout the years that have made me the scientist that I am that I couldn't possibly name them all, but I thank them because I am better for these collaborations. However, I acknowledge the ones who are most related to the work covered in this book: Marc Mermelstein, Jean-Christophe Bouteiller, and Khush Brar. I especially thank J.-C. Bouteiller for reading through Chapter 7. I also thank my mother, Lynette Headley, who taught me through example that hard work does pay off. Thanks also to my sons Jelani and Gabriel who bring balance to my life and lift me up with their youthful joy. Finally, I acknowledge my beautiful wife, who has supported throughout this process and has been the wind beneath my wings. After many a late night, our home is always a joy to come home to. Finally, thank you God for all that you do for me.

**Govind Agrawal** (*Editor/Author, Chapters 1 and 2*)

A large number of persons have contributed, either directly or indirectly, to the work reported in Chapter 2. It is impossible to mention all of them by name. I thank my graduate students who helped improve my understanding through their sharp questions and comments. I acknowledge the help of Y. Emori and S. Namiki through e-mails on several issues discussed in this chapter. I am also grateful to my colleagues at the Institute of Optics for creating a relaxing atmosphere.

**Karsten Rottwitt** (*Author, Chapter 3*)

Much of the work reported in this chapter was done at Bell Laboratories, Lucent Technologies. Colleagues at the Fiber Research Department are greatly acknowledged and especially Andrew Stentz, Jake Bromage, and Mei Du are thanked—without them I would not have been able to write this chapter. I also thank my former colleagues at Tyco Submarine Systems, especially Howard Kidorf, with whom I did much initial work on Raman amplification.

**Shu Namiki, Yoshiro Emori, and Atsushi Oguri** (*Authors, Chapter 4*)

We acknowledge Drs. M. Sakano and H. Yanagawa for their support.

# Chapter 1

# Introduction

**Clifford Headley** and **Govind P. Agrawal**

The first fiber-optical telecommunication systems emerged with the engineering of low loss optical fiber [1]. Even though the complexity of the system has increased, the basic elements remain the same. They consist of an optical source, a means of modulating the source, the transmission medium (i.e., the optical fiber), and a detector at the output end of the fiber. Fiber loss is one limitation to the transmission distance of this system. In the early days of fiber-optic communications the loss of the fiber was compensated for in long spans by using electrical regenerators. As their name implies, these devices detected the signal, converted it to an electrical signal, and using a new laser transmitted a new version of the signal. Electrical regenerators were expensive and also limited the rate at which data could be transmitted as time for the much slower electrical processing to occur had to be built into the system.

In order to overcome the limitations imposed by electrical regeneration, a means of optical amplification was sought. Two competing technologies emerged: the first was erbium-doped fiber amplifiers (EDFA) [2, 3] and the second Raman amplification [4–7]. In the first deployed systems EDFA emerged as the preferred approach. One reason was that the optical pump powers required for Raman amplification were significantly higher than that for EDFA, and the pump laser technology could not reliably deliver the required powers. However, with the improvement of pump laser technology Raman amplification is now an important means of expanding span transmission reach and capacity.

This chapter provides an overview of this book. It begins with a brief description of optical fibers and is followed by an explanation of stimulated Raman scattering (SRS), the nonlinear phenomenon responsible for Raman amplification. The third section discusses the advantages of a Raman amplified system. The next section focuses on the problems associated with Raman amplification, while Section 1.5 points out some countermeasures to these problems. The next two sections provide a brief explanation of the pump sources used for Raman amplification, semiconductor diodes and Raman fiber lasers. Finally, the rest of the book is summarized.

## 1.1  Optical Fibers

Optical fibers are able to guide light by using the principle of total internal reflection [8]. Consider the interface of two dielectric media in which the refractive index of the incident medium is greater than that of the medium in which the light is refracted. As the angle of incidence, $\theta_i$, is increased eventually an angle of incidence, $\theta_c$, is reached beyond which all the incoming energy is totally reflected. This is illustrated in Figure 1.1, which is a cross-section of an optical fiber, with a core refractive index $n_{co}$, and a cladding index $n_{cl}$, such that $n_{co} > n_{cl}$. As light propagates down the fiber, it can be totally internally reflected and hence guided by the fiber.

In practice the cladding of the fiber is generally made of silica. The choice of silica glass, formed by fusing $SiO_2$ molecules, satisfies many criteria for a good fiber material [9]. It has low optical loss properties, and its refractive index can be controlled precisely in the radial direction with low fluctuations in the longitudinal direction. In addition it is chemically stable and mechanically durable. The core of the fiber is usually formed by adding dopants to the core such as germanium, phosphorus, and aluminum, which raise the index of the silica. Finally, a polymer jacket is placed on the silica cladding to provide the mechanical protection. The index of the polymer is higher than that of the cladding to prevent guiding in the cladding.

The sources of loss in optical fibers can be put into two categories, extrinsic and intrinsic. Extrinsic losses arise from fiber fabrication processes such as impurity absorption and structural imperfections. With state-of-the-art fiber fabrication facilities, most of the extrinsic losses can

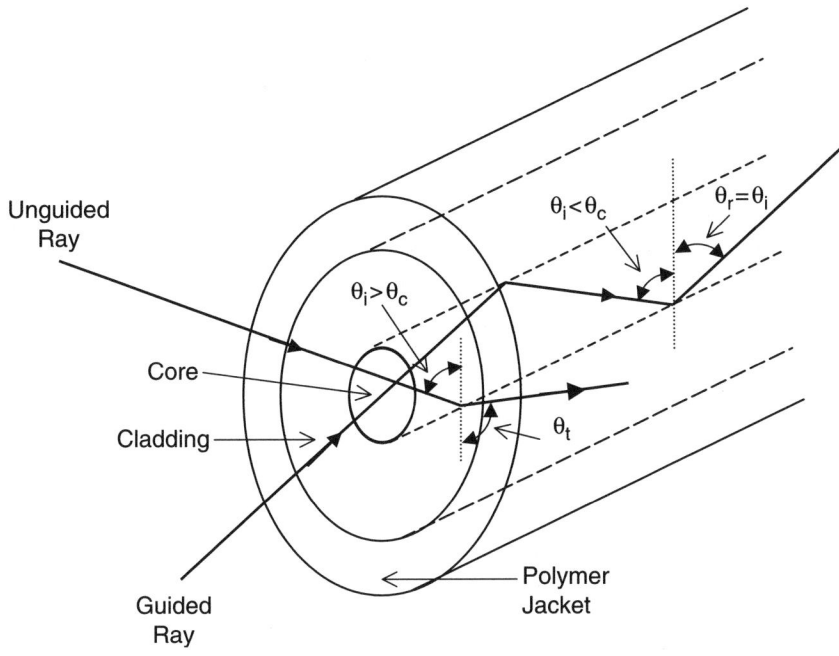

**Figure 1.1:** A schematic of the cross-section of an optical fiber showing the principle of total internal reflection. $\theta_i$, $\theta_r$, and $\theta_t$ are the angles of incidence, reflection, and transmission, respectively. $\theta_c$ is the critical angle.

be eliminated, even to the point of elimination of the OH-absorption peak, as seen in Figure 1.2. However, an intrinsic fundamental loss mechanism is Rayleigh scattering loss [10], which arises from microscopic fluctuations in the density of the core that are frozen in during manufacture. The density fluctuations lead to fluctuations in the refractive index of the core, on a scale that is smaller than the optical wavelength. The addition of dopants in the core increases the amount of Rayleigh scattering loss as the material becomes more inhomogeneous. A good fit to the measured data of the Rayleigh scattering coefficient $\alpha_s$ is given by [11]

$$\alpha_s(\text{dB/km}) = \frac{(0.75 + 66\Delta n_{\text{Ge}})}{\lambda^4}, \tag{1.1}$$

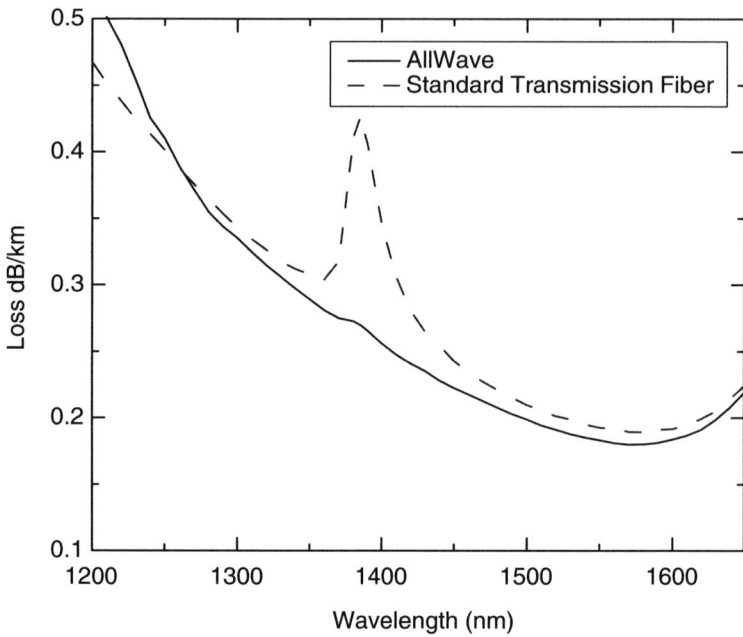

**Figure 1.2:** The loss of typical transmission fiber and one with the water peak removed as a function of wavelength. (Courtesy of Robert Lingle, Jr. OFS Fitel.)

where $\lambda$ is in $\mu$m, and $\Delta n_{Ge}$ is the index difference due to germanium in the fiber. Typical values of $\alpha_s$ are 0.12–0.16 dB/km at $\lambda = 1.55$ $\mu$m. This loss ultimately limits the distance over which signals can propagate and necessitates optical amplification.

## 1.2   Raman Amplification

Raman scattering is a nonlinear effect [12]. Intuition into nonlinear effects can be gained by considering a simple spring. If a small load is attached to a spring, the extension of the spring is linearly related to the load. However, as the load is increased, the dependence of the extension of the spring on the load becomes nonlinearly related to the applied load. Likewise the response of a dielectric medium, such as an optical fiber, to an intense

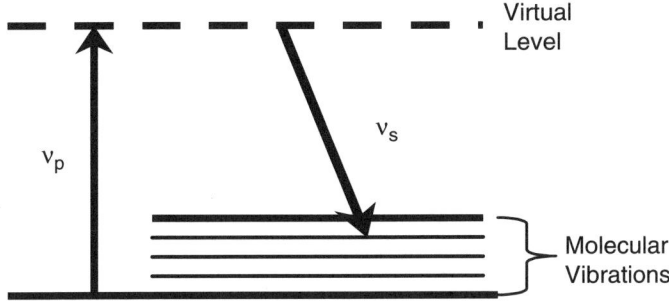

**Figure 1.3:** Schematic of the quantum mechanical process taking place during Raman scattering.

amount of light is nonlinear, and Raman scattering is the result of such a nonlinear process.

During Raman scattering, light incident on a medium is converted to a lower frequency [13]. This is shown schematically in Figure 1.3. A pump photon, $\nu_p$, excites a molecule up to a virtual level (nonresonant state). The molecule quickly decays to a lower energy level emitting a signal photon $\nu_s$ in the process. The difference in energy between the pump and signal photons is dissipated by the molecular vibrations of the host material. These vibrational levels determine the frequency shift and shape of the Raman gain curve. Due to the amorphous nature of silica the Raman gain curve is fairly broad in optical fibers. Figure 1.4 shows the Raman gain spectrum in three types of optical fibers [14]. The frequency (or wavelength) difference between the pump and the signal photon ($\nu_p - \nu_s$) is called the Stokes shift, and in standard transmission fibers with a Ge-doped core, the peak of this frequency shift is about 13.2 THz.

For high enough pump powers, the scattered light can grow rapidly with most of the pump energy converted into scattered light. This process is called SRS, and it is the gain mechanism in Raman amplification. Three important points are (i) SRS can occur in any fiber; (ii) because the pump photon is excited to a virtual level, Raman gain can occur at any signal wavelength by proper choice of the pump wavelength; and (iii) the Raman gain process is very fast. This differs from EDFA, in that (i) they require a specially fabricated fiber, (ii) the pump and signal wavelengths are determined by the resonant levels of the erbium, and (iii) because there

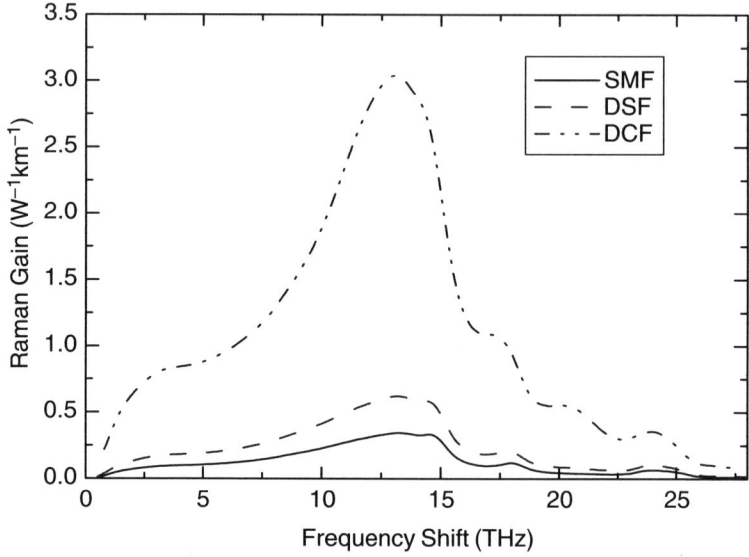

**Figure 1.4:** Raman gain profiles for a 1510-nm pump in three different fiber types. SMF, standard single mode fiber; DSF, dispersion shifted fiber; DCF, dispersion compensating fiber. (Courtesy of authors of Ref. 14.)

is energy stored in the upper level of the erbium the transfer of energy is much slower. The advantages and disadvantages of these differences will be discussed shortly.

A schematic of an optical telecommunication system employing Raman amplification is shown in Figure 1.5. The signal propagates from the transmitter (Tx) to the receiver (Rx). The pump traveling in the same direction as the signal is called the co- or forward pump, and the pump traveling in the opposite direction of the signal is called the counter- or backward pump. When the fiber being pumped is the actual transmission span that links two points, this setup is referred to as a distributed Raman amplifier. If the amplifier is contained in a box at the transmitter or receiver end of the system it is called a discrete Raman amplifier. Another distinctive feature between distributed and discrete Raman amplifiers tends to be the length of the fiber used. Typically distributed Raman amplifiers have lengths greater than 40 km whereas discrete Raman amplifiers have lengths around 5 km. This difference can lead to some slightly different considerations.

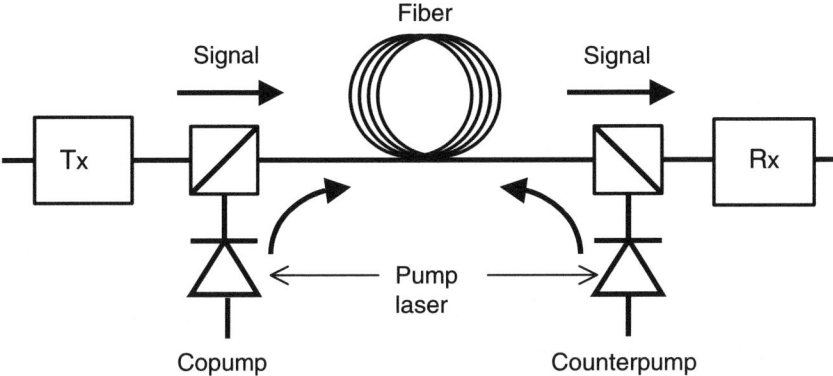

**Figure 1.5:** Schematic of an optical communication system employing Raman amplification.

The evolution of the pump, $P_p$, and signal, $P_s$, powers along the longitudinal axis of the fiber $z$ in a Raman amplified system can be expressed by the following equations [12]:

$$\frac{dP_s}{dz} = g_R P_p P_s - \alpha_s P_s, \tag{1.2}$$

and

$$\pm\frac{dP_p}{dz} = -\frac{\omega_p}{\omega_s} g_R P_p P_s - \alpha_p P_p, \tag{1.3}$$

where $g_R (W^{-1}m^{-1})$ is the Raman gain coefficient of the fiber normalized with respect to the effective area of the fiber $A_{\text{eff}}$, $\alpha_{s/p}$ are the attenuation coefficient at the pump and signal wavelength, and $\omega_{s/p}$ are the angular frequencies of the pump and signal. The $\pm$ signs represent a co- and counterpropagating pump wave, respectively. The first term on the right-hand side of Eq. (1.2) (Eq. (1.3)) represents the signal gain (pump depletion) due to SRS; the second term represents the intrinsic signal (pump) loss. If the depletion of the pump by the signal is ignored, Eq. (1.3) can be solved for the counterpropagating case to give, $P_p(z) = P_0 e^{-\alpha_p(L-z)}$, where $P_0$ is the input pump power and $L$ is the fiber length. This result is substituted into

Eq. (1.2), and the resulting differential equation can be solved analytically to yield

$$P_s(L) = P_s(0) \exp\left(g_R P_0 L_{\text{eff}} - \alpha_s L\right) \equiv G_N(L) P_s(0) \qquad (1.4)$$

where

$$L_{\text{eff}} = \left[1 - \exp\left(-\alpha_p L\right)\right] / \alpha_p, \qquad (1.5)$$

and $G_N$ is the net gain. Equation (1.4) is a first-order approximation of the signal evolution in the fiber. In typical transmission fibers, the loss is approximately 0.25 dB/km at 1455 nm; therefore, $L_{\text{eff}}$ for the pump light is approximately 17 km for spans greater than 60 km.

The effect of Raman amplification on the evolution of the signal power in an 80-km distributed Raman amplifier (i) without Raman amplification and (ii) with counterpumping is shown in Figure 1.6 [15]. The signal power is larger at the receiver in the span containing Raman amplification. This intuitively implies that the system will perform better. It will be shown

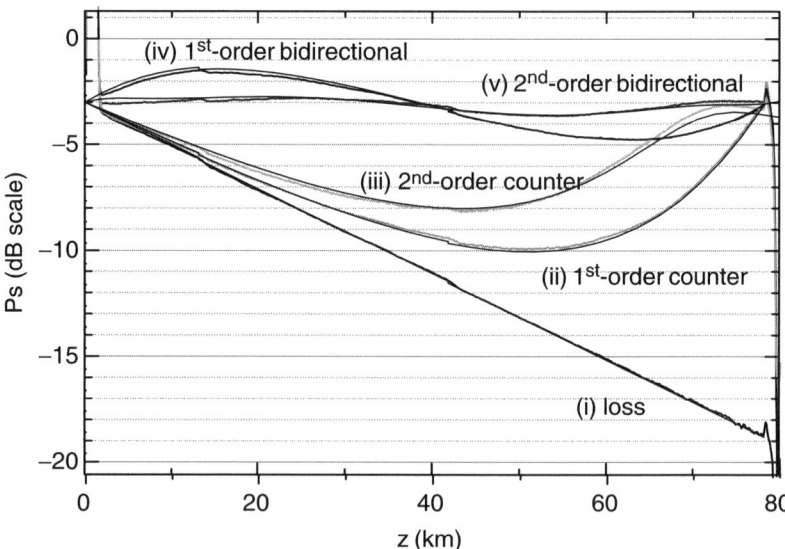

**Figure 1.6:** Experimental measurements of the signal evolution in a transmission fiber for several different pumping configurations. (Courtesy of authors of Ref. 15.)

next that this leads to a better signal-to-noise ratio (SNR) than if a discrete amplifier was used at the receiver of an unpumped transmission span.

## 1.3 Advantages of Raman Amplification

In the following sections, two advantages of Raman amplification in a transmission system are considered. The first is how it can be used to improve the noise figure of a system; the second is how a flat gain profile can be obtained. It should be borne in mind that these advantages can be applied to any transmission fiber.

### 1.3.1 Improved Noise Figure

The noise figure (NF) of an amplifier is the ratio of the SNR of the input signal to SNR of the output signal. It is a measure of how much the amplifier degrades the signal. In a Raman amplified system, the equivalent noise figure, ($NF_{eq}$), represents the noise figure an amplifier placed at the receiver end of the transmission span would need, in the absence of Raman amplification, to provide the same SNR as that obtained using distributed Raman amplification [13]. The two equivalent systems are shown schematically in Figure 1.7. The loss in the span in Figure 1.7b is ($\alpha_s L$); hence the gain is $G = (\alpha_s L)^{-1}$, and the noise figure of the unpumped span is $\alpha_s L$ (no noise added so NF $= P_{in}/P_{out}$). A well-known expression [3] for the noise figure for two cascaded amplifiers is given as $NF_{sys} = NF_1 + (NF_2 - 1)/G_1$. Where $NF_1$ ($NF_2$) is the noise figure of the first (second) amplifier, and $G_1$ is the gain of the first amplifier. For the equivalent system in Figure 1.7b,

$$NF_{sys} = NF_{eq}\alpha_s L. \tag{1.6}$$

Equating the noise figure of the Raman amplified system to that of the equivalent system, it is seen that

$$NF_{eq} = \frac{NF_R}{\alpha_s L} \quad \text{or} \quad NF_{eq}^{dB} = NF_R^{dB} - (\alpha_s L)^{dB}, \tag{1.7}$$

where the superscript dB indicates the variable expressed in dB. From Eq. (1.7) it is seen that $NF_{eq}^{dB}$ can be less than zero. Such an amplifier

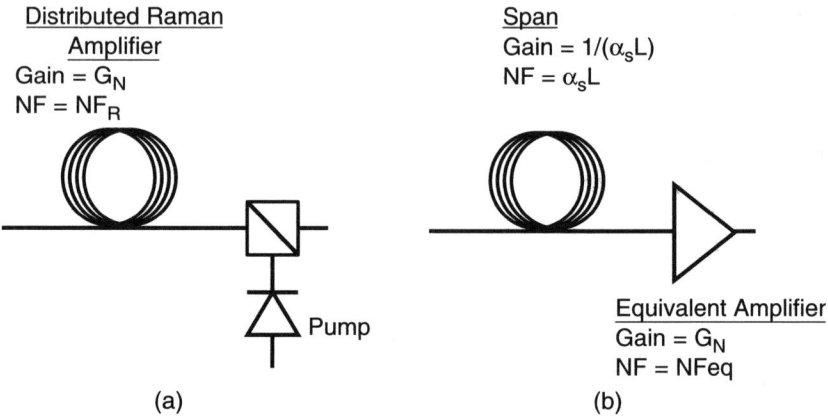

**Figure 1.7:** Schematic of a distributed Raman amplified system (a) and the equivalent system of a transmission span and a discrete Er-doped fiber amplifier (b).

is not physically realizable, but is indicative of the superior performance provided by the distributed Raman amplification, which cannot be matched by a discrete amplifier placed after the span. An intuitive if not rigorous explanation is that amplification always adds noise to the signal, degrading its SNR. In the best case, if the signal propagates along the fiber with no loss and with no amplification its SNR would be equal to its input value and the NF equal to one. The worst case is if the signal experiences the full loss of the span and then is amplified. This is the worst case because the gain required from the amplifier at the end of the span has increased; because more pump power is required, more amplified spontaneous emission (ASE) is generated in the amplifier. In addition the input signal power to the amplifier has decreased. The lower signal power means that the ASE can more successfully compete with the signal for gain in the amplifier. These two factors combine to lower the output SNR and increase the NF. If the transmission span is considered to be a series of discrete amplifiers, then the more evenly the gain is distributed along the fiber the less gain is required from each of the individual amplifiers and the higher the signal power into each of these amplifiers. This is why distributed amplification provides improved performance compared to discrete amplification. In addition it also explains why even when doing distributed Raman amplification, the more evenly gain is distributed along the fiber length the larger the improved performance provided by the distributed amplification scheme. In many of the discussions that follow the focus will be on raising the gain by more evenly distributing it along the fiber.

## 1.3.2 Improved Gain Flatness

In a multiple wavelength telecom system it is important that all signal wavelengths have similar optical powers. The variation in the gain provided to different wavelengths after passing through an amplifier is referred to as the gain flatness. If the signal at one wavelength is disproportionately amplified, as it passes through several amplifiers, it will grow superlinearly relative to the other channels reducing the gain to other channels. The system, however, will still be limited by the channel with the lowest gain. As a result, after each amplifier the gain spectrum generally is flattened. One approach is to insert wavelength-dependent lossy elements, within the amplifier, with the appropriate spectral profile. Raman amplification offers the ability to achieve this without lossy elements.

In Raman amplification a flat spectral profile can be obtained by using multiple pump wavelengths [14, 16–19]. For a given fiber the location of the Raman gain is only dependent on the wavelength of the pump, the magnitude of the gain is proportional to the pump power, and the shape of the gain curve is independent of the pump wavelength. Therefore, if multiple pumps are used a flat spectral gain profile can be obtained.

The required pump wavelengths and the gain required at each wavelength can be predicted by summing the logarithmic gain profiles at the individual pump wavelengths [14]. Figure 1.8 shows the individual pump gain profiles along with their logarithmic superposition. It is seen that a significant amount of the gain comes from the longest wavelength pump. However, the required pump powers cannot be obtained simply from calculating the amount of pump power required to produce a given value of Raman gain. This is because of interpump Raman amplification in which the short wavelength pumps amplify the long wavelength pumps. This effect is also shown in Figure 1.8, in which the curve marked "pump–pump" interactions shows a more than 3-dB tilt in the gain spectrum when the interpump Raman amplification is considered.

Pump–pump interaction also affects pump power evolution as shown in Figure 1.9. It shows the decrease of the pump power in a 25-km span

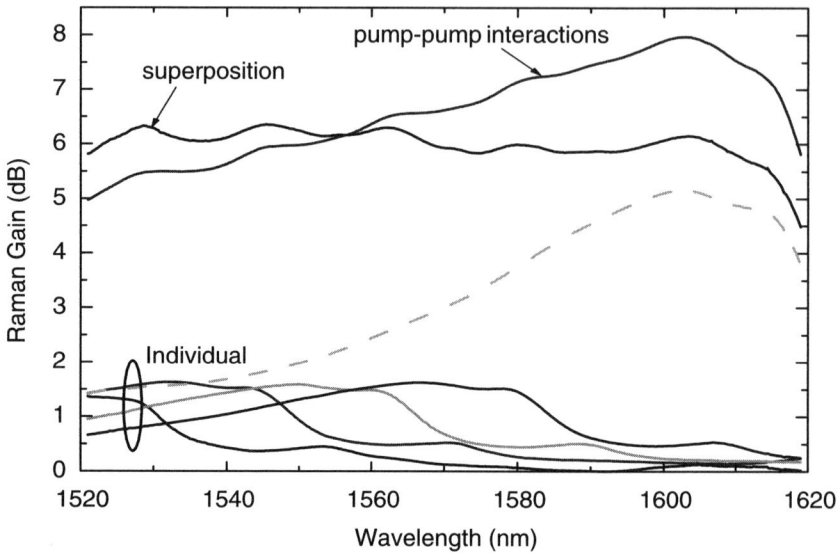

**Figure 1.8:** The gain contribution from individual pump wavelengths (1420, 1435, 1450, 1465, and 1480 nm), their logarithmic superposition, and the resultant gain curve with pump-to-pump interactions for a 25-km span of dispersion shifted fiber. Launched pump powers were 61, 55, 48, 47, and 142 mW, respectively. (Courtesy of authors of Ref. 14.)

of dispersion shifted fiber when 100 mW of power at each wavelength is launched. Initially, the power at the longest pump wavelength increases as it receives Raman gain from the other pump wavelengths [14]. Therefore a typical launched pump power distribution looks similar to that in Figure 1.10 [14]. Even though most of the gain is provided by the longest wavelength, the power launched at this wavelength is relatively modest. For the pump wavelengths and powers shown in Figure 1.10, 12 dB of gain can be obtained with a gain variation of less than 0.5 dB from 1525 to 1595 nm [14].

The nonlinear pump–pump interaction means numerical simulations have to be used to find the required pump powers. Several algorithms have emerged for determining the required number of pumps, their wavelengths, and their powers [20–22].

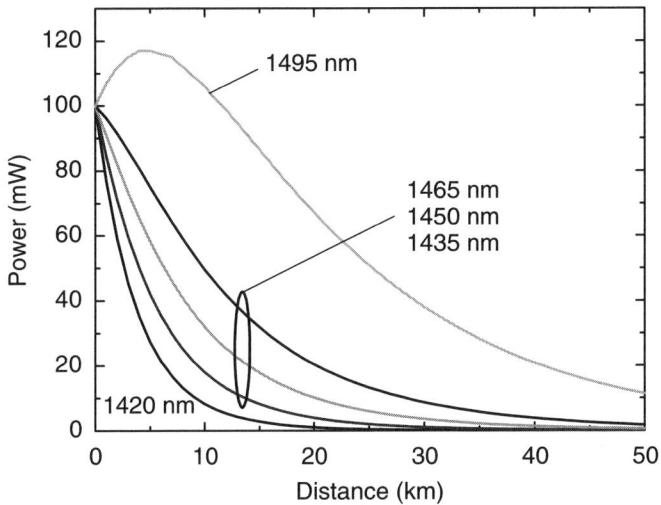

**Figure 1.9:** Simulated pump power evolution in a Raman amplified system. (Courtesy of authors of Ref. 14.)

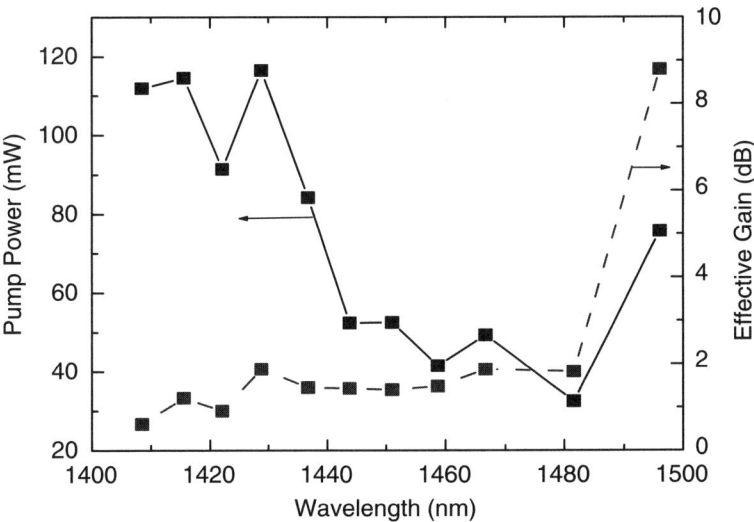

**Figure 1.10:** Comparison between launched pump power and gain provided by each pump wavelength. (Courtesy of authors of Ref. 14.)

## 1.4   Concerns in Raman Amplification

The benefits of Raman amplification are now clear. However, additional issues arise in Raman amplification that are not present in EDFA. These include multipath interference, pump noise transfer, and noise figure tilt and are examined in the following sections.

### 1.4.1   Multipath Interference

Light can be scattered twice in a fiber due to Rayleigh scattering [10, 23–28]. The first scattering event takes place in a direction counterpropagating with the signal; this scattered light can undergo a second scattering event and copropagate with the signal. This double Rayleigh backscattered (DRB) light will pass through the amplifier twice, in some circumstances experiencing significant gain, and appear at the receiver as a noise source. This is illustrated in Figure 1.11. The DRB light is problematic. Because the reflections are occurring all along the fiber length, the phase of the DRB light will fluctuate at the fiber output. In addition, if the path difference between the signal DRB light exceeds the coherence length of the signal

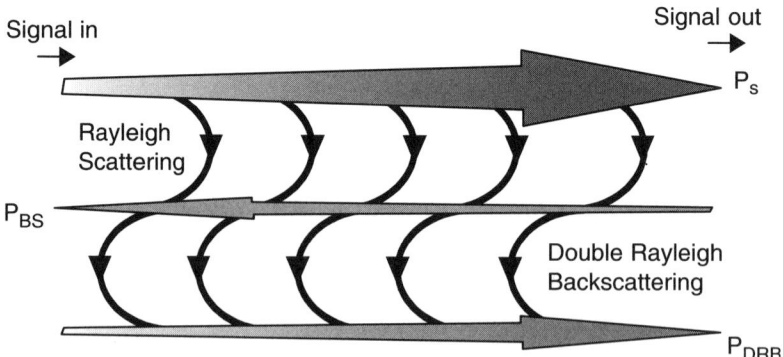

**Figure 1.11:** Schematic showing the growth of double Rayleigh backscattered light in a fiber. $P_{BS}$ is power in the backscattered light and $P_{DRB}$ is power of the double Rayleigh backscattered light.

source, phase changes occur due to phase changes in the source. In either case, the interference of the many different signals leads to intensity fluctuations. This phenomenon is also referred to as multiple path interference (MPI).

The growth of the Rayleigh backscattered coefficient is mathematically described by adding the term $\alpha_R S P_s$ describing signal loss due to Rayleigh backscattering to Eq. (1.2) and the following coupled equations terms representing the backward, $P_{BS}$, and DRB, $P_{DRB}$, light [13, 14, 23]

$$-\frac{dP_{BS}}{dz} = \frac{\omega_p}{\omega_s} g_p P_p P_{BS} - \alpha_s P_{BS} + \alpha_R S (P_s - P_{DRB}) \qquad (1.8)$$

$$\frac{dP_{DRB}}{dz} = \frac{\omega_p}{\omega_s} g_p P_p P_{DRB} - \alpha_s P_{DRB} + \alpha_R S P_{BS}, \qquad (1.9)$$

where $\alpha_R$ is the Rayleigh scattering coefficient, and $S$ is the Rayleigh capture coefficient describing what fraction of the backscattered light is captured, and is proportional to $1/A_{eff}$ [26, 27]. The first two terms describe the gain and loss of the backscattered and DRB signal. The next term describes the growth of $P_{BS}$ and $P_{DRB}$ through Rayleigh scattering. The last term in Eq. (1.8) represents the loss of backscattered light to DRB through Rayleigh scattering. Typical values for $\alpha_R S$ range from $0.6(10^{-4})$ to $5(10^{-4})$ km$^{-1}$ depending on the effective area of the fiber [13]. The long lengths of fiber used for distributed Raman amplification make DRB light an issue. As the gain in the Raman amplifier increases, MPI will eventually limit the achievable gain.

## 1.4.2  Pump Noise Transfer to the Signal

As pointed out in Section 1.2, the Raman gain process is extremely fast, $<1$ ps. Therefore any fluctuations in the pump at frequencies $<1$ THz can cause fluctuations in the Raman gain and therefore fluctuations in the signal power [29–31]. This phenomenon is generally referred to as pump to signal relative intensity-to-noise (RIN) transfer. The RIN is a standard measure of the noise in a laser.

The amount of noise transferred between the pump and the signal will depend on the pump configuration and the noise frequency. Figure 1.12 is

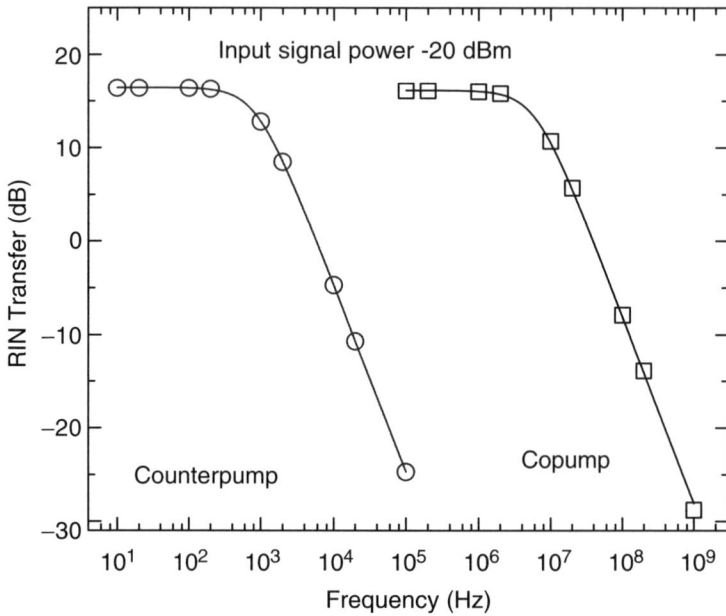

**Figure 1.12:** Calculated RIN transfer spectra for co- and counterpumped configurations. $G_N = 10$ dB, $L = 100$ km, $\alpha_p = 0.235$ dB/km, $\alpha_s = 0.189$ dB/km, $g_s = 1.4$ W$^{-1}$km$^{-1}$, $D = 15$ ps/km/nm. (Courtesy of authors of Ref. 30.)

a plot of the RIN transfer function, which measures how much of the pump modulation is transferred to the signal, as a function of the noise frequency. The curves shown are for the case of co- and counterpumping. For very low frequencies the noise on the pump is amplified (RIN transfer $>0$ dB) as it is impressed on the signal, but it is attenuated at higher frequencies. Less transfer occurs at higher frequencies since the averaging effect due to walkoff between the pump and the signal is more significant. The RIN transfer between the pump and the signal is reduced in a counterpumped configuration, since any pump noise will be averaged over the length of the fiber. For the case of a copropagating pump and signal, the only averaging effect is from the walkoff due to chromatic dispersion. This leads to more stringent requirements on the noise properties of lasers used for copumping versus counterpumping.

In addition to the pumping configuration, the amount of noise transferred will depend on the gain and the length of fiber used. It is intuitive that a larger gain will increase the noise transferred onto the signal, and indeed this result holds up. For longer lengths of fiber, more averaging of the pump noise can take place. However, for typical transmission spans where $L_{\text{eff}}$ is 17 km, the actual length of the span has little effect. In discrete Raman amplifiers in which the effective length is 5 km, the fiber length becomes important.

### 1.4.3 Noise Figure Tilt

In Section 1.3.2 it was pointed out that there is significant Raman interaction between the different pump wavelengths. This results in amplification of the longest wavelength pump in the fiber span as shown in Figure 1.9. As a consequence of this the longer wavelength signal channels, which receive more gain from the longest wavelength pump (see Figure 1.8), will have a gain that is more evenly distributed along the fiber length and hence a better NF than the shorter wavelengths as pointed out in Section 1.3.1 [32–34].

A second effect that contributes to NF tilt is the dependence of the amount of generated ASE on the frequency difference between the pump and the signal [32]. As this difference decreases, more ASE is generated. Therefore, the shorter signal wavelengths, which still experience some gain from the longest wavelength pump (see Figure 1.8), also generate more ASE.

Figure 1.13 shows the experimentally measured Raman gain spectrum obtained by counterpumping an 80-km span of standard single mode fiber in order to obtain 9.5 dB of gain. Also shown is the corresponding $\text{NF}_R^{\text{dB}}$ for the counterpumped case. It is seen that there is almost a 1.5-dB change in $\text{NF}_R^{\text{dB}}$ in moving from 1530 to 1590 nm.

## 1.5 Advanced Concepts in Raman Amplification

Thus far the primary focus has been on a counterpumped Raman amplified system and the problems associated with it. Now we examine more complex schemes that are proposed to address some of these problems.

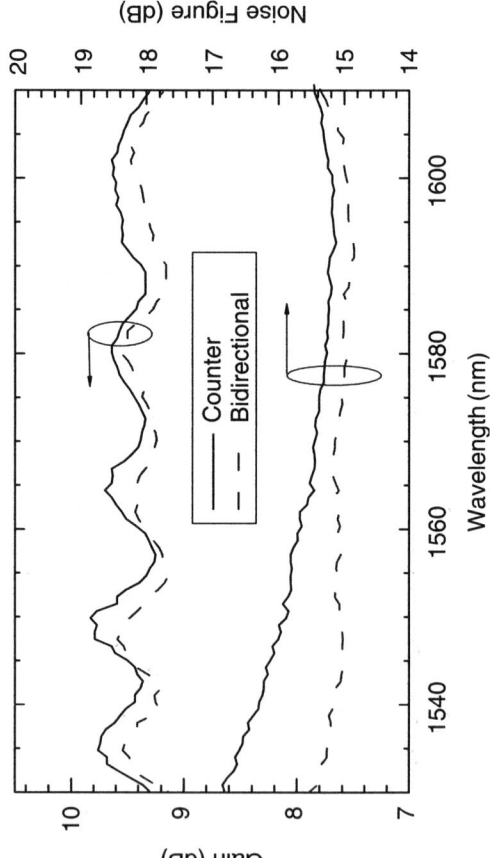

**Figure 1.13:** Measured gain and noise figure versus wavelength for 76 km of standard single mode fiber. Solid line is for counterpumping and dashed curve is for bidirectional pumping. (Courtesy of authors of Ref. 33.)

## 1.5.1  Bidirectional Pumping

In Section 1.3.2 when the NF was considered it was made clear that having gain closer to the input end of the fiber improves the NF. However, copumping by itself limits the amplifier gain since large amplification of the signals near the input end of the fiber raises the path averaged power in the span and could produce deleterious nonlinear effects. This points to the need for bidirectionally pumping the amplifier to achieve a low noise figure (copumping) and high gain (counterpumping) [34–37].

One problem that had to be overcome in bidirectional pumping is the noise transfer between the pump and the signal light. The first demonstrated bidirectionally pumped system used an approach called second-order pumping, which will be described in the next section, to overcome this problem [34]. However, now diode lasers with low enough RIN values for copumping are available [36].

An example of signal power evolution in bidirectional pumping is shown in Figure 1.6, curve (iv). The small signal variations compared to curve (ii) are remarkable. In this example, equal amounts of pump power were launched at the input and output ends of the fiber. Noise figure measurements corresponding to the signal power evolution shown in Figure 1.6 (iv) show a >4-dB improvement in the NF over most of the signal wavelength range compared to first-order counterpumping only [15].

In addition to the improvement in the absolute value of the NF, bidirectional pumping also reduces the gain tilt in the fiber. This is shown in Figure 1.13 where there is approximately a 1-dB improvement in the NF tilt [14]. The improvement in the NF results because in a sense the selection of the copump wavelengths determines the spectral shape of the NF, while the counterpropagating pumps primarily determine the shape of the Raman gain profile.

Finally, bidirectional pumping also reduces the MPI penalty, by reducing the amount of DRB light [24, 37]. This result is not especially intuitive, but results from a more even distribution of the gain inside the fiber.

## 1.5.2  Higher Order Pumping

In the amplification schemes discussed so far, the signal light is approximately one Stokes shift away from the pump light. This is referred to as first-order Raman amplification. In higher order pumping schemes, one or more pumps that are two or more Stokes shifts away from the signal are used to primarily amplify the first-order pumps, which in turn amplify the signal [15, 35, 38–41]. When the pump is two Stokes shifts away from the signal it is referred to as second-order Raman amplification and so forth for higher order pumping schemes. Higher order pumping reduces the NF of the amplifier by more evenly distributing the gain across the length of the fiber.

The first bidirectionally pumped system used second-order pumping to overcome the problems associated with the transfer of pump RIN to the signal [34]. In this scheme a first-order counterpump was used with a second-order copump. The second-order pump amplified the first-order pump at the signal input. The noise transfer was reduced compared to first-order pumping since the second-order pump only provided a small amount of direct gain to the signal and was counterpropagating to the first-order pump. Since this initial demonstration second-order counterpumping, bidirectional amplification, and even third-order amplification have been demonstrated.

Examples of second-order counter- and bidirectional second-order pumping are shown in Figure 1.6 as curves (iii) and (v), respectively. The more even distribution of the gains along the fiber length, compared to first-order pumping, are evident. The improvement in NF as a result of higher order pumping is illustrated in Figure 1.14. It is seen that second-order counterpumping provides approximately a 1-dB improvement in NF compared to first-order pumping. Second-order bidirectional pumping provides about a 0.5-dB improvement compared to first-order pumping, for the conditions indicated in Figure 1.14. Note that in Figure 1.14 the improvement in NF tilt is not as evident as it is in Figure 1.13 because of the narrower signal bandwidth under consideration and the fact the pump wavelengths were not optimized for minimizing NF tilt.

## 1.5.3   Frequency Modulated Pumps

Two other noteworthy pumping schemes that recently emerged are time varying and broadband Raman pumping. The problem of NF tilt was

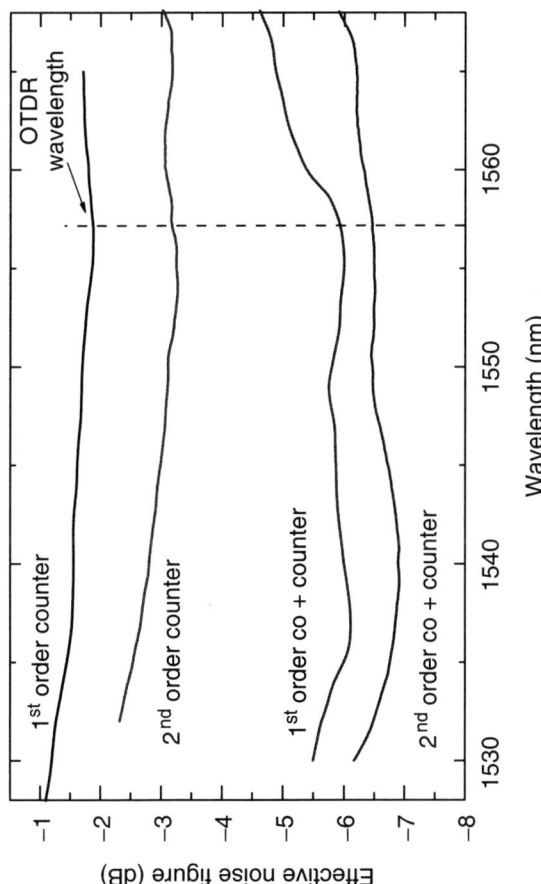

**Figure 1.14:** Measured noise figure for the four different amplifier configurations. (Courtesy of authors of Ref. 15.)

outlined in Section 1.4.3. This was primarily due to pump-to-pump inter-actions. In addition to this problem, another detrimental nonlinear effect, four-wave mixing, between pumps may arise. One way to avoid these problems is to not have the Raman pump wavelengths spatially overlap in the fiber. This can be done in a counterpumped amplifier by temporally modulating the pump wavelength.

There are two ways of doing this. One is to alternatively turn on and off several multiplexed diodes, and the other is to have a source whose frequency can be varied in time [42–45]. An additional benefit of this approach is that by varying the amount of time a given pump diode is on (or the dwell time at a certain frequency), a way of actively varying the gain profile can be realized.

Two problems with this approach are that it can only be used in a counterpumped configuration in which the temporal variations in the pump wavelength can be averaged out during counterpropagation with the signal. In addition there can be an enhancement of both amplified spontaneous emission and DRB light in these schemes [46, 47].

### 1.5.4   Broadband Spectral Width Pump Sources

It was also recently pointed out that the use of spectrally broad Raman pumps can broaden the Raman gain curve [48, 49]. The Raman gain profile produced by a pump is the convolution of the pump spectrum and the Raman gain curve. If a broadband pump source is used by properly shaping the pump spectrum, a flatter Raman gain profile can be obtained. This can reduce the number of pump diodes needed in a system.

## 1.6   Pump Sources

A key enabler of Raman amplification is the relatively high power sources needed to accomplish Raman amplification. For telecommunication wave-lengths in the 1500–1600-nm region, the required pump wavelengths for first-order pumping are in the 1400-nm region. Two competing pump sources have emerged. These are semiconductor diode lasers and Raman fiber lasers (RFL).

## 1.6.1  Diode Lasers

Although the basic technology for fabricating InGaAsP lasers operating in
the wavelength range extending from 1200 to 1600 nm has been available
since the early 1980s, power levels emitted from most InGaAsP lasers were
typically below 10 mW. Such power levels are adequate for diode lasers
used as an optical source within the optical transmitter. In contrast, diode
lasers employed for pumping of Raman amplifiers should provide fiber-
coupled power in excess of 100 mW, and power levels as high as 400 mW
are often desirable. Moreover, they should operate in the wavelength range
1400–1500 nm if the Raman amplifier is designed to amplify wavelength-
division-multiplexed signals in the C and L bands, covering the wavelength
region from 1530 to 1620 nm. High-power InGaAsP diode lasers operating
in this wavelength range were developed during the 1990s for the purpose
of pumping Raman amplifiers.

   Diode lasers used for Raman amplifiers are grown on an InP substrate
with an active region consisting of multiple quantum wells (MQWs) made
of $In_{1-x}Ga_xAs_yP_{1-y}$ material. The values of fractions $x$ and $y$ can be
tailored to fabricate a diode laser operating at the desired wavelength. In
the buried-heterostructure (BH) design, the active region is surrounded
on all sides by a lower refractive-index material to ensure that the laser
operates in a single transverse mode in a stable fashion. The entire structure is grown using an epitaxial growth technique such as metal-organic
chemical vapor deposition (MOCVD) or molecular beam epitaxy (MBE).
Both laser facets are coated with suitable dielectric layers to enhance the
power output from the laser. The coating reduces the reflectivity of the
front facet (used for power output), while it enhances the reflectivity of
the rear facet. Power levels in excess of 700 mW can be coupled into an
optical fiber when such lasers are designed with a relatively long cavity
length of 3 mm [1].

   Several features are unique to pump modules used for Raman amplifiers. First, it is common to employ several diode lasers operating at
suitable wavelengths within a single pump module in order to realize
a uniform Raman gain over a wide bandwidth [2]. Second, the wavelength of each pump laser is stabilized through a fiber Bragg grating
fabricated within the pigtail that is used to couple the laser output into
a single-mode fiber. Third, the polarization dependence of the Raman
gain is reduced either by employing two orthogonally polarized diode

lasers at each wavelength or by inserting a depolarizer within the pump module [2].

## 1.6.2   Raman Fiber Lasers

Conceptually, a complete RFL consist of three parts [52]. The first portion is a set of multimode diodes that are the optical pumps. The next section is a rare-earth-doped cladding-pumped fiber laser (CPFL), which converts the multimode diode light into single mode light at another wavelength. Finally, this single mode light is converted to the desired wavelength by a cascaded Raman resonator.

Multimode diodes capable of emitting 1–2W in the 9xx-nm region, from a 100-$\mu$m-diameter fiber core with 0.22 NA (numerical aperture) are typically used as the optical pumps to an RFL [53]. The larger area of the multimode diode pigtail fiber, compared to single mode diodes, allows more light to be coupled into the fiber from the chip.

The light from the multimode diode is next coupled into a CPFL [54–58]. Cladding pumped fiber (CPF) or double clad fiber consists of a rare-earth-doped core surrounded by a silica glass cladding [59]. What differentiates the fiber from typical fiber is that surrounding the glass is a polymer whose index is lower than that of silica. This allows light to be guided by the silica cladding as well as the core. Light from a multimode diode is transmitted along the cladding of the fiber. As the light propagates down the core it is absorbed by the rare earth dopant. The light emitted by the rare-earth-doped ions is solely guided in the single mode core. By placing Bragg gratings with the appropriate reflectivity at either end of the single mode core of the CPF, a CPFL is formed. The CPFL acts as a brightness and wavelength converter, coupling the high NA multimode light from the diodes into a small area low NA fiber. Typically, the wavelengths that are chosen for the CPFL range from 1064 to 1117 nm in Yb-doped fibers.

Rare earth dopants for the single mode core of the CPF that allow efficient lasing in the 14xx region are not available. Therefore, a cascaded Raman resonator is used to shift the light at 11xx nm out to 14xx nm. The RFL consists of two sets of gratings separated by a fiber with an enhanced Raman gain coefficient [60]. In an exemplary 1480-nm RFL, light from the 1117-nm CPFL enters the cavity. As it propagates down the fiber it is converted into light at the next Stokes shift at 1175 nm. Any 1117-nm light that is not converted will be reflected by a high reflector (HR) ($\sim$100%) on the output grating set. The light at 1175 nm is confined in the cavity by two HR gratings on either end of the (highly nonlinear fiber) HNLF. The 1175-nm light is converted to 1240-nm light. The process continues in a similar manner with nested pairs of HR at all intermediate Stokes shifts to aid the efficiency of the process. When the desired output wavelength is reached the output grating set contains a grating whose reflectivity is less than 100% so that light is coupled out of the cavity. This grating is called the output coupler (OC). By selecting an OC with a center wavelength in the 14xx-nm region a pump for Raman amplification in the 1550-nm region can be formed.

## 1.7 Summary of Chapters

This book contains six more chapters dealing with various aspects of Raman amplifiers and covering both the physics and the engineering issues involved while designing such amplifiers. Each chapter is written by one or more experts in their subfield and is intentionally kept relatively long so that it can provide in-depth coverage and act as a tutorial.

Chapter 2 is devoted to the basic theory behind Raman amplifiers. It introduces in Section 2.1 a set of two coupled pump and signal equations that govern power transfer from the pump to signal leading to signal amplification. To provide physical insight, these equations are first solved approximately in an analytic form in the case in which a single pump laser provides Raman gain. The more realistic case of multiple pump lasers is then addressed to illustrate how such a scheme can provide gain over a wide signal bandwidth. Section 2.2 is devoted to several practical issues that must be addressed for Raman amplifiers. More specifically, it considers the impact of spontaneous Raman scattering, double Rayleigh scattering, and various kinds of cross talk. The impact of polarization-mode dispersion is discussed in a separate section as this phenomenon can affect the performance of a Raman amplifier considerably by inducing fluctuations in the amplified signal. The last section is devoted to the amplification of ultrashort optical pulses to discuss the effects of group-velocity dispersion and fiber nonlinearities.

In Chapter 3 the performance of distributed amplifiers is considered. The benefits of this scheme are discussed including improved noise performance, upgradeability of existing systems, amplification at any wavelength, broadband amplification, and shaping of the gain curve and improved noise figure. This is followed by a discussion of the challenges of Raman amplification Rayleigh scattering, nonlinear impairments resulting from amplification, pump–signal cross talk, and signal–pump–signal cross talk. Finally advanced pumping schemes such as co- or forward, bidirectional higher order pumping are discussed, and low noise pumps are discussed.

In Chapter 4 the focus switches to discrete amplifiers. In discrete amplifiers there is more flexibility in the choice of gain fiber as compared to distributed amplifiers; therefore the Raman properties of fibers with different dopants are discussed. This is followed by several design issues affecting the performance of the amplifier including fiber length, pump mediated noise, ASE, nonlinear effects, and DRBS noise. Special attention is paid to dispersion compensating Raman amplifiers (DCRA), which provide both amplification and dispersion compensation simultaneously. The chapter closes with a discussion on wideband operation.

In Chapter 5 two impairment mechanisms that are strongly emphasized by or specific to Raman based systems, transfer of RIN from pumps to signals and double Rayleigh backscattering, are analyzed. The RIN transfer significantly limits the usefulness of forward pumping unless special low-RIN line-broadened pump lasers are used. Double Rayleigh scattering can be a significant source of multipath interference in Raman-based systems and ultimately limits the useful Raman gain and number of spans. Simple analytical models describing these impairments and their system impact are presented, as well as methods for experimental characterization.

Chapter 6 describes the technology for fabricating high-power laser diodes that operate in the wavelength range of 1400–1500 nm and are used almost universally for pumping modern Raman amplifiers. After providing the relevant fabrication details, the chapter focuses on the characteristics of such high-power laser diodes with particular attention paid to their noise, stimulated Brillouin scattering (SBS) characteristics, and polarization properties. The stabilization of the pump-laser wavelength through a fiber Bragg grating is also discussed.

In the final chapter, Raman fiber lasers, which are used as pumps for Raman amplifiers, are discussed. The constituent components of these devices are discussed including the pump diodes, gain fiber, and Bragg gratings. This is followed by a description on optimizing the fiber laser. Finally, single cavities capable of providing multiple wavelengths or of providing light at multiple Stokes orders are discussed.

# References

[1] F. P. Kapron, D. B. Keck, and R. D. Maurer, *Appl. Phys. Lett.* **17,** 423 (1970).

[2] E. Desurvire, *Erbium-Doped Fiber Amplifiers* (Wiley, New York, 1994).

[3] P. C. Becker, N. A. Olsson, and J. R. Simpson, *Erbium-Doped Fiber Amplifiers Fundamentals and Technology* (Academic Press, San Diego, CA, 1999).

[4] R. H. Stolen, E. P. Ippen, and A. R. Tynes, *Appl. Phys. Lett.* **20,** 276 (1972).

[5] R. H. Stolen and E. P. Ippen, *Appl. Phys. Lett.* **22,** 276 (1973).

[6] L. Eskildsen, P. B. Hansen, S. G. Grubb, A. J. Stentz, T. A. Strasser, J. Judkins, J. J. DeMarco, R. Pedrazzani, and D. J. DiGiovanni, Capacity upgrade of transmission systems by Raman amplification (Optical Amplifiers and Their Applications, Monterey, CA, 1996), Paper ThB4.

[7] P. B. Hansen, L. Eskildsen, S. G. Grubb, A. J. Stentz, T. A. Strasser, J. Judkins, J. J. DeMarco, R. Pedrazzani, and D. J. DiGiovanni, *IEEE Photon. Technol. Lett.* **9,** 262 (1997).

[8] G. P. Agrawal, *Fiber-Optic Communication Systems* (Wiley, New York, 1997).

[9] T. Izawa and S. Sudo, *Optical Fibers: Materials and Fabrication* (KTK Scientific, Tokyo, 1987).

[10] P. Wan and J. Conradi, *J. Lightwave Technol.* **14,** 288 (1996).

[11] L. B. Jeunhome, *Single-Mode Fiber Optics Principles and Applications* (Dekker, New York, 1990), 100.

[12] G. P. Agrawal, *Nonlinear Fiber Optics* (Academic Press, San Diego, 1995), Chap. 8.

[13] J. Bromage, Raman amplification for fiber communication systems (Optical Fiber Communication Conference, Atlanta, GA, 2003), Paper TuC1.

[14]  Y. Emori and S. Namiki, *IEICE Trans. Commun.* **E84-B,** 1219 (2001).

[15]  J.-C. Bouteiller, K. Brar, and C. Headley, Quasi-constant signal power transmission, ECOC 2002 Paper, S3.04.

[16]  K. Rottwitt and H. D. Kidorf, A 92 nm bandwidth Raman amplifier (Optical Fiber Communication Conference, San Jose, CA, 1998), Paper PD6.

[17]  Y. Emori, K. Tanaka, and S. Namiki, *Electron. Lett.* **35,** 1355 (1999).

[18]  C. R. S. Fludger, V. Handerek, and R. J. Mears, Ultra-wide bandwidth Raman amplifiers (Optical Fiber Communication Conference, Anaheim, CA, 2002), Paper TuJ3.

[19]  H. Kidorf, K. Rottwitt, M. Nissov, M. Ma, and E. Rabarijaona, *IEEE Photon. Technol. Lett.* **11,** 530 (1999).

[20]  X. Zhou, C. Lu, P. Shum, and T. H. Cheng, *IEEE Photon. Technol. Lett.* **13,** 945 (2001).

[21]  M. Yan, J. Chen, W. Jiang, J. Li, J. Chen, and X. Li, *IEEE Photon. Technol. Lett.* **13,** 948 (2001).

[22]  V. E. Perlin and H. G. Winful, *J. Lightwave Technol.* **20,** 250 (2002).

[23]  P. B. Hansen, L. Eskildsen, A. J. Stentz, T. A. Strasser, J. Judkins, J. J. DeMarco, R. Pedrazzani, and D. J. DiGiovanni, *IEEE Photon. Technol. Lett.* **10,** 159 (1998).

[24]  M. Nissov, K. Rottwitt, H. D. Kidorf, and M. X. Ma, *Electron. Lett.* **35,** 997 (1999).

[25]  C. R. S. Fludger and R. J. Mears, *J. Lightwave Technol.* **19,** 536 (2001).

[26]  E. Brinkmeyer, *J. Opt. Soc. Am.* **70,** 1010 (1980).

[27]  A. H. Hartog and M. P. Gold, *J. Lightwave Technol.* **2,** 76 (1984).

[28]  S. A. E. Lewis, S. V. Chernikov, and J. R. Taylor, *IEEE Photon. Technol. Lett.* **12,** 528 (2000).

[29]  C. R. S. Fludger, V. Handerek, and R. J. Mears, *J. Lightwave Technol.* **19,** 1140 (2001).

[30]  M. D. Mermelstein, C. Headley, and J.-C. Bouteiller, *Electron. Lett.* **38,** 403 (2002).

[31]  M. Yan, J. Chen, W. Jiang, J. Li, J. Chen, and X. Li, *IEEE Photon. Technol. Lett.* **13,** 651 (2001).

[32]  C. R. Fludger, V. Handerek, and R. J. Mears, Fundamental noise limits in broadband Raman amplifiers (Optical Fiber Communication Conference, Anaheim, CA, 2001), Paper MA5.

[33]  S. Kado, Y. Emori, and S. Namiki, Gain and noise tilt control in multi-wavelength bi-directionally pumped Raman amplifier, OFC, Anaheim, CA, 2002, Paper TuJ4.

[34] K. Rottwitt, A. Stentz, T. Nielsen, P. B. Hansen, K. Feder, and K. Walker, Transparent 80 km bi-directionally pumped distributed Raman amplifier with second order pumping, ECOC, Nice, France, 1999, Paper II-144.

[35] S. Kodo, Y. Emori, S. Namiki, N. Tsukiji, J. Yoshida, and T. Kimura, Broadband flat-noise Raman amplifier using low noise bi-directionally pumping sources, ECOC, Amsterdam, Netherlands, 2001, Paper PD.F.1.8.

[36] R.-J. Essiambre, P. Winzer, J. Bromage, and C. H. Kim, *IEEE Photon. Technol. Lett.* **14,** 914 (2002).

[37] V. Dominic, A. Mathur, and M. Ziari, Second-order distributed Raman amplification with a high-power 1370 nm laser diode (Optical Amplifiers and Their Applications, 2001), Paper OMC6.

[38] L. Labrunie, F. Boubal, E. Brandon, L. Buet, N. Darbois, D. Dufournet, V. Havard, P. La Roux, M. Mesic, L. Piriou, A. Tran, and J. P. Blondel, 1.6 Terabit/s (160 × 10.66 Gbit/s) unrepeated transmission over 321 km using second order pumping distributed Raman amplification (Optical Amplifers and Their Applications, 2001), Paper PD3.

[39] J.-C. Bouteiller, K. Brar, S. Radic, and C. Headley, *IEEE Photon. Technol. Lett.* **15,** 212 (2003).

[40] S. B. Papernyi, V. I. Karpov, and W. R. L. Clements, Third-Order Cascaded Raman Amplification (Optical Fiber Communication Conference, Anaheim, CA, 2002), Paper FB4.

[41] C. Martinelli, D. Mongardien, J. C. Antona, C. Simonneau, and D. Bayart, Analysis of bidirectional and second-order pumping in long-haul systems with distributed Raman amplification (European Conference on Optical Communication, Copenhagen, Denmark, 2002), Paper P3.30.

[42] L. F. Mollenauer, A. R. Grant, and P. V. Mamyshev, *Opt. Lett.* **27,** 592 (2002).

[43] C. R. S. Fludger, V. Handerek, and N. Jolley, Novel ultra-broadband high performance distributed Raman amplifier employing pump modulation (Optical Fiber Communication Conference, Anaheim, CA, 2002), Paper WB4.

[44] P. J. Winzer, K. Sherman, and M. Zirngibl, Experimental demonstration of time division multiplexed Raman pumping (Optical Fiber Communication Conference, Anaheim, CA, 2002), Paper WB5.

[45] J. Nicholson, J. Fini, J. C. Bouteiller, J. Bromage, and K. Brar, A swept-wavelength Raman pump with 69 MHz repetition rate (Optical Fiber Communication Conference, Atlanta, GA, 2003), Paper PD46.

[46] A. Artamonov, V. Smokovdin, M. Kleshov, S. A. E. Lewis, and S. V. Chernikov, Enhancement of double Rayleigh scattering by pump intensity noise in a fiber Raman amplifier (Optical Fiber Communication Conference, Anaheim, CA, 2002), Paper WB6.

[47] J. Bromage, P. J. Winzer, L. E. Nelson, M. D. Mermelstein, C. Horn, and C. Headley, *IEEE Photon. Technol. Lett.* **15,** 667 (2003).

[48] T. J. Ellingham, L. M. Gleeson, and N. J. Doran, Enhanced Raman amplifier performance using non-linear pump broadening (European Conference on Optical Communication, Copenhagen, Denmark, 2002), Paper 4.1.3.

[49] D. Vakhshoori, M. Azimi, P. Chen, B. Han, M. Jiang, K. J. Knopp, C. C. Lu, Y. Shen, G. Vander Rhodes, S. Vote, P. D. Wang, and X. Zhu, Raman amplification using high-power incoherent semiconductor pump sources (Optical Fiber Conference, Atlanta, GA, 2003), Paper PD47.

[50] D. Garbuzov, R. Menna, A. Komissarov, M. Maiorov, V. Khalfin, A. Tsekoun, S. Todorov, and J. Connolly, 1400–1480nm ridge-waveguide pump lasers with 1 watt CW output power for EDFA and Raman amplification (Optical Fiber Communication Conference, Anaheim, CA, 2001), Paper PD18.

[51] S. Namiki and Y. Emori, *J. Sel. Top. Quantum Electron.* **7,** 3 (2001).

[52] S. Grubb, T. Erdogan, V. Mizrahi, T. Strasser, W. Y. Cheung, W. A. Reed, P. J. Lemaire, A. E. Miller, S. G. Kosinski, G. Nykolak, and P. C. Becker, 1.3 $\mu$m cascaded Raman amplifier in germanosilicate fibers (Proc. Optical Amplifiers and Their Applications, Davos, Switzerland, 1994), Paper PD3–1, p. 187.

[53] See, for example, Alfalight, Boston Lasr Inc., IRE-Polus Group, JDSU, LaserTel, SLI, Spectra Physics Products List.

[54] L. Goldberg, B. Cole, and E. Snitzer, *Electron. Lett.* **33,** 2127 (1997).

[55] L. Goldberg and J. Koplow, *Electron. Lett.* **34,** 2027 (1998).

[56] A. B. Grudin, J. Nilsson, P. W. Turner, C. C. Renaud, W. A. Clarkson, and D. N. Payne, Single clad coiled optical fibre for high power lasers and amplifiers (Conf. on Lasers and Electro-Optics, 1999), Paper CPD26-1.

[57] C. Codermard, K. Yla-Jarkko, J. Singleton, P. W. Turner, I. Godfrey, S.-U. Alam, J. Nilsson, J. Sahu, and A. B. Grudinin, Low noise, intelligent cladding pumped L-band EDFA (European Conference on Optical Communication, Anaheim, CA, 2002), Paper PD1.6.

[58] D. J. DiGiovanni and A. J. Stentz, Tapered fiber bundles for coupling light into and out of cladding-pumped fiber devices (U.S. Patent 5 864 644, January 26, 1999).

[59] E. Snitzer, H. Po, F. Hakimi, R. Tumminelli, and B. C. McCollum, Double-clad, offset core Nd fiber laser (Proc. Optical Fiber Communication Conf. 1998), Paper PD5.

[60] Y. Qian, J. H. Povlsen, S. N. Knudsen, and L. Grüner-Nielsen, On Rayleigh backscattering and nonlinear effects evaluations and Raman amplification characterizations of single-mode fibers (Proc. Optical Amplifiers and Their Applications Conference, Quebec, Canada, 2000), Paper OMD18.

# Chapter 2

# Theory of Raman Amplifiers

## Govind P. Agrawal

Stimulated Raman scattering (SRS) is the fundamental nonlinear process that turns optical fibers into broadband Raman amplifiers. Although Raman amplification in optical fibers was observed as early as 1972, until recently SRS was mainly viewed as a harmful nonlinear effect because it can also severely limit the performance of multichannel lightwave systems by transferring energy from one channel to its neighboring channels. This chapter focuses on the SRS phenomenon from the standpoint of Raman amplifiers. Section 2.1 discusses the basic physical mechanism behind SRS, with emphasis on the Raman gain spectrum. It also presents the coupled pump and Stokes equations that one needs to solve for predicting the performance of Raman amplifiers. The simplest case in which both the pump and Stokes are in the form of continuous-wave (CW) beams is considered in this section for both the single-pump and the multiple-pump situations. This section also introduces forward, backward, and bidirectional pumping configurations. Section 2.2 focuses on several practical issues that are relevant for Raman amplifiers. The topics covered include spontaneous Raman scattering, double Rayleigh backscattering, and pump-noise transfer. Section 2.3 is devoted to the effects of polarization-mode dispersion (PMD) on Raman amplification. A vector theory of the SRS process is used to show that PMD introduces fluctuations in the amplified signal whose magnitude depends not only on the PMD parameter but also on the pumping configuration. More specifically, fluctuations are reduced significantly when a backward-pumping configuration is used. The amplification of short pulses is discussed in Section 2.4 with emphasis on the role of group-velocity mismatch and dispersion.

## 2.1   Pump and Signal Equations

In any molecular medium, spontaneous Raman scattering can transfer a small fraction (typically $<10^{-6}$) of power from one optical field to another field whose frequency is downshifted by an amount determined by the vibrational modes of the medium. This phenomenon was discovered by Raman in 1928 and is known as the Raman effect [1]. As shown schematically in Figure 2.1, it can be viewed quantum-mechanically as scattering of a photon of energy $\hbar\omega_p$ by one of the molecules to a lower-frequency photon of energy $\hbar\omega_s$. The energy difference is used by an optical phonon that is generated during this process, that is, the molecule makes a transition to an excited vibrational state. From a practical standpoint, the incident light acts as a pump for generating the red-shifted radiation called the Stokes line. A blue-shifted component, known as the anti-Stokes line, is also generated but its intensity is much weaker than that of the Stokes line because the anti-Stokes process requires the vibrational state to be initially populated

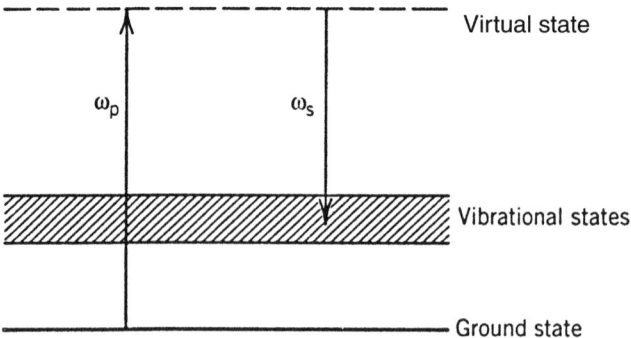

**Figure 2.1:** Schematic illustration of the Raman-scattering process from a quantum-mechanical viewpoint. A Stokes photon of reduced energy $\hbar\omega_s$ is created spontaneously when a pump photon of energy $\hbar\omega_p$ is lifted to a virtual level shown as a dashed line.

with a phonon of right energy and momentum. In what follows we ignore the anti-Stokes process as it plays virtually no role in fiber amplifiers.

Although spontaneous Raman scattering takes place in any molecular medium, it is weak enough that it can be ignored when an optical beam propagates through an optical fiber. It was observed in 1962 that, for intense optical fields, the nonlinear phenomenon of SRS can occur in which the Stokes wave grows rapidly inside the medium such that most of the power of the pump beam is transferred to it [2]. Since 1962, SRS has been studied extensively in a variety of molecular media and has found a number of applications [3–9]. SRS was observed in silica fibers in 1972; soon after, losses of such fibers were reduced to acceptable levels [10]. Since then, the properties of the Raman scattering process have been quantified for many optical glasses, in both the bulk and the fiber form [11–18].

From a practical standpoint, SRS is not easy to observe in optical fibers using CW pump beams because of its relatively high threshold ($\sim 1$ W). However, if a Stokes beam of right frequency is launched together with the pump beam as shown in Figure 2.2, it can be amplified significantly using a CW pump beam with power levels $\sim 100$ mW. The pump and signal can even be launched in the opposite directions because of the nearly isotropic nature of SRS. In fact, as will become clear later, the backward-pumping configuration is often preferred in practice because it leads to better performance of Raman amplifiers. Although fiber-based Raman amplifiers attracted considerable attention during the 1980s [19–37], it was only with the availability of appropriate pump lasers in the late 1990s that their development matured for telecommunication applications [38–62].

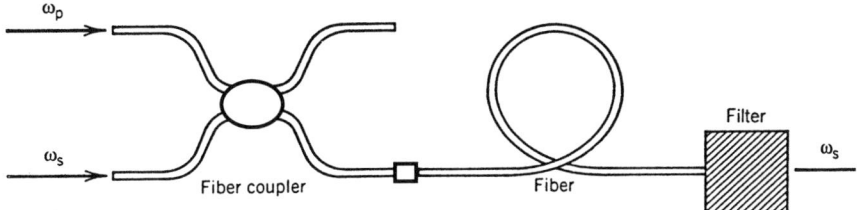

**Figure 2.2:** Schematic of a fiber-based Raman amplifier in the forward-pumping configuration. The optical filter passes the signal beam but blocks the residual pump.

## 2.1.1  Raman Gain Spectrum

The most important parameter characterizing Raman amplifiers is the Raman gain coefficient $g_R$, which is related to the cross section of spontaneous Raman scattering [7]. It describes how the Stokes power grows as pump power is transferred to it through SRS. On a more fundamental level, $g_R$ is related to the imaginary part of the third-order nonlinear susceptibility [8]. In a simple approach, valid under the CW or quasi-CW conditions, the initial growth of a weak optical signal is governed by

$$\frac{dI_s}{dz} = \gamma_R\left(\Omega\right) I_p I_s, \tag{2.1.1}$$

where $\gamma_R\left(\Omega\right)$ is related to $g_R$, $\Omega \equiv \omega_p - \omega_s$ represents the Raman shift, and $\omega_p$ and $\omega_s$ are the optical frequencies associated with the pump and signal fields having intensities $I_p$ and $I_s$, respectively.

The Raman gain spectrum has been measured for silica glasses as well as silica-based fibers [10–18]. Figure 2.3 shows the Raman gain coefficient for bulk silica as a function of the frequency shift $\Omega$ when the pump and signal are copolarized (solid curve) or orthogonally polarized (dotted curve). The peak gain is normalized to 1 in the copolarized case so that the same curves can be used for any pump wavelength $\lambda_p$. The peak value scales inversely with $\lambda_p$ and is about $6 \times 10^{-14}$ m/W for a pump near 1.5 $\mu$m.

The most significant feature of the Raman gain spectrum for silica fibers is that the gain exists over a large frequency range (up to 40 THz) with a broad peak located near 13.2 THz. This behavior is due to the noncrystalline nature of silica glasses. In amorphous materials such as fused silica, molecular vibrational frequencies spread out into bands that overlap and create a continuum. As a result, in contrast to most molecular media, for which the Raman gain occurs at specific well-defined frequencies, it extends continuously over a broad range in silica fibers. Optical fibers can act as broadband Raman amplifiers because of this feature. Another important feature of Figure 2.3 is the polarization dependence of the Raman gain; the gain nearly vanishes when pump and signal are orthogonally polarized. As discussed in Section 2.3, the polarization-dependence of the Raman gain affects the performance of Raman amplifiers in several different ways.

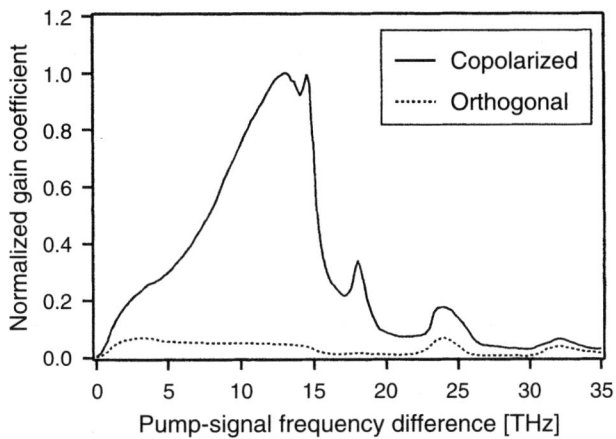

**Figure 2.3:** Raman gain spectrum for bulk silica measured when the pump and signal are copolarized (solid curve) or orthogonally polarized (dotted curve). The peak gain is normalized to 1 in the copolarized case. (After Ref. [62]; © 2004 IEEE.)

In single-mode fibers, the spatial profile of both the pump and the signal beams is dictated by the fiber design and does not change along the entire fiber length. For this reason, one often deals with the total optical power defined as

$$P_j(z) = \int \int_{-\infty}^{\infty} I_j(x, y, z)\, dx dy, \qquad (2.1.2)$$

where $j = p$ or $s$. Equation (2.1.1) can be written in terms of optical powers as

$$\frac{dP_s}{dz} = (\gamma_R / A_{\text{eff}})\, P_p P_s \equiv g_R P_p P_s, \qquad (2.1.3)$$

where the effective core area is defined as

$$A_{\text{eff}} = \frac{\left(\int \int_{-\infty}^{\infty} I_p(x, y, z)\, dx dy\right) \left(\int \int_{-\infty}^{\infty} I_s(x, y, z)\, dx dy\right)}{\int \int_{-\infty}^{\infty} I_p(x, y, z) I_s(x, y, z)\, dx dy} \qquad (2.1.4)$$

This complicated expression simplifies considerably if we assume that the field-mode profile $F(x,y)$ is nearly the same for both the pump and the Stokes. In terms of this mode profile, $A_{\text{eff}}$ can be written as

$$A_{\text{eff}} = \frac{\left(\int \int_{-\infty}^{\infty} |F(x, y)|^2 \, dx dy\right)^2}{\int \int_{-\infty}^{\infty} |F(x, y)|^4 \, dx dy}. \tag{2.1.5}$$

If we further approximate the mode profile by a Gaussian function of the form $F(x, y) = \exp\left[-\left(x^2 + y^2\right)/w^2\right]$, where $w$ is the field-mode radius [9], and perform the integrations in Eq. (2.1.5), we obtain the simple result $A_{\text{eff}} \approx \pi w^2$. Since the field-mode radius $w$ is specified for any fiber, $A_{\text{eff}}$ is a known parameter whose values can range from 10 to 100 $\mu$m$^2$ depending on fiber design; low values of $A_{\text{eff}}$ occur for dispersion-compensating fibers (DCFs) for which the core diameter is relatively small.

Figure 2.4 shows $g_R \equiv \gamma_R/A_{\text{eff}}$ (sometimes called the Raman gain efficiency) for a DCF, a nonzero dispersion fiber (NZDF), and a superlarge area (SLA) fiber with $A_{\text{eff}} = 15$, 55, and 105 $\mu$m$^2$, respectively. In all cases, the fiber was pumped at 1.45 $\mu$m and provided gain near 1.55 $\mu$m. The main point to note is that a DCF is nearly 10 times more efficient

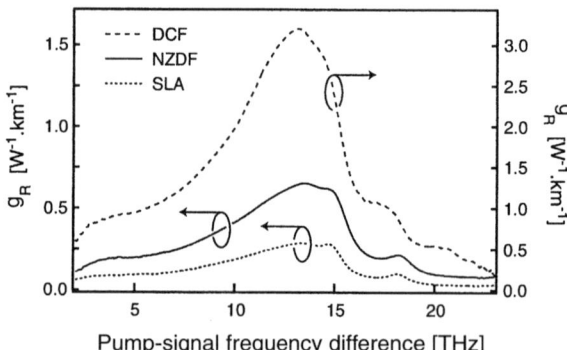

**Figure 2.4:** Measured Raman gain spectra for three kinds of fibers pumped at 1.45 $\mu$m. Both the effective core area and the GeO$_2$ doping levels are different for three fibers. (After Ref. [62]; © 2004 IEEE.)

for Raman amplification. An increase by a factor of 7 is expected from its reduced effective core area. The remaining increase is due to a higher doping level of germania in DCFs ($GeO_2$ molecules exhibit a larger Raman gain peaking near 13.1 THz). Spectral changes seen in Figure 2.4 for three fibers can be attributed to $GeO_2$ doping levels.

It is evident from Figure 2.4 that, when a pump beam is launched into the fiber together with a weak signal beam, it will be amplified because of the Raman gain as long as the frequency difference $\Omega = \omega_p - \omega_s$ lies within the bandwidth of the Raman gain spectrum. The signal gain depends considerably on the frequency difference $\Omega$ and is maximum when the signal beam is downshifted from the pump frequency by 13.2 THz (about 100 nm in the 1.5-$\mu$m region). The Raman gain exists in all spectral regions; that is, optical fibers can be used to amplify any signal provided an appropriate pump source is used. This remarkable feature of Raman amplifiers is quite different from erbium-doped fiber amplifiers, which can amplify only signals whose wavelength is close to the atomic transition wavelength occurring near 1.53 $\mu$m.

The nonuniform nature of the Raman gain spectrum in Figure 2.4 is of concern for wavelength-division-multiplexed (WDM) lightwave systems because different channels will be amplified by different amounts. This problem is solved in practice by using multiple pumps at slightly different wavelengths. Each pump provides nonuniform gain but the gain spectra associated with different pumps overlap partially. With a suitable choice of wavelengths and powers for each pump laser, it is possible to realize nearly flat gain profile over a considerably wide wavelength range. We discuss the single-pump scheme first, as it allows us to introduce the basic concepts in a simple manner, and then focus on the multiple-pump configuration of Raman amplifiers.

## 2.1.2 Single-Pump Raman Amplification

Consider the simplest situation in which a single CW pump beam is launched into an optical fiber used to amplify a CW signal. Even in this case, Eq (2.1.3) should be modified to include fiber losses before it can be used. Moreover, the pump power does not remain constant along the fiber. When these effects are included, the Raman-amplification process is governed by the following set of two coupled equations,

$$\frac{dP_s}{dz} = g_R P_p P_s - \alpha_s P_s, \tag{2.1.6}$$

$$\xi \frac{dP_p}{dz} = \frac{\omega_p}{\omega_s} g_R P_p P_s - \alpha_p P_p, \tag{2.1.7}$$

where $\alpha_s$ and $\alpha_p$ account for fiber losses at the Stokes and pump wavelengths, respectively. The parameter $\xi$ takes values $\pm 1$ depending on the pumping configuration; the minus sign should be used in the backward-pumping case.

Equations (2.1.6) and (2.1.7) can be derived rigorously from Maxwell's equations. They can also be written phenomenologically by considering the processes through which photons appear in or disappear from each beam. The frequency ratio $\omega_p/\omega_s$ appears in Eq. (2.1.7) because the pump and signal photons have different energies. One can readily verify that, in the absence of losses,

$$\frac{d}{dz}\left(\frac{P_s}{\omega_s} + \xi \frac{P_p}{\omega_p}\right) = 0 \tag{2.1.8}$$

Noting that $P_j/\omega_j$ is related to photon flux at the frequency $\omega_j$, this equation merely represents the conservation of total number of photons during the SRS process.

Equations (2.1.6) and (2.1.7) are not easy to solve analytically because of their nonlinear nature. In many practical situations, pump power is so large compared with the signal power that pump depletion can be neglected by setting $g_R = 0$ in Eq. (2.1.7), which is then easily solved. As an example, $P_p(z) = P_0 \exp(-\alpha_p z)$ in the forward-pumping case ($\xi = 1$), where $P_0$ is the input pump power at $z = 0$. If we substitute this solution in Eq. (2.1.6), we obtain

$$\frac{dP_s}{dz} = g_R P_0 \exp\left(-\alpha_p z\right) P_s - \alpha_s P_s. \tag{2.1.9}$$

This equation can be easily integrated to obtain

$$P_s(L) = P_s(0) \exp\left(g_R P_0 L_{\text{eff}} - \alpha_s L\right) \equiv G(L) P_s(0), \tag{2.1.10}$$

where $G(L)$ is the net signal gain, $L$ is the amplifier length, and $L_{\text{eff}}$ is an effective length defined as

$$L_{\text{eff}} = \left[1 - \exp\left(-\alpha_p L\right)\right] / \alpha_p. \qquad (2.1.11)$$

The solution (2.1.10) shows that, because of pump absorption, the effective amplification length is reduced from $L$ to $L_{\text{eff}}$.

The backward-pumping case can be considered in a similar fashion. In this case, Eq. (2.1.7) should be solved with $g_R = 0$ and $\xi = -1$ using the boundary condition $P_p(L) = P_0$; the result is $P_p(z) = P_0 \exp\left[-\alpha_p(L - z)\right]$. The integration of Eq. (2.1.6) yields the same solution given in Eq. (2.1.10), indicating that the amplified signal power at a given pumping level is the same in both the forward- and backward-pumping configurations.

The case of bidirectional pumping is slightly more complicated because two pump lasers are located at the opposite fiber ends. The pump power in Eq. (2.1.6) now represents the sum $P_p = P_f + P_b$, where $P_f$ and $P_b$ are obtained by solving (still ignoring pump depletion)

$$dP_f/dz = -\alpha_p P_f, \qquad dP_b/dz = \alpha_p P_b. \qquad (2.1.12)$$

Solving these equations, we obtain total pump power $P_p(z)$ at a distance $z$ in the form

$$P_p(z) = P_0 \left\{ r_f \exp\left(-\alpha_p z\right) + \left(1 - r_f\right) \exp\left[-\alpha_p(L - z)\right] \right\}, \qquad (2.1.13)$$

where $P_0$ is the total pump power and $r_f = P_L / P_R$ is the fraction of pump power launched in the forward direction. The integration of Eq. (2.1.6) yields the signal gain

$$G(z) = \frac{P_s(z)}{P_s(0)} = \exp\left(g_R \int_0^z P_p(z)\, dz - \alpha_s z\right). \qquad (2.1.14)$$

Figure 2.5 shows how the signal power changes along a 100-km-long distributed Raman amplifier as $r_f$ is varied from 0 to 1. In all cases, the total pump power is chosen such that the Raman gain is just sufficient to compensate for fiber losses, that is, $G(L) = 1$.

One may ask which pumping configuration is the best from the system standpoint. The answer is not so simple as it depends on many factors.

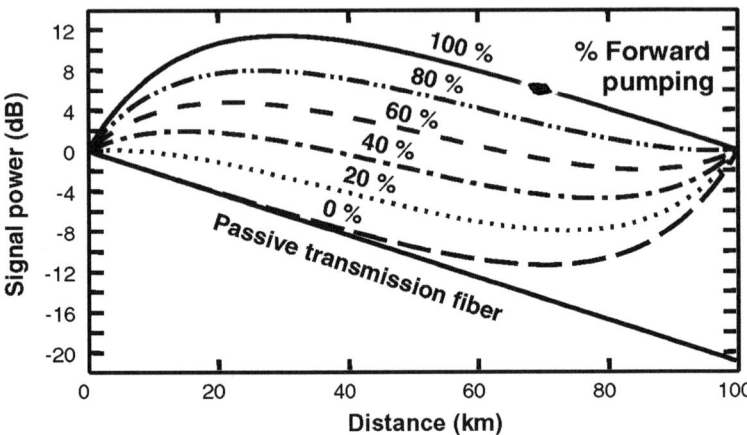

**Figure 2.5:** Evolution of signal power in a bidirectionally pumped, 100-km-long Raman amplifier as the contribution of forward pumping is varied from 0 to 100%. The straight line shows for comparison the case of a passive fiber with no Raman gain. (After Ref. [57]; © 2003 Springer.)

As discussed in Section 2.2, forward pumping is superior from the noise viewpoint. However, for a long-haul system limited by fiber nonlinearities, backward pumping may offer better performance because the signal power is the smallest throughout the link length in this case. The total accumulated nonlinear phase shift induced by self-phase modulation (SPM) can be obtained from [9]

$$\phi_{NL} = \gamma \int_0^L P_s(z)dz = \gamma P_s(0) \int_0^L G(z)\, dz. \qquad (2.1.15)$$

where $\gamma = 2\pi n_2/(\lambda_s A_{eff})$ is the nonlinear parameter responsible for SPM. The increase in the nonlinear phase shift occurring because of Raman amplification can be quantified through the ratio [57]

$$R_{NL} = \frac{\phi_{NL}\,(\text{pump on})}{\phi_{NL}\,(\text{pump off})} = L_{eff}^{-1} \int_0^L G(z)\, dz. \qquad (2.1.16)$$

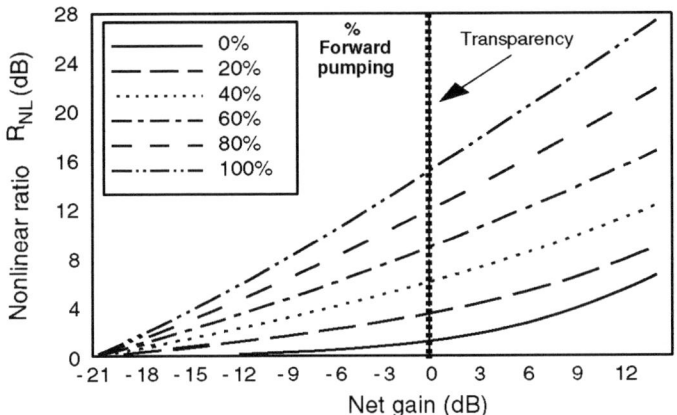

**Figure 2.6:** Enhancement in the nonlinear effects as a function of net gain in a 100-km-long, bidirectionally pumped, distributed Raman amplifier as the contribution of forward pumping is varied from 0 to 100%. The vertical line shows the case in which Raman gain just compensates total fiber losses. (After Ref. [57]; © 2003 Springer.)

Figure 2.6 shows how this ratio changes as a function of the net gain $G(L)$ for a 100-km-long distributed Raman amplifier for different combinations of forward and backward pumping. Clearly, the nonlinear effects are the least in the case of backward pumping and become enhanced by more than 10 dB when forward pumping is used.

The quantity $G(L)$ represents the net signal gain and can be even <1 (net loss) if the Raman gain is not sufficient to overcome fiber losses. It is useful to introduce the concept of the on–off Raman gain using the definition

$$G_A = \frac{P_s\,(L) \text{ with pump on}}{P_s\,(L) \text{ with pump off}} = \exp\left(g_R P_0 L_{\text{eff}}\right). \tag{2.1.17}$$

Clearly, $G_A$ represents the total amplifier gain distributed over a length $L_{\text{eff}}$. If we use a typical value of $g_R = 3$ W$^{-1}$/km for a DCF from Figure 2.4 together with $L_{\text{eff}} = 1$ km, the signal can be amplified by 20 dB for $P_0 \approx 1.5$ W. Figure 2.7 shows the variation of $G_A$ with $P_0$ observed in a 1981 experiment in which a 1.3-km-long fiber was used to amplify the

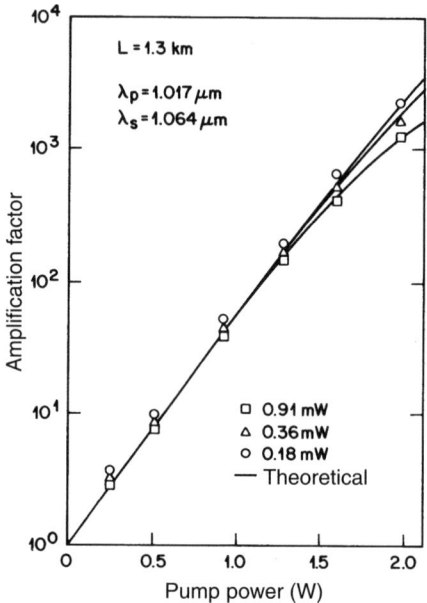

**Figure 2.7:** Variation of amplifier gain $G_A$ with pump power $P_0$. Different symbols show the experimental data for three values of input signal power. Solid curves show the theoretical prediction. (After Ref. [19]; © 1981 Elsevier.)

1.064-$\mu$m signal by using a 1.017-$\mu$m pump [19]. The amplification factor $G_A$ increases exponentially with $P_0$ initially, as predicted by Eq. (2.1.17), but starts to deviate for $P_0 > 1$ W. This is due to gain saturation occurring because of pump depletion. The solid lines in Figure 2.7 are obtained by solving Eqs. (2.1.6) and (2.1.7) numerically to include pump depletion. The numerical results are in excellent agreement with the data and serve to validate the use of Eqs. (2.1.6) and (2.1.7) for modeling Raman amplifiers.

An approximate expression for the saturated gain $G_s$ in Raman amplifiers can be obtained by solving Eqs. (2.1.6) and (2.1.7) analytically [9] with the assumption $\alpha_s = \alpha_p \equiv \alpha$. This approximation is not always valid but can be justified for optical fibers in the 1.55-$\mu$m region. Assuming forward pumping ($\xi = 1$) and making the transformation $P_j = \omega_j F_j \exp(-\alpha z)$

with $j = s$ or $p$, we obtain two simple equations:

$$\frac{dF_s}{dz} = \omega_p g_R F_p F_s, \qquad \frac{dF_p}{dz} = -\omega_p g_R F_p F_s. \qquad (2.1.18)$$

Noting that $F_p(z) + F_s(z) = C$, where $C$ is a constant, the differential equation for $F_s$ can be integrated over the amplifier length to obtain the following result:

$$G_s = \frac{F_s(L)}{F_s(0)} = \left(\frac{C - F_s(L)}{C - F_s(0)}\right) \exp\left(\omega_p g_R C L_{\text{eff}}\right). \qquad (2.1.19)$$

Using $C = F_p(0) + F_s(0)$ in the preceding equation, the saturated gain of the amplifier is given by

$$G_s = \frac{(1 + r_0) G_A^{1+r_0}}{1 + r_0 G_A^{1+r_0}}, \qquad (2.1.20)$$

where $r_0$ is related to the signal-to-pump power ratio at the fiber input as

$$r_0 = \frac{F_s(0)}{F_p(0)} = \frac{\omega_p}{\omega_s} \frac{P_s(0)}{P_p(0)} \qquad (2.1.21)$$

and $G_A = \exp(g_R P_0 L_{\text{eff}})$ is the unsaturated gain introduced in Eq. (2.1.17). Typically, $P_s(0) \ll P_p(0)$. For example, $r_0 < 10^{-3}$ when $P_s(0) < 1$ mW while $P_p(0) \sim 1$ W. Under such conditions, the saturated gain of the amplifier can be approximated as

$$G_s = \frac{G_A}{1 + r_0 G_A}. \qquad (2.1.22)$$

The gain is reduced by a factor of 2 or 3 dB when the Raman amplifier is pumped hard enough that $r_0 G_A = 1$. This can happen for $r_0 = 10^{-3}$ when the on–off Raman gain approaches 30 dB. This is precisely what we observe in Figure 2.7.

Figure 2.8 shows the saturation characteristics by plotting $G_s/G_A$ as a function of $G_A r_0$ for several values of $G_A$. The saturated gain is reduced by a factor of two when $G_A r_0 \approx 1$. This condition is satisfied when the power in the amplified signal starts to approach the input pump power $P_0$.

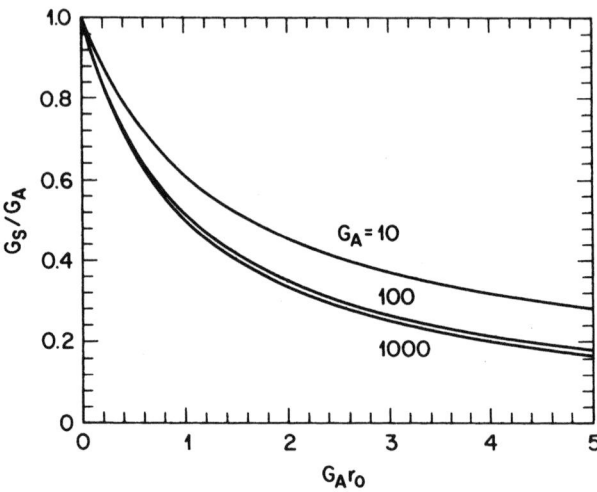

**Figure 2.8:** Gain-saturation characteristics of Raman amplifiers for several values of the unsaturated amplifier gain $G_A$.

In fact, $P_0$ is a good measure of the saturation power of Raman amplifiers. As typically $P_0 > 1$ W, the saturation power of Raman amplifiers is much larger compared with that of erbium-doped fiber amplifiers.

As seen in Figure 2.7, Raman amplifiers can amplify an input signal by a factor of 1000 (30-dB gain) when the pump power exceeds 1 W [19]. Most of the early experiments used for pumping a Nd:YAG laser operated at 1.06 μm because it can provide such CW power levels. This laser can also operate at 1.32 μm. In a 1983 experiment [21], a 1.4-μm signal was amplified using such a laser, and gain levels of up to 21 dB were obtained at a pump power of 1 W. The amplifier gain was nearly the same in both the forward- and the backward-pumping configurations. Signal wavelengths of most interest from the standpoint of optical fiber communications are near 1.5 μm. A Nd:YAG laser can still be used if a higher-order Stokes line is used as a pump. For example, the first-order Stokes line at 1.4 μm from a 1.32-μm laser can act as a pump to amplify an optical signal near 1.5 μm. As early as 1984, amplification factors of more than 20 dB were realized by using such schemes [24–26]. These experiments also indicated the importance of matching the polarization directions of the pump and signal waves as SRS nearly ceases to occur in the case of orthogonal polarizations. The use of a polarization-preserving fiber with a high-germania core resulted in 20-dB gain at 1.52 μm when such a fiber was pumped with 3.7 W of pump power.

The main drawback of Raman amplifiers from the standpoint of lightwave system applications is that they require a high-power CW laser for pumping. Most experiments performed in the 1980s in the 1.55-μm spectral region used tunable color-center lasers as a pump; such lasers are too bulky for telecommunication applications. For this reason, with the advent of erbium-doped fiber amplifiers around 1989, Raman amplifiers were rarely used in the 1.55-μm wavelength region.

The situation changed with the availability of compact high-power semiconductor and fiber lasers. Indeed, the development of Raman amplifiers underwent a virtual renaissance during the 1990s [38–56]. In one approach, three pairs of fiber gratings are inserted within the fiber used for Raman amplification [39]. The Bragg wavelengths of these gratings are chosen such that they form three cavities for three Raman lasers operating at wavelengths of 1.117, 1.175, and 1.24 μm that correspond to the first-, second-, and third-order Stokes line of a 1.06-μm pump. All three lasers are pumped through cascaded SRS using a single, diode-pumped, Nd-fiber laser. The 1.24-μm laser then pumps the Raman amplifier to provide signal amplification in the 1.3-μm region. The same idea of cascaded SRS was used to obtain 39-dB gain at 1.3 μm by using WDM couplers in place of fiber gratings [38]. In a different approach, the core of silica fiber is doped heavily with germania. Such a fiber can be pumped to provide 30-dB gain at a pump power of only 350 mW [40], power levels that can be obtained using one or more semiconductor lasers. A dual-stage configuration has also been used in which a 2-km-long germania-doped fiber is placed in series with a 6-km-long dispersion-shifted fiber in a ring geometry [46]. Such a Raman amplifier, when pumped with a 1.24-μm Raman laser, provided 22-dB gain in the 1.3-μm wavelength region with a noise figure of about 4 dB.

### 2.1.3 Multiple-Pump Raman Amplification

Starting in 1998, the use of multiple pumps for Raman amplification was pursued for developing broadband optical amplifiers required for WDM lightwave systems operating in the 1.55-μm region [47–56]. Massive WDM systems (80 or more channels) typically require optical amplifiers capable of providing uniform gain over a 70- to 80-nm wavelength range. In a simple approach, hybrid amplifiers made by combining erbium doping with Raman gain were used. In one implementation of this idea [49], nearly 80-nm bandwidth was realized by combining an erbium-doped fiber amplifier with two Raman amplifiers, pumped simultaneously at three different wavelengths (1471, 1495, and 1503 nm) using four pump modules, each module launching more than 150 mW of power into the fiber. The combined gain of 30 dB was nearly uniform over the wavelength range 1.53–1.61 μm.

Broadband amplification over 80 nm or more can also be realized by using a pure Raman-amplification scheme. In this case, a relatively long span (typically >5 km) of a fiber with a relatively narrow core (such as a DCF) is pumped using multiple-pump lasers. Alternatively, one can use the transmission fiber itself as the Raman gain medium. In the latter scheme, the entire long-haul fiber link is divided into multiple segments (60 to 100 km long), each one pumped backward using a pump module consisting of multiple-pump lasers. The Raman gain accumulated over the entire segment length compensates for fiber losses of that segment in a distributed manner.

Multiple-pump Raman amplifiers make use of the fact that the Raman gain exists at any wavelength as long as the pump wavelength is suitably chosen. Thus, even though the gain spectrum of a single pump is not very wide and is flat only over a few nanometers (see Figure 2.4), it can be broadened and flattened considerably by using several pumps of different wavelengths. Each pump creates a gain profile that mimics the spectrum shown in Figure 2.4. Superposition of several such spectra can produce relatively constant gain over a wide spectral region when pump wavelengths and power levels are chosen judiciously. Figure 2.9 shows a numerical example when six pump lasers operating at wavelengths in the range of 1420–1500 nm are employed [62]. The individual pump powers (vertical bars) are chosen to provide individual gain spectra (dashed curves) such that the total Raman gain of 18 dB is nearly flat over a 80-nm bandwidth (solid trace). Pump powers range from 40 to 200 mW and are larger for shorter wavelength pumps because all pumps interact through SRS, and some power is transferred to longer wavelength pumps within the amplifier. This technique can provide a gain bandwidth of more than 100 nm with

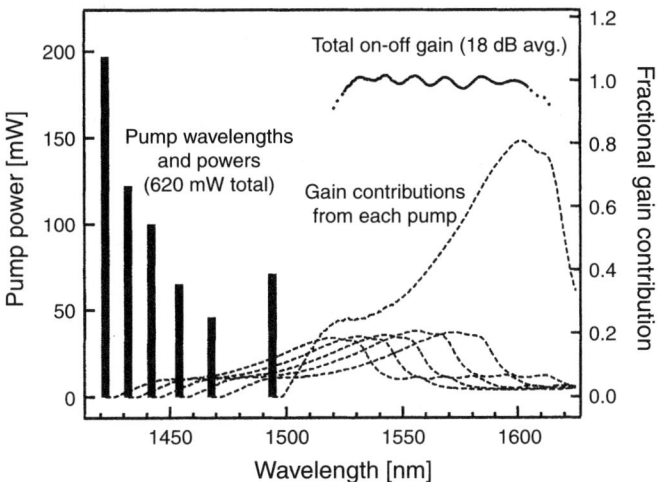

**Figure 2.9:** Numerically simulated composite Raman gain (solid trace) of a Raman amplifier pumped with six lasers with different wavelengths and input powers (vertical bars). Dashed curves show the Raman gain provided by individual pumps. (After Ref. [62]; © 2004 IEEE.)

a suitable design [50–54]. In a 2000 demonstration, 100 WDM channels with 25-GHz channel spacing, each operating at a bit rate of 10 Gb/s, were transmitted over 320 km [53]. All channels were amplified simultaneously by pumping each 80-km fiber span in the backward direction using four semiconductor lasers. Such a distributed Raman amplifier provided 15-dB gain at a total pump power of 450 mW.

As seen in Figure 2.3, the Raman gain is polarization sensitive. This creates a problem in practice since signal polarization is unpredictable in most lightwave systems. The polarization problem can be solved by pumping a Raman amplifier such that two orthogonally polarized lasers are used at each pump wavelength or by depolarizing the output of each pump laser. It should be stressed that the state of polarization of the pump and signal fields changes randomly in any realistic fiber because of birefringence variations along the fiber length. This issue is discussed in Section 2.3. Other issues that must be addressed are related to double Rayleigh backscattering and noise induced by spontaneous Raman scattering.

Broadband Raman amplifiers are designed using a numerical model that includes pump–pump interactions, Rayleigh backscattering, and spontaneous Raman scattering. Such a model considers each frequency component separately and requires the solution of a large set of coupled equations of the form [52]

$$
\begin{aligned}
\frac{dP_f\,(\nu)}{dz} = {} & \int_{\mu>\nu} g_R\,(\mu,\nu)\left[P_f\,(\mu) + P_b\,(\mu)\right] \\
& \times \left[P_f\,(\nu) + 2h\nu n_{\text{sp}}\,(\mu - \nu)\right] d\mu \\
& - \int_{\mu>\nu} g_R\,(\nu,\mu)\left[P_f\,(\mu) + P_b\,(\mu)\right] \\
& \times \left[P_f\,(\nu) + 4h\nu n_{\text{sp}}\,(\nu - \mu)\right] d\mu, \\
& - \alpha\,(\nu)\,P_f\,(\nu) + f_r\alpha_r P_b\,(\nu),
\end{aligned}
\tag{2.1.23}
$$

where $\mu$ and $\nu$ denote optical frequencies and the subscripts $f$ and $b$ denote forward- and backward-propagating waves, respectively. The parameter $n_{\text{sp}}$ is defined as

$$
n_{\text{sp}}\,(\Omega) = \left[1 - \exp\left(-h\Omega/k_B T\right)\right]^{-1},
\tag{2.1.24}
$$

where $\Omega = |\mu - \nu|$ is the Raman shift and $T$ denotes absolute temperature of the amplifier. In Eq. (2.1.23), the first and second terms account for the Raman-induced power transfer into and out of each frequency band. The factor of 2 in the first term accounts for the two polarization modes of the fiber. An additional factor of 2 in the second term includes spontaneous emission in both the forward and the backward directions [54]. Fiber losses and Rayleigh backscattering are included through the last two terms and are governed by the parameters $\alpha$ and $\alpha_r$, respectively; $f_r$ represents the fraction of backscattered power that is recaptured by the fiber mode. A similar equation holds for the backward-propagating waves.

To design broadband Raman amplifiers, the entire set of such equations is solved numerically to find the channel gains, and input pump powers are adjusted until the gain is nearly the same for all channels (see Figure 2.9). Figure 2.10 shows an experimentally measured gain spectrum for a Raman amplifier made by pumping a 25-km-long dispersion-shifted fiber with 12 diode lasers. The frequencies and powers of pump lasers are

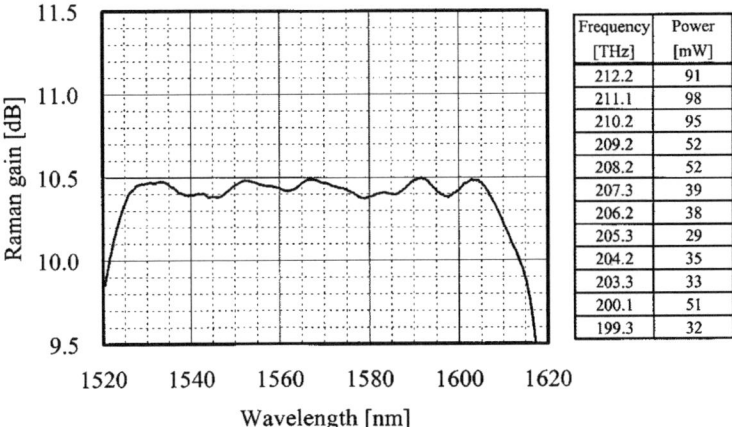

**Figure 2.10:** Measured Raman gain as a function of signal wavelength for a 25-km-long amplifier pumped with 12 lasers. Pump frequencies and power levels used are indicated on the right. (After Ref. [54]; © 2001 IEEE.)

also indicated in a tabular form. Notice that all powers are under 100 mW. The amplifier provided about 10.5-dB gain over an 80-nm bandwidth with a ripple of less than 0.1 dB. Such amplifiers are suitable for dense WDM systems covering both the C and L bands. Several experiments have used broadband Raman amplifiers to demonstrate transmission over long distances at high bit rates. In a 2001 experiment, 77 channels, each operating at 42.7 Gb/s, were transmitted over 1200 km by using the C and L bands simultaneously [55]. Since then, many demonstrations have employed Raman amplification for a wide variety of WDM systems [58–61].

## 2.2 Performance Limiting Factors

The performance of modern Raman amplifiers is affected by several factors that need to be controlled. In this section we focus on spontaneous Raman scattering, double Rayleigh backscattering, and pump-noise transfer. The impact of PMD on the performance of Raman amplifier is considered in Section 2.3.

## 2.2.1   Spontaneous Raman Scattering

Spontaneous Raman scattering adds to the amplified signal and appears as a noise because of random phases associated with all spontaneously generated photons. This noise mechanism is similar to the spontaneous emission that affects the performance of erbium-doped fiber amplifiers except that, in the Raman case, it depends on the phonon population in the vibrational state, which in turn depends on the temperature of the Raman amplifier. On a more fundamental level, one should consider the evolution of signal with the noise added by spontaneous Raman scattering. However, since noise photons are not in phase with the signal photons, we cannot use the rate equations (2.1.6) and (2.1.7) satisfied by the signal and pump powers. Rather, we should write equations for the two optical fields. If we neglect pump depletion, it is sufficient to replace Eq. (2.1.6) with

$$\frac{dA_s}{dz} = \frac{g_R}{2} P_p(z) A_s - \frac{\alpha_s}{2} A_s + f_n(z, t), \qquad (2.2.1)$$

where $A_s$ is the signal field defined such that $P_s = |A_s|^2$, $P_p$ is the pump power, and the Langevin noise source $f_n(z, t)$ takes into account the noise added through spontaneous Raman scattering. Since each scattering event is independent of others, this noise can be modeled as a Markovian stochastic process with Gaussian statistics such that $\langle f_n(z, t) \rangle = 0$ and its second moment is given by [63]

$$\langle f_n(z, t) f_n(z', t') \rangle = n_{sp} h \nu_0 g_R P_p(z) \delta(z - z') \delta(t - t'), \qquad (2.2.2)$$

where $n_{sp}$ is the spontaneous-scattering factor introduced earlier and $h\nu_0$ is the average photon energy. The two delta functions ensure that all spontaneous events are independent of each other.

Equation (2.2.1) can be easily integrated to obtain $A_s(L) = \sqrt{G(L)} A_s(0) + a_{ASE}(t)$, where $G(L)$ is the amplification factor defined earlier in Eq. (2.1.10) and the total accumulated noise from spontaneous

Raman scattering is given by

$$a_{ASE}(t) = \sqrt{G(L)} \int_0^L \frac{f_n(z,t)}{\sqrt{G(z)}} dz,$$

$$G(z) = \exp\left(\int_0^z [g_R P_p(z') - a_s] dz'\right). \qquad (2.2.3)$$

This noise is often referred to as amplified spontaneous emission (ASE) because of its amplification by the distributed Raman gain. It is easy to show that it vanishes on average ($\langle a_{ASE}(t)\rangle = 0$) and its second moment is given by

$$\langle a_{ASE}(t)a_{ASE}(t')\rangle = G_L \int_0^L dz \int_0^L dz' \frac{\langle f_n(z,t) f_n(z',t')\rangle}{\sqrt{G(z) G(z')}}$$

$$= S_{ASE}\delta(t-t'), \qquad (2.2.4)$$

where $G_L \equiv G(L)$ and the ASE spectral density is defined as

$$S_{ASE} = n_{sp} h\nu_0 g_R G_L \int_0^L \frac{P_p(z)}{G(z)} dz. \qquad (2.2.5)$$

The presence of the delta function in Eq. (2.2.4) is due to the Markovian assumption implying that $S_{ASE}$ is constant and exists at all frequencies (white noise). In practice, the noise exists only over the amplifier bandwidth and can be further reduced by placing an optical filter at the amplifier output. Assuming this to be the case, we can calculate the total ASE power after the amplifier using

$$P_{ASE} = 2\int_{-\infty}^{\infty} S_{ASE} H_f(\nu) d\nu = 2S_{ASE} B_{opt}, \qquad (2.2.6)$$

where $B_{opt}$ is the bandwidth of the optical filter. The factor of 2 in this equation accounts for the two polarization modes of the fiber. Indeed, ASE can be reduced by 50% if a polarizer is placed after the amplifier. Assuming that a polarizer is not used, the optical signal-to-noise ratio (SNR) of the

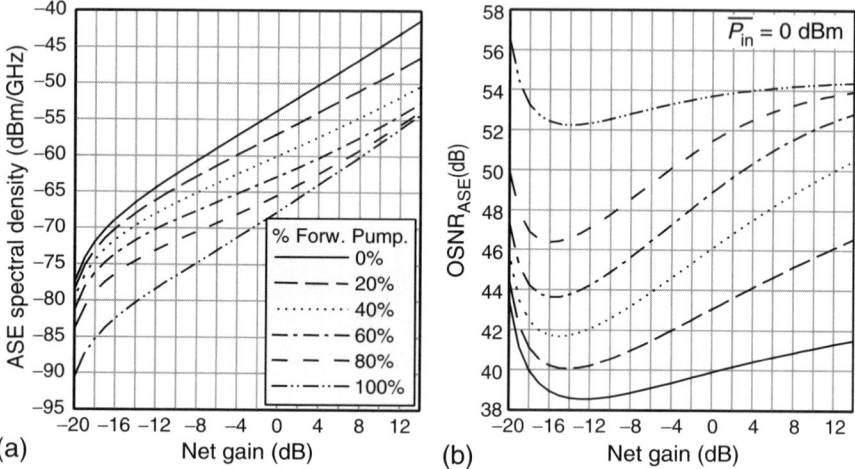

**Figure 2.11:** (a) Spontaneous spectral density and (b) optical SNR as a function of net gain $G(L)$ at the output of a 100-km-long, bidirectionally pumped, distributed Raman amplifier assuming $P_{in} = 1$ mW. (After Ref. [57]; © 2003 Springer.)

amplified signal is given by

$$\text{SNR}_o = \frac{P_s(L)}{P_{\text{ASE}}} = \frac{G_L P_{in}}{P_{\text{ASE}}}. \tag{2.2.7}$$

It is evident from Eq. (2.2.5) that both $P_{\text{ASE}}$ and $\text{SNR}_o$ depend on the pumping scheme through pump-power variations $P_p(z)$ occurring inside the Raman amplifier. As an example, Figure 2.11 shows how the spontaneous power per unit bandwidth, $P_{\text{ASE}}/B_{\text{opt}}$, and the optical SNR vary with the net gain $G(L)$ for several different pumping schemes assuming that a 1-mW input signal is amplified by a 100-km-long, bidirectionally pumped, distributed Raman amplifier. The fraction of forward pumping varies from 0 to 100%. The other parameters were chosen to be $\alpha_s = 0.21$ dB/km, $\alpha_p = 0.26$ dB/km, $n_{\text{sp}} = 1.13$, $h\nu_0 = 0.8$ eV, and $g_R = 0.68$ W$^{-1}$/km. The optical SNR is highest in the case of purely forward pumping (about 54 dB or so) but degrades by as much as 15 dB as the fraction of backward pumping is increased from 0 to 100%. This can be understood by noting that the spontaneous noise generated near the input end experiences losses over the full length of the fiber in the case of forward pumping, whereas it experiences only a fraction of such losses in the case of backward pumping. Mathematically, $G(z)$ in the denominator in Eq. (2.2.5) is larger in the forward-pumping case, resulting in reduced $S_{ASE}$.

The preceding discussion shows that spontaneous Raman scattering degrades the SNR of the signal amplified by a Raman amplifier. The extent of SNR degradation is generally quantified through a parameter $F_n$, called the *amplifier noise figure*, and defined as [64–66]

$$F_n = \frac{(\text{SNR})_{\text{in}}}{(\text{SNR})_{\text{out}}}. \tag{2.2.8}$$

In this equation, SNR is not the optical SNR but refers to the *electric power* generated when the optical signal is converted into an electric current. In general, $F_n$ depends on several detector parameters that govern thermal noise associated with the detector. A simple expression for $F_n$ can be obtained by considering an ideal detector whose performance is limited by shot noise only. The electrical SNR of the input signal is then given by [67]

$$(\text{SNR})_{\text{in}} = \frac{\langle I_d \rangle^2}{\sigma_s^2} = \frac{\langle R_d P_{\text{in}} \rangle^2}{\sigma_s^2}, \tag{2.2.9}$$

where $I_d$ is the current and $R_d = q/h\nu_0$ is the responsivity of a detector with 100% quantum efficiency. The variance of shot noise over the receiver bandwidth $\Delta f$ can be written as $\sigma^2 = 2q (R_d P_{\text{in}}) \Delta f$, resulting in an input SNR of $P_{\text{in}}/(2h\nu_0\Delta f)$.

To calculate the electrical SNR of the amplified signal, we should add the contribution of ASE to the receiver noise. When all noise sources are included, the detector current takes the form

$$I_d = R_d \left[ \left| \sqrt{G_L}\, E_s + E_{\text{cp}} \right|^2 + \left| E_{\text{op}} \right|^2 \right] + i_s + i_T, \tag{2.2.10}$$

where $i_s$ and $i_T$ are current fluctuations induced by shot and thermal noises, respectively, $E_s$ is the signal field, $E_{\text{cp}}$ is the part of ASE copolarized with the signal, and $E_{\text{op}}$ is its orthogonally polarized part. It is necessary to separate the ASE into two parts because only its copolarized part can beat with the signal. Since ASE occurs over a broader bandwidth than the signal bandwidth $\Delta v_s$, it is common to divide the ASE bandwidth $B_{opt}$ into $M$ bins, each of bandwidth $\Delta v_s$, and write $E_{cp}$ in the form [64–66]

$$E_{cp} = \sum_{m=1}^{M} \sqrt{S_{ASE}\Delta v_s} \exp{(\phi_m - i\omega_m t)}, \qquad (2.2.11)$$

where $\phi_m$ is the phase of the noise component at the frequency $\omega_m = \omega_f + m(2\pi\Delta v_s)$. An identical form should be used for $E_{op}$.

Using $E_s = \sqrt{P_{in}} \exp{(\phi_s - i\omega_s t)}$ in Eq. (2.2.10) and including all beating terms, the current $I_d$ can be written in the form

$$I_d = R_d G_L P_{in} + i_b + i_{ASE} + i_s + i_T, \qquad (2.2.12)$$

where $i_b$ and $i_{ASE}$ represent current fluctuations resulting from signal–ASE and ASE–ASE beating, respectively, and are given by

$$i_b = 2R_d (G_L P_{in} S_{ASE}\Delta v_s)^{1/2} \sum_{m=1}^{M} \cos{[(\omega_s - \omega_m)t + \phi_m - \phi_s]}, \qquad (2.2.13)$$

$$i_{ASE} = 2R_d S_{ASE}\Delta v_s \sum_{m=1}^{M} \sum_{n=1}^{M} \cos{[(\omega_n - \omega_m)t + \phi_m - \phi_n]}. \qquad (2.2.14)$$

Since these two noise terms fluctuate with time, we need to find their variances. As details are available in several texts, we write the final result directly [64–66]:

$$\sigma_b^2 = 4R_d^2 G_L P_{in} S_{ASE}\Delta f, \qquad (2.2.15)$$

$$\sigma_{ASE}^2 = 4R_d^2 S_{ASE}^2 \Delta f \left(B_{opt} - \Delta f/2\right). \qquad (2.2.16)$$

The total variance of current fluctuations can be written from Eq. (2.2.12) as

$$\sigma^2 = \sigma_b^2 + \sigma_{ASE}^2 + G_L\sigma_s^2 + \sigma_T^2. \qquad (2.2.17)$$

We can neglect the thermal-noise contribution $\sigma_T^2$ as it is relatively small. The $\sigma_{ASE}^2$ term is also small in comparison with $\sigma_b^2$. For this reason, the electrical SNR of the amplified signal is approximately given by

$$(\text{SNR})_{\text{out}} = \frac{(R_d G_L P_{\text{in}})^2}{G_L \sigma_s^2 + \sigma_b^2}. \tag{2.2.18}$$

The noise figure can now be obtained by substituting Eqs. (2.2.9) and (2.2.18) in Eq. (2.2.8). The result is found to be

$$F_n = \frac{1}{G_L} \left( 1 + \frac{\sigma_b^2}{G_L \sigma_s^2} \right). \tag{2.2.19}$$

Using $\sigma_b^2$ from Eq. (2.2.15) and $S_{\text{ASE}}$ from Eq. (2.2.5), the noise figure becomes

$$F_n = 2n_{\text{sp}} g_R \int_0^L \frac{P_p(z)}{G(z)} dz + \frac{1}{G_L}. \tag{2.2.20}$$

This equation shows that the noise figure of a Raman amplifier depends on the pumping scheme. It provides reasonably small noise figures for "lumped" Raman amplifiers for which fiber length is $\sim 1$ km and the net signal gain exceeds 10 dB.

When the fiber within the transmission link itself is used for distributed amplification, the length of the fiber section typically exceeds 50 km, and pumping is such that net gain $G(z) < 1$ throughout the fiber length. In this case, $F_n$ predicted by Eq. (2.2.20) can be very large and exceed 15 dB depending on the span length. This does not mean distributed amplifiers are noisier than lumped amplifiers. To understand this apparent contradiction, consider a 100-km-long fiber span with a loss of 0.2 dB/km. The 20-dB span loss is compensated using a hybrid scheme in which a lumped amplifier with 5-dB noise figure is combined with the Raman amplification through backward pumping. The on–off gain $G_A$ of the Raman amplifier can be varied in the range of 0–20 dB by adjusting the pump power. Clearly, $G_A = 0$ and 20 dB correspond to the cases of pure lumped and distributed amplifications, respectively.

The solid line in Figure 2.12 shows how the noise figure of such a hybrid amplifier changes as $G_A$ is varied from 0 to 20 dB [62]. When $G_A = 0$, Eq. (2.2.20) shows that the passive fiber has a noise figure of 20 dB. This is not surprising since any fiber loss reduces signal power and thus degrades

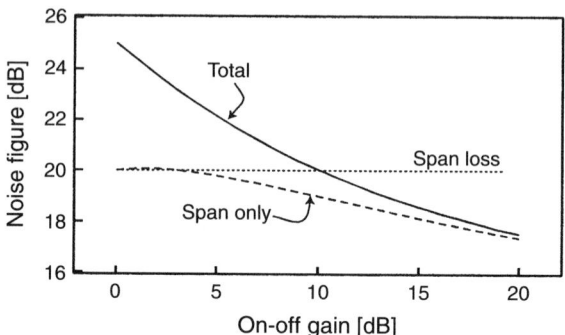

**Figure 2.12:** Total noise figure as a function of the on–off gain of a Raman amplifier when a 20-dB loss of a 100-km-long fiber span is compensated using a hybrid-amplification scheme. Dashed line shows the noise figure of Raman-pumped fiber span alone. Dotted line shows 20-dB span loss. (After Ref. [62]; © 2004 IEEE.)

the SNR [56]. When the signal is amplified by the lumped amplifier, an additional 5-dB degradation occurs, resulting in a total noise figure of 25 dB in Figure 2.12. This value decreases as $G_A$ increases, reaching a level of about 17.5 dB for $G_A = 20$ dB (no lumped amplification). The dashed line shows the noise figure of the Raman-pumped fiber span alone as predicted by Eq. (2.2.20). The total noise figure is higher because the lumped amplifier adds some noise if span losses are only partially compensated by the Raman amplifier. The important point is that total noise figure drops below the 20-dB level (dotted line) when the Raman gain exceeds a certain value.

To emphasize the noise advantage of distributed amplifiers, it is common to introduce the concept of an *effective* noise figure using the definition $F_{eff} = F_o \exp(-\alpha L)$, where $\alpha$ is the fiber-loss parameter at the signal wavelength. In decibel units, $F_{eff} = F_n - \mathcal{L}$, where $\mathcal{L}$ is the span loss in dB. As seen from Figure 2.12, $F_{eff} < 1$ (or negative on the decibel scale) by definition. It is this feature of distributed amplification that makes it so attractive for long-haul WDM lightwave systems. For the example shown in Figure 2.12, $F_{eff} \approx -2.5$ dB when pure distributed amplification is employed. Note, however, the noise advantage is almost 7.5 dB when the noise figures are compared for lumped and distributed amplification schemes.

As seen from Eq. (2.2.20), the noise figure of a Raman amplifier depends on the pumping scheme used because, as discussed in Section 2.1, $P_p(z)$ can be quite different for forward, backward, and bidirectional pumping. In general, forward pumping provides the highest SNR, and the smallest noise figure, because most of the Raman gain is then concentrated toward the input end of the fiber where power levels are high. However, backward pumping is often employed in practice because of other considerations such as the transfer of pump noise to signal.

## 2.2.2 Rayleigh Backscattering

The phenomenon that limits the performance of distributed Raman amplifiers most turned out to be Rayleigh backscattering [68–73]. Rayleigh backscattering occurs in all fibers and is the fundamental loss mechanism for them. Although most of the scattered light escapes through the cladding, a part of backscattered light can couple into the core mode supported by a single-mode fiber. Normally, this backward-propagating noise is negligible because its power level is smaller by more than 40 dB compared with the forward-propagating signal power. However, it can be amplified over long lengths in fibers with distributed Raman gain.

Rayleigh backscattering affects the performance of Raman amplifiers in two ways. First, a part of backward-propagating ASE can appear in the forward direction, enhancing the overall noise. This noise is relatively small and is not of much concern for Raman amplifiers. Second, *double Rayleigh backscattering* of the signal creates a cross-talk component in the forward direction that has nearly the same spectral range as the signal (in-band cross talk). It is this Rayleigh-induced noise, amplified by the distributed Raman gain, that becomes the major source of power penalty in Raman-amplified lightwave systems. Figure 2.13 shows schematically the phenomenon of double Rayleigh backscattering of the signal. Density fluctuations at the location $z_2$ inside the transmission fiber reflect a small portion of the signal through Rayleigh backscattering. This backward-propagating field is reflected a second time by density fluctuations at the location $z_1$ and thus ends up propagating in the same direction as the signal. Of course, $z_1$ and $z_2$ can vary over the entire length of the fiber, and the total noise is obtained by summing over all possible paths. For this reason, the Rayleigh noise is also referred to as originating from *multiple-path interference* [57].

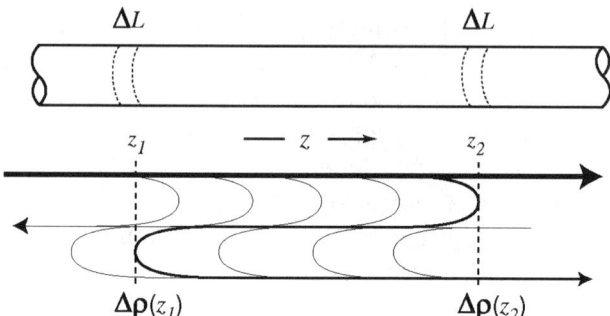

**Figure 2.13:** Schematic illustration of double Rayleigh backscattering in distributed Raman amplifiers. The signal is first reflected at $z_2$ and then again at $z_1$ through local density fluctuations $\Delta\rho$ at these two locations. (After Ref. [57]; © 2003 Springer.)

The calculation of the noise produced by double Rayleigh backscattering is somewhat complicated because it should include not only the statistical nature of density fluctuations but also any depolarizing effects induced by birefringence fluctuations. If we ignore the depolarizing effects, the Rayleigh-noise field can be written as [57]

$$a_r = \int_0^L z_2 \int_0^{z_2} dz_1 r(z_1) r(z_2) G(z_1, z_2)$$
$$\times A_s \left[ L, t - 2(z_2 - z_1)/v_g \right] \exp \left[ 2ik_s(z_2 - z_1) \right], \qquad (2.2.21)$$

where $v_g$ is the group velocity, $k_s$ is the propagation constant at the signal frequency, and $G(z_1, z_2)$ is the gain experienced in the fiber section between $z_1$ and $z_2$. From Eq. (2.2.3), this gain can be written as

$$G(z_1, z_2) = \exp \left( \int_{z_1}^{z_2} \left[ g_R P_p(z) - \alpha_s \right] dz \right). \qquad (2.2.22)$$

The random variable $r(z)$ in Eq. (2.2.21) represents the Rayleigh backscatter coefficient at the location $z$ and is related to density fluctuations occurring within the fiber. It is common to assume that $r(z)$ is a Markovian Gaussian stochastic process whose odd moments vanish. Moreover, its second moments can be expressed in terms of the second moment given by

$$\langle r^*(z_1)\, r(z_2) \rangle = \alpha_r f_r \delta(z_1 - z_2), \qquad \langle r(z_1)\, r(z_2) \rangle = 0 \qquad (2.2.23)$$

where $\alpha_r$ is the Rayleigh scattering loss and $f_r$ is the fraction that is recaptured by the fiber mode.

Even though the statistical properties of Rayleigh-induced noise require the calculation of the moments of $a_r$ in Eq. (2.2.21), it is relatively easy to calculate the fraction of *average* signal power that ends up propagating in the forward direction after double Rayleigh scattering. In a simple approach, one assumes that depletion of the signal through Rayleigh scattering is negligible and supplements the signal equation (2.1.9) with two more equations of the form

$$-\frac{dP_1}{dz} = \left[ g_R P_p(z) - \alpha_s \right] P_1 + f_r \alpha_r P_s, \qquad (2.2.24)$$

$$\frac{dP_2}{dz} = \left[ g_R P_p(z) - \alpha_s \right] P_2 + f_r \alpha_r P_1, \qquad (2.2.25)$$

where $P_1$ and $P_2$ represent the average power levels of the noise components created through single and double Rayleigh backscattering, respectively.

The three equations, Eqs. (2.1.9), (2.2.24), and (2.2.25), need to be integrated over the fiber length to find the fraction $f_{DRS}$ that ends up coming out with the signal at the output end at $z = L$. The signal equation has already been integrated and has the solution $P_s(z) = G(z) P_s(0)$, where $G(z)$ is given in Eq. (2.2.3). If we introduce the new variables using $P_1 = Q_1/G(z)$ and $P_2 = G(z) Q_2$, Eqs. (2.2.24) and (2.2.25) reduce to

$$-\frac{dQ_1}{dz} = f_r \alpha_r G^2(z) P_s(0), \qquad \frac{dQ_2}{dz} = f_r \alpha_r G^{-2}(z) Q_1(z). \quad (2.2.26)$$

These equations can be readily integrated to obtain

$$Q_2(L) = f_r \alpha_r \int_0^L G^{-2}(z) Q_1(z)\, dz, \qquad (2.2.27)$$

$$Q_1(z) = f_r \alpha_r \int_z^L G^2(z) P_s(0)\, dz, \qquad (2.2.28)$$

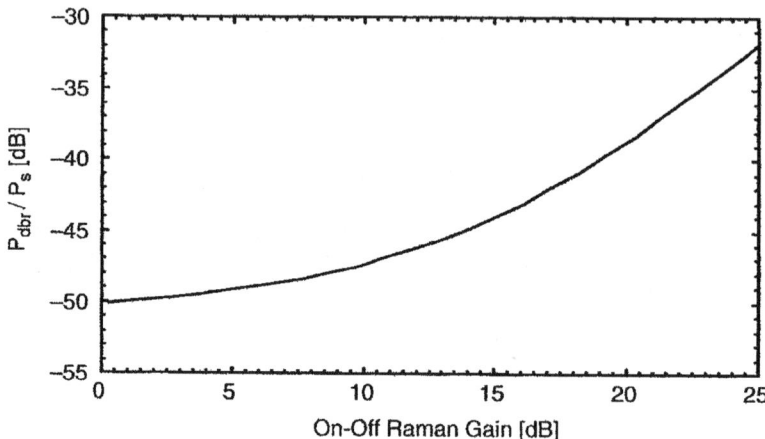

**Figure 2.14:** Fraction of signal power that is converted to noise by double Rayleigh backscattering plotted as a function of Raman gain for a 100-km-long backward-pumped distributed Raman amplifier. (After Ref. [56]; © 2002 Elsevier.)

The fraction of input power that ends up coming out with the signal at the output end is thus given by [70]

$$f_{\text{DRS}} = \frac{P_2(L)}{P_s(L)} = (f_r \alpha_r)^2 \int_0^L dz\, G^{-2}(z) \int_z^L G^2\left(z'\right) dz'. \quad (2.2.29)$$

Figure 2.14 shows how $f_{\text{DRS}}$ increases with the on–off Raman gain $G_A$ using $f_r \alpha_r = 10^{-4}$, $g_R = 0.7$ W$^{-1}$/km, $\alpha_s = 0.2$ dB/km, $\alpha_p = 0.25$ dB/km, and backward pumping for a 100-km-long Raman amplifier. The crosstalk begins to exceed the $-35$-dB level for the 20-dB Raman gain needed for compensating fiber losses. Since this crosstalk accumulates over multiple amplifiers, it can lead to large power penalties in long-haul lightwave systems.

A simple way to estimate the degradation induced by double Rayleigh backscattering is to consider how the noise figure of a Raman amplifier increases because of it. The noise field of power $P_2(L)$ is incident on the photodetector together with the signal field with power $P_s(L)$. Similar to the case of spontaneous Raman scattering, the two fields beat and produce a noise component in the receiver current of the form

$$\Delta I_r = 2R_d \, (G_L P_{in}) \, f_{DRS}^{1/2} \cos \theta, \qquad (2.2.30)$$

and $\theta$ is a rapidly varying random phase. If $\sigma_r^2$ represents the variance of this current noise, we should add it to the other noise sources and replace Eq. (2.2.18) with

$$(SNR)_{out} = \frac{(R_d G_L P_{in})^2}{\left( G_L \sigma_s^2 + \sigma_b^2 + \sigma_r^2 \right)}, \qquad (2.2.31)$$

where $\sigma_r^2 = 2 f_{DRS} \, (R_d G_L P_{in})^2$ after using $\langle \cos^2 \theta \rangle = \frac{1}{2}$.

The noise figure of a Raman amplifier can now be calculated following the procedure discussed earlier. In place of Eq. (2.2.19), we obtain

$$F_n = \frac{1}{G_L} \left( 1 + \frac{\sigma_b^2}{G_L \sigma_s^2} + \frac{\sigma_r^2}{G_L \sigma_s^2} \right), \qquad (2.2.32)$$

where the last term represents the contribution of double Rayleigh backscattering. This contribution has been calculated including the depolarization effects as well as the statistical nature of Rayleigh backscattering [57]. In the case of a Gaussian pulse, Gaussian-shaped filters, and a relatively large optical bandwidth, it is found to be

$$\frac{\sigma_r^2}{G_L \sigma_s^2} = \frac{5 P_{in} f_{DRS}}{9 h \nu_o G_L \Delta f}, \qquad (2.2.33)$$

where $P_{in}$ is the input channel power, $f_{DRS}$ is the fraction given in Eq. (2.2.29), and we assumed that signal bandwidth $\Delta \nu_s \approx \Delta f$. As an example, for $P_{in} = 1$ mW, $f_{DRS} = 10^{-4}$, and $\Delta f = 40$ GHz, the factor of 8 is large enough to increase the noise figure.

## 2.2.3 Pump-Noise Transfer

All lasers exhibit some intensity fluctuations. The situation is worse for semiconductor lasers used for pumping a Raman amplifier because the level of power fluctuations in such lasers can be relatively high owing to their relatively small size and a large rate of spontaneous emission. As seen from Eq. (2.1.14), the gain of a Raman amplifier depends on the pump power exponentially. It is intuitively expected from this equation that any fluctuation in the pump power would be magnified and result in even larger fluctuations in the amplified signal power. This is indeed the case, and this source of noise is known as pump-noise transfer. Details of the noise transfer depend on many factors including amplifier length, pumping scheme, and the dispersion characteristics of the fiber used for making the Raman amplifier.

Power fluctuations of a semiconductor laser are quantified through a frequency-dependent quantity called the *relative intensity noise* (RIN). It represents the spectrum of intensity or power fluctuations and is defined as

$$\frac{\sigma_p^2}{\langle P_0 \rangle^2} = \int_0^\infty \text{RIN}_p \, (f) \, df, \tag{2.2.34}$$

where $\sigma_p^2$ is the variance of pump-power fluctuations and $\langle P_0 \rangle$ is the average pump power. When pump noise is included, the amplified signal also exhibits fluctuations, and one can introduce the RIN of the amplified signal as

$$\frac{\sigma_s^2}{\langle P_s(L) \rangle^2} = \int_0^\infty \text{RIN}_s \, (f) \, df. \tag{2.2.35}$$

The pump-noise transfer function represents the enhancement in the signal noise at a specific frequency $f$ and is defined as

$$H \, (f) = \text{RIN}_s \, (f) \, / \text{RIN}_p \, (f) . \tag{2.2.36}$$

To calculate the noise transfer function $H(f)$, one must include the fact that the pump and signal do not travel along the fiber at the same speed because of their different wavelengths. The speed difference depends on the dispersion of the fiber. For this reason, the pump-noise transfer process depends on the pumping scheme, the amplifier length, and the dispersion parameter $D$ of the fiber used to make the amplifier. Mathematically, Eqs. (2.1.6)

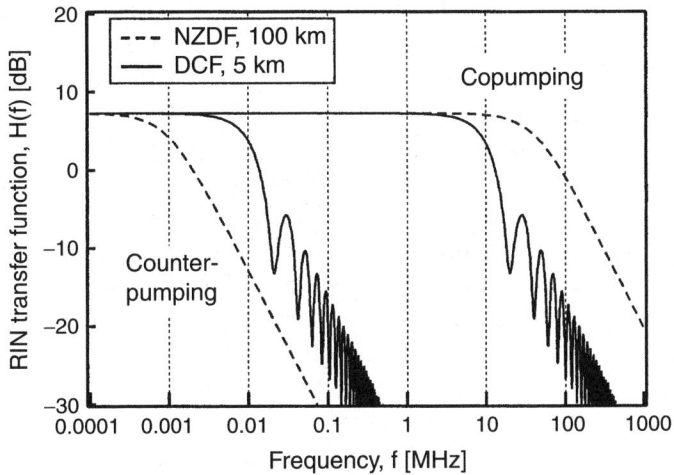

**Figure 2.15:** Calculated RIN transfer function for two types of Raman amplifiers when pump and signal copropagate or counterpropagate. (After Ref. [62]; © 2004 IEEE.)

and (2.1.7) are modified to include an additional term and take the form

$$\frac{\partial P_s}{\partial z} + \frac{1}{v_{gs}}\frac{\partial P_s}{\partial t} = g_R P_p P_s - \alpha_s P_s, \tag{2.2.37}$$

$$\xi\frac{\partial P_p}{\partial z} + \frac{1}{v_{gp}}\frac{\partial P_p}{\partial t} = -\frac{\omega_p}{\omega_s}g_R P_p P_s - \alpha_p P_p, \tag{2.2.38}$$

where $v_{gs}$ and $v_{gp}$ are the group velocities for the signal and pump, respectively. These equations can be solved with some reasonable approximations [74] to calculate the noise transfer function $H(f)$.

Figure 2.15 compares the calculated frequency dependence of $H(f)$ in the forward- and backward-pumping configurations for two types of Raman amplifiers pumped at 1450 nm to provide gain near 1550 nm [62]. Solid curves are for a discrete Raman amplifier made with a 5-km-long DCF with $\alpha_p = 0.5$ dB/km and a dispersion of $D = -100$ ps/(km-nm) at 1.55 μm. Dashed curves are for a distributed Raman amplifier made with a 100-km-long NZDF with $\alpha_p = 0.25$ dB/km and a dispersion of 4.5 ps/(km-nm) at 1.55 μm. The on–off Raman gain is 10 dB in all cases.

In the case of forward pumping, RIN is enhanced by a factor of nearly 6 in the frequency range of 0–5 MHz for the DCF and the frequency range increases to 50 MHz for the NZDF because of its relatively low value of $D$. However, when backward pumping is used, RIN is actually reduced at all frequencies except for frequencies below a few kilohertz. Notice that $H(f)$ exhibits an oscillatory structure in the case of a DCF. This feature is related to a relatively short length of discrete amplifiers.

In general, the noise enhancement is relatively small in the backward-pumping configuration and for large values of $D$ [74]. This behavior can be understood physically as follows. The distributed Raman gain builds up as the signal propagates inside the fiber. In the case of forward pumping and low dispersion, pump and signal travel at nearly the same speed. As a result, any fluctuation in the pump power stays in the same temporal window of the signal. In contrast, when dispersion is large, the signal moves out of the temporal window associated with the fluctuation and sees a somewhat averaged gain. The averaging is much stronger in the case of backward pumping because the relative speed is extremely large (twice that of the signal group velocity). In this configuration, the effects of pump-power fluctuations are smoothed out so much that almost no RIN enhancement occurs. For this reason, backward pumping is often used in practice even though the noise figure is larger for this configuration. Forward pumping can only be used if fiber dispersion is relatively large and pump lasers with low RIN are employed.

## 2.3   Effects of Polarization-Mode Dispersion

In the scalar approach used so far, it has been implicitly assumed that both pump and signal are copolarized and maintain their state of polarization (SOP) inside the fiber. However, unless a special kind of polarization-maintaining fiber is used for making Raman amplifiers, residual fluctuating birefringence of most fibers changes the SOP of any optical field in a random fashion and also leads to PMD, a phenomenon that has been studied extensively in recent years [75–79]. Indeed, the effects of PMD on Raman amplification have been observed in several experiments [80–83]. In this section we develop a vector theory of Raman amplification capable of including the PMD effects [84–86]. It turns out that the amplified signal fluctuates over a wide range if PMD changes with time, and the average gain is significantly lower than that expected in the absence of PMD.

## 2.3.1 Vector Theory of Raman Amplification

The third-order nonlinear polarization induced in a medium such as silica glass in response to an optical field $E(r, t)$ can be written in the form [87]

$$P^{(3)}(r, t) = \frac{\varepsilon_0}{2} \sigma [E(r, t) \cdot E(r, t)] E(r, t)$$

$$+ \varepsilon_0 E(r, t) \int_{-\infty}^{t} a(t - \tau) E(r, \tau) d\tau$$

$$+ \varepsilon_0 E(r, t) \cdot \int_{-\infty}^{t} b(t - \tau) E(r, \tau) d\tau, \qquad (2.3.1)$$

where $a(t)$ and $b(t)$ govern the Raman response (related to nuclear motion) while $\sigma$ accounts for the instantaneous electronic response of the nonlinear medium. Also, $\varepsilon_0$ is the vacuum permittivity.

In a Raman amplifier, the pump and signal waves propagate simultaneously, and the total field is given by

$$E = \mathrm{Re}\left[ E_p \exp(-i\omega_p t) + E_s \exp(-i\omega_s t) \right], \qquad (2.3.2)$$

where $E_p$ and $E_s$ are the slowly varying envelope for the pump and signal fields oscillating at frequencies $\omega_p$ and $\omega_s$, respectively. Writing $P^{(3)}$ in the same form as

$$P^{(3)} = \mathrm{Re}\left[ P_p^{(3)} \exp(-i\omega_p t) + P_s^{(3)} \exp(-i\omega_s t) \right], \qquad (2.3.3)$$

and using Maxwell's equations, the slowly varying fields are found to satisfy

$$\left( \nabla^2 + \frac{\omega_j^2}{c^2} \overleftrightarrow{\varepsilon}_j \right) E_j = \frac{1}{\varepsilon_0 c^2} P_j^{(3)}, \qquad (2.3.4)$$

where $j = p$ or $s$ and $\overleftrightarrow{\varepsilon}$ represents the linear part of the dielectric constant.

Birefringence fluctuations responsible for PMD are included through the tensorial nature of $\overset{\leftrightarrow}{\varepsilon}$. Writing it in the basis of the Pauli matrices as [77]

$$\left(\omega_j^2/c^2\right) \overset{\leftrightarrow}{\varepsilon}_j = \left(k_j + i\alpha_j/2\right)^2 \sigma_0 - k_j\omega_j\boldsymbol{\beta} \cdot \boldsymbol{\sigma}, \qquad (2.3.5)$$

where $\alpha_j$ accounts for fiber losses, $\sigma_0$ is a unit matrix and the vector $\boldsymbol{\beta}$ governs the local value of fiber birefringence. The vector $\boldsymbol{\sigma}$ is formed using the Pauli matrices as $\boldsymbol{\sigma} = \sigma_1\hat{e}_1 + \sigma_2\hat{e}_2 + \sigma_3\hat{e}_3$, where $\hat{e}_1$, $\hat{e}_2$, and $\hat{e}_3$ are the three unit vectors in the Stokes space and

$$\sigma_1 = \begin{pmatrix} 1 & 0 \\ 0 & -1 \end{pmatrix}, \quad \sigma_2 = \begin{pmatrix} 0 & 1 \\ 1 & 0 \end{pmatrix}, \quad \sigma_3 = \begin{pmatrix} 0 & -i \\ i & 0 \end{pmatrix}. \qquad (2.3.6)$$

In the Jones-matrix notation, the pump and signal fields at any point $r$ inside the fiber can be written as

$$E_p(r) = F_p(x, y)|A_p\rangle e^{ik_p z}, \quad E_s(r) = F_s(x, y)|A_s\rangle e^{ik_s z}, \qquad (2.3.7)$$

where $F_p(x, y)$ and $F_s(x, y)$ represent the fiber-mode profile, $k_p$ and $k_s$ are propagation constants, and the Jones vectors $|A_p\rangle$ and $|A_s\rangle$ are two-dimensional column vectors representing the two components of the electric field in the $x-y$ plane. Since $F_p$ and $F_s$ do not change with $z$, we only need to consider the evolution of $|A_p\rangle$ and $|A_s\rangle$ along the fiber. After some algebra, they are found to satisfy the following vector equation [86]:

$$\frac{d|A_j\rangle}{dz} = -\frac{\alpha_j}{2}|A_j\rangle - \frac{i}{2}\omega_j\boldsymbol{\beta} \cdot \boldsymbol{\sigma}|A_j\rangle$$
$$+ \frac{i\gamma_{jj}}{3}\left[2\langle A_j \mid A_j\rangle + (\kappa_b/\kappa_a)|A_j^*\rangle\langle A_j^*|\right]|A_j\rangle$$
$$+ \frac{2i\gamma_{jm}}{3}\left[(1 + \delta_b)\langle A_m \mid A_m\rangle + (1 + \delta_a)|A_m\rangle\langle A_m|\right.$$
$$+ (\kappa_b/\kappa_a + \delta_b)|A_m^*\rangle\langle A_m^*|\right]|A_j\rangle$$
$$+ \frac{\zeta}{2}g_R\left[|A_m\rangle\langle A_m| + \mu\left(\langle A_m \mid A_m\rangle + |A_m^*\rangle\langle A_m^*|\right)\right]|A_j\rangle, \qquad (2.3.8)$$

where $j, m = p$ or $s$ with $m \neq j$ and $\zeta = 1$ when $j = s$ but $\zeta = -\omega_p/\omega_s$ when $j = p$. The PMD effects are included through the birefringence vector $\boldsymbol{\beta}$.

The Raman gain $g_R$ and the parameter $\mu$ in Eq. (2.3.8) are defined as [87]

$$g_R = \frac{\omega_s^2 \omega_p \text{Im}\left[\tilde{a}\left(\Omega\right)\right]}{c^4 k_p k_s A_{\text{eff}}}, \quad \mu = \frac{\text{Im}\left[\tilde{b}\left(\Omega\right)\right]}{\text{Im}\left[\tilde{a}\left(\Omega\right)\right]}. \tag{2.3.9}$$

Physically, $\mu$ represents the ratio of the Raman gain for copolarized and orthogonally polarized pumps, respectively. As seen in Figure 2.3, $\mu$ is much smaller than 1 for optical fibers [17]. It has a value of about 0.012 for silica fibers near the Raman gain peak [87].

The nonlinear effects in Eq. (2.3.8) are governed by the parameters

$$\gamma_{jm} = 3\omega_j^2 \omega_m \kappa_a / \left(8c^4 k_j k_m A_{\text{eff}}\right), \tag{2.3.10}$$

$$\kappa_a = \sigma + 2\tilde{a}\left(0\right) + \tilde{b}\left(0\right), \quad \kappa_b = \sigma + 2\tilde{b}\left(0\right), \tag{2.3.11}$$

$$\delta_a = 2\left\{\text{Re}\left[\tilde{a}\left(\Omega_R\right)\right] - \tilde{a}\left(0\right)\right\}/\kappa_a, \tag{2.3.12}$$

$$\delta_b = \left\{\text{Re}\left[\tilde{b}\left(\Omega_R\right)\right] - \tilde{b}\left(0\right)\right\}/\kappa_a, \tag{2.3.13}$$

where $\Omega = \omega_p - \omega_s$ is the Raman shift and $\tilde{a}\left(\omega\right)$ and $\tilde{b}\left(\omega\right)$ are the Fourier transforms of the Raman response functions $a\left(\tau\right)$ and $b\left(\tau\right)$, respectively. The nonlinear parameters $\gamma_{\text{ps}}$ and $\gamma_{\text{sp}}$ are responsible for cross-phase modulation (XPM).

Equation (2.3.8) looks complicated in the Jones-matrix formalism. It can be simplified considerably by writing it in the Stokes space [77]. After introducing the Stokes vectors for the pump and signal as

$$P = \langle A_p | \sigma | A_p \rangle \equiv P_1 \hat{e}_1 + P_2 \hat{e}_2 + P_3 \hat{e}_3, \tag{2.3.14}$$

$$S = \langle A_s | \sigma | A_s \rangle \equiv S_1 \hat{e}_1 + S_2 \hat{e}_2 + S_3 \hat{e}_3, \tag{2.3.15}$$

we obtain the following two vector equations governing the dynamics of $P$ and $S$:

$$\xi \frac{dP}{dz} = -\frac{\omega_p g_R}{2\omega_s}\left[(1 + 3\mu)\, P_s P + (1 + \mu)\, P_p S - 2\mu P_p S_3\right]$$
$$- \alpha_p P + \left(\omega_p \beta + W_p\right) \times P, \tag{2.3.16}$$

$$\frac{dS}{dz} = -\frac{g_R}{2} \left[ (1 + 3\mu) \, P_p S + (1 + \mu) \, P_s P - 2\mu P_s S_3 \right]$$

$$- \alpha_s S + (\omega_s \boldsymbol{\beta} + W_s) \times S, \qquad (2.3.17)$$

where $P_p = |P|$ and $P_s = |S|$ are the pump and signal powers and

$$W_p = \frac{2}{3} \left[ \gamma_{pp} P_3 + 2\gamma_{ps} \, (1 + \delta_b) \, S_3 - \gamma_{ps} \, (2 + \delta_a + \delta_b) \, S \right], \qquad (2.3.18)$$

$$W_s = \frac{2}{3} \left[ \gamma_{ss} S_3 + 2\gamma_{sp} \, (1 + \delta_b) \, P_3 - \gamma_{sp} \, (2 + \delta_a + \delta_b) \, P \right]. \qquad (2.3.19)$$

The vectors $W_p$ and $W_s$ account for the SPM- and XPM-induced nonlinear polarization rotation. We have also added $\xi \equiv \pm 1$ to the pump equation to allow for the possibility of backward pumping.

Because of fiber birefringence, both $P$ and $S$ rotate on the Poincaré sphere around the same axis but with different rates. However, the SRS process depends only on the relative orientation of $P$ and $S$. To simplify the following analysis, we choose to work in a rotating frame in which the pump Stokes vector $P$ is not affected by birefringence. Moreover, the beat length of residual birefringence ($\sim 1$ m) and its correlation length ($\sim 10$ m) are typically much smaller than the nonlinear length scale. As a result, rotations of Stokes vectors induced by fiber birefringence are so fast that one can average over them. The averaged equations are found to be [86]

$$\xi \frac{dP}{dz} = -\frac{\omega_p g_R}{2\omega_s} \left[ (1 + 3\mu) \, P_s P + (1 + \mu/3) \, P_p S \right]$$

$$- \alpha_p P - \varepsilon_{ps} S \times P, \qquad (2.3.20)$$

$$\frac{dS}{dz} = \frac{g_R}{2} \left[ (1 + 3\mu) \, P_p S + (1 + \mu/3) \, P_s P \right]$$

$$- \alpha_s S - \left( \Omega B + \varepsilon_{sp} P \right) \times S, \qquad (2.3.21)$$

where $B$ is related to $\boldsymbol{\beta}$ through a rotation on the Poincaré sphere and

$$\varepsilon_{jm} = 2\gamma_{jm} \, (4 + 3\delta_a + \delta_b) \, /9, \qquad \Omega = \xi \omega_p - \omega_s. \qquad (2.3.22)$$

Equations (2.3.20) and (2.3.21) describe Raman amplification under quite general conditions. We make two further simplifications and neglect both the pump depletion and the signal-induced XPM because the pump power is much larger than the signal power in practice. The pump equation (2.3.20) then contains only the loss term and can be easily integrated. The effect of fiber losses is to reduce the magnitude of $\boldsymbol{P}$ but the direction of $\boldsymbol{P}$ remains fixed in the rotating frame. In the case of forward pumping, the solution is

$$\boldsymbol{P}(z) = \hat{\boldsymbol{p}} P_{\text{in}} \exp(-\alpha_p z) \equiv \hat{\boldsymbol{p}} P_p(z), \qquad (2.3.23)$$

where $\hat{\boldsymbol{p}}$ is the input SOP of the pump.

The last term in Eq. (2.3.21) accounts for the pump-induced nonlinear polarization rotation and does not affect the Raman gain because of its deterministic nature. We can eliminate this term with a simple transformation to obtain

$$\frac{d\boldsymbol{S}}{dz} = \frac{g_R}{2} \left[ (1+3\mu) P_p \boldsymbol{S} + (1+\mu/3) P_s \boldsymbol{P} \right] - a_s \boldsymbol{S} - \Omega \boldsymbol{b} \times \boldsymbol{S}, \quad (2.3.24)$$

where $\boldsymbol{b}$ is related to $\boldsymbol{B}$ by a deterministic rotation. As optical fibers used for Raman amplification are much longer than the birefringence correlation length, $\boldsymbol{b}(z)$ can be modeled as a three-dimensional stochastic process whose first-order and second-order moments are given by

$$\langle \boldsymbol{b}\,(z) \rangle = 0, \quad \langle \boldsymbol{b}\,(z_1)\,\boldsymbol{b}\,(z_2) \rangle = \frac{1}{3} D_p^2 \overset{\leftrightarrow}{I} \delta\,(z_2 - z_1), \qquad (2.3.25)$$

where angle brackets denote an ensemble average, $\overset{\leftrightarrow}{I}$ is a unit tensor, and $D_p$ is the PMD parameter of the fiber. On physical grounds, we treat all stochastic differential equations in the Stratonovich sense [88].

Equation (2.3.24) can be further simplified by noting that two terms on its right side do not change the direction of $\boldsymbol{S}$ and can be removed with the transformation

$$\boldsymbol{S} = \boldsymbol{s} \exp \left\{ \int_0^z \left[ \frac{g_R}{2} (1+3\mu) P_p\,(z) - \alpha_s \right] dz \right\}. \qquad (2.3.26)$$

The Stokes vector $\boldsymbol{s}$ of the signal is found to satisfy [86]

$$\frac{d\boldsymbol{s}}{dz} = \frac{g_R}{2} (1+\mu/3) P_p\,(z)\, s_0 \hat{\boldsymbol{p}} - \Omega_R \boldsymbol{b} \times \boldsymbol{s}, \qquad (2.3.27)$$

where $s_0 = |\boldsymbol{s}|$ is the magnitude of vector $\boldsymbol{s}$.

Equation (2.3.27) applies for both the forward and the backward pumping schemes, but the $z$ dependence of $P_p(z)$ and the magnitude of $\Omega$ depend on the pumping configuration. More specifically, $\Omega = \omega_p - \omega_s$ in the case of forward pumping but it is replaced with $\Omega = -(\omega_p + \omega_s)$ in the case of backward pumping. In the absence of birefringence ($b = 0$), $s$ remains oriented along $\hat{p}$, and we recover the scalar case.

## 2.3.2   Average Raman Gain and Signal Fluctuations

Equation (2.3.27) can be used to calculate the power $P_s$ of the amplified signal as well as its SOP at any distance within the amplifier. In the presence of PMD, $b(z)$ fluctuates with time. As a result, the amplified signal $P_s(L)$ at the output of an amplifier also fluctuates. Such fluctuations would affect the performance of any Raman amplifier. In this section we calculate the average and the variance of such PMD-induced fluctuations. We focus on the forward-pumping case for definiteness.

The signal gain is defined as $G(L) = P_s(L)/P_s(0)$. Its average value, $G_{av}$, and variance of signal power fluctuations, $\sigma_s^2$, can be calculated using

$$G_{av} = \frac{\langle P_s(L) \rangle}{P_s(0)}, \qquad \sigma_s^2 = \frac{\langle P_s^2(L) \rangle}{\langle P_s(L) \rangle^2} - 1. \qquad (2.3.28)$$

To calculate the average signal power $\langle P_s(L) \rangle$ at the end of a Raman amplifier of length $L$, we introduce an angle $\theta$ as the relative angle between the pump-signal SOPs with the definition $s_0 \cos(\theta) = s \cdot \hat{p}$ and use (2.3.27) and a well-known technique [88] to obtain the following two coupled but deterministic equations [86]:

$$\frac{d \langle s_0 \rangle}{dz} = \frac{g_R}{2}(1 + \mu/3)P_p(z)\langle s_0 \cos \theta \rangle, \qquad (2.3.29)$$

$$\frac{d \langle s_0 \cos \theta \rangle}{dz} = \frac{g_R}{2}(1 + \mu/3)P_p(z)\langle s_0 \rangle - \eta \langle s_0 \cos \theta \rangle, \qquad (2.3.30)$$

where $\eta = 1/L_d = D_p^2 \Omega^2/3$, $L_d$ is the diffusion length, and $\theta$ is the angle between $P$ and $S$. The diffusion length is a measure of the distance after which the SOPs of the two optical fields, separated in frequency by $\Omega$, become decorrelated.

Equations (2.3.29) and (2.3.30) are two linear first-order differential equations that can be easily integrated. When PMD effects are large and $L_d \ll L$, $\langle s_0 \cos \theta \rangle$ reduces to zero over a short fiber section. The average gain in this case is given by

$$G_{\text{av}} = \exp \left[ \frac{1}{2} (1 + 3\mu) \, g_R \, P_{\text{in}} L_{\text{eff}} - \alpha_s L \right]. \tag{2.3.31}$$

Compared to the scalar case discussed in Section 2.1, PMD reduces the Raman gain coefficient by a factor of $(1 + 3\mu)/2$, exactly the average Raman gain coefficients in the copolarized and orthogonally polarized cases [12]. Since $\mu \ll 1$, $g_R$ is reduced by about a factor of 2. This is expected on physical grounds. It should however be stressed that the on–off Raman gain $G_A$ is reduced by a large factor when $g_R$ is halved because of the exponential relation between the two.

The variance of signal fluctuations requires the second-order moment $\langle P_s^2(L) \rangle$ of the amplified signal. Following the averaging procedure discussed in Ref. [86], Eq. (2.3.27) leads to the following set of three linear equations [88]:

$$\frac{d \langle s_0^2 \rangle}{dz} = g_R (1 + \mu/3) \, P_p(z) \langle s_0^2 \cos \theta \rangle, \tag{2.3.32}$$

$$\frac{d \langle s_0^2 \cos \theta \rangle}{dz} = -\eta \langle s_0^2 \cos \theta \rangle + \frac{g_R}{2} (1 + \mu/3) \, P_p(z) \left[ \langle s_0^2 \rangle + \langle s_0^2 \cos^2 \theta \rangle \right], \tag{2.3.33}$$

$$\frac{d \langle s_0^2 \cos^2 \theta \rangle}{dz} = -3\eta \langle s_0^2 \cos^2 \theta \rangle + \eta \langle s_0^2 \rangle + g_R (1 + \mu/3) \, P_p(z) \langle s_0^2 \cos \theta \rangle. \tag{2.3.34}$$

These equations show that signal fluctuations have their origin in fluctuations of the angle $\theta$ between the Stokes vectors associated with the pump and signal.

To illustrate the impact of PMD on the performance of Raman amplifiers, we focus on a 10-km-long Raman amplifier pumped with 1 W of power using a single 1.45-μm laser. The Raman gain coefficient $g_R = 0.6 \text{ W}^{-1}/\text{km}$ while $\mu = 0.012$ near the signal wavelength 1.55 μm. Fiber losses are taken to be 0.273 dB/km and 0.2 dB/km at the pump and signal wavelengths, respectively. Figure 2.16 shows how the average gain and $\sigma_s$ change with the PMD parameter $D_p$ when the input signal is copolarized (solid curves) or orthogonally polarized (dashed curves) with respect to the pump. The curves are shown for both the forward- and the backward-pumping schemes. When $D_p$ is zero, the two beams maintain their SOP, and the copolarized signal experiences a maximum gain of 17.6 dB while the orthogonally polarized signal has 1.7-dB loss, irrespective of the pumping configuration. The loss is not exactly 2 dB because a small gain exists for the orthogonally polarized input signal. As the PMD parameter increases, the gain difference between the copolarized and the orthogonally polarized cases decreases and disappears eventually.

The level of signal fluctuations in Figure 2.16 increases quickly with the PMD parameter, reaches a peak, and then decreases slowly to zero with further increase in $D_p$. The location of the peak depends on the pumping scheme as well as on the initial polarization of pump. The noise level can exceed 20% for $D_p = 0.05$ ps/$\sqrt{\text{km}}$ in the case of forward pumping. If a fiber with low PMD is used, the noise level can exceed 70% under some conditions. These results suggest that forward-pumped Raman amplifiers will perform better if a fiber with $D_p > 0.1$ ps/$\sqrt{\text{km}}$ is used. The curves for backward pumping are similar to those for forward pumping but shift to smaller $D_p$ values and have a higher peak. In spite of an enhanced peak, the backward pumping produces the smallest amount of fluctuations for all fibers for which $D_p > 0.01$ ps/$\sqrt{\text{km}}$. Note that the curves in the case of backward pumping are nearly identical to those for forward pumping except that they are shifted to left. As a result, the solid and dashed curves merge at a value of $D_p$ that is smaller by about a factor of 30. This difference is related to the definition of $\Omega = \xi\omega_p - \omega_s$ in Eq. (2.3.27). In the case of backward pumping, $|\Omega| = \omega_p + \omega_s$ is about 30 times larger than the value of $\Omega = \omega_p - \omega_s$ in the forward-pumping case.

In practice, fibers used to make a Raman amplifier have a constant value of $D_p$. Figure 2.17 shows the average Raman gain and $\sigma_s$ as a function of amplifier length for a fiber with $D_p = 0.05$ ps/$\sqrt{\text{km}}$. All other parameters are the same as in Figure 2.16. The solid and dashed lines correspond to copolarized and orthogonally polarized cases, respectively (the two lines are indistinguishable in the case of backward pumping). Physically, it takes some distance for the orthogonally polarized signal to adjust its SOP through PMD before it can experience the full Raman gain. Within the PMD

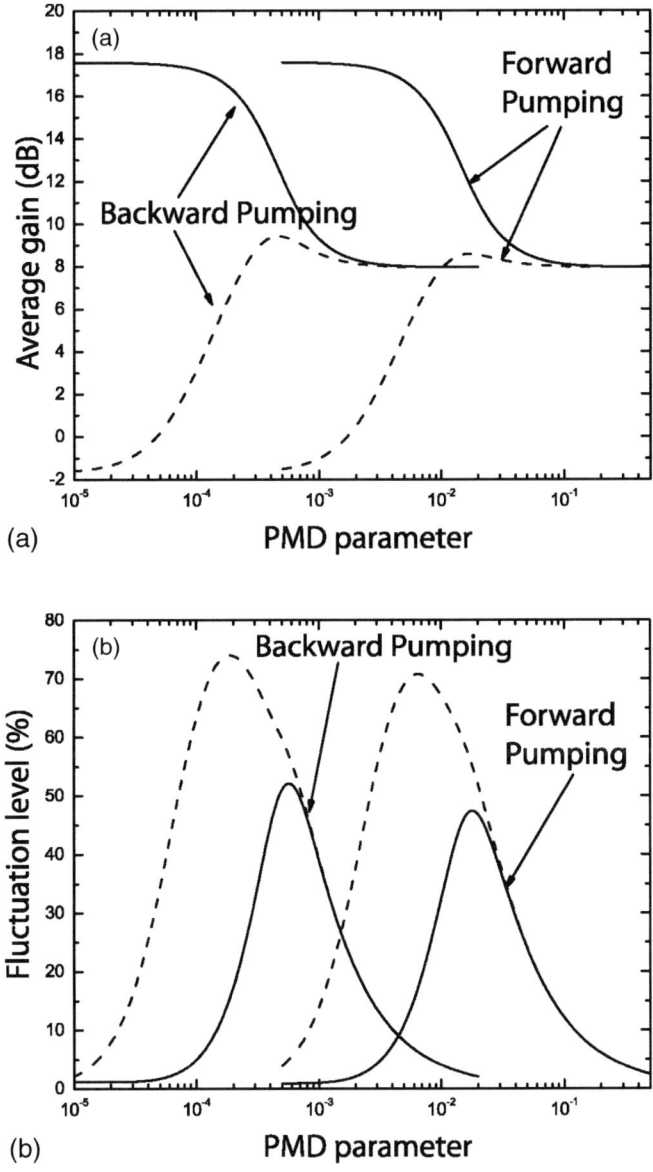

**Figure 2.16:** (a) Average gain and (b) standard deviation of signal fluctuations at the output of a Raman amplifier as a function of PMD parameter in the cases of forward and backward pumping. The solid and dashed curves correspond to the cases of copolarized and orthogonally polarized signals, respectively. (After Ref. [86]; © 2003 OSA.)

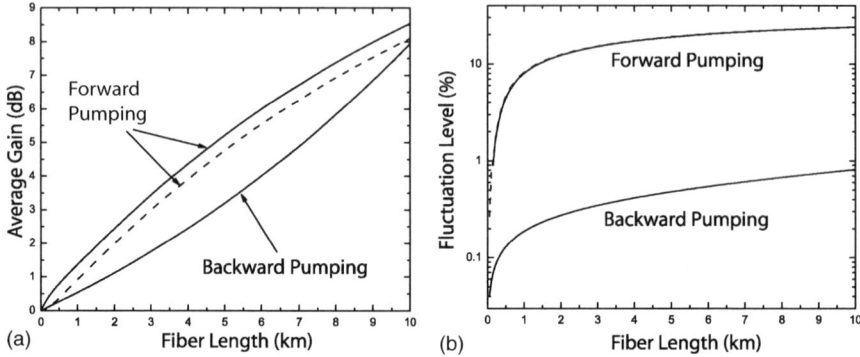

(a) Fiber Length (km)                    (b) Fiber Length (km)

**Figure 2.17:** (a) Average gain and (b) level of signal fluctuations as a function of amplifier length for a fiber with $D_p = 0.05 \text{ ps}/\sqrt{\text{km}}$. The solid and dashed curves correspond to the cases of copolarized and orthogonally polarized signal, respectively. The two curves nearly coincide in the case of backward pumping. (After Ref. [86]; © 2003 OSA.)

diffusion length (around 175 m in this case of forward pumping), fiber loss dominates and the signal power decreases; beyond the diffusion length, Raman gain dominates and the signal power increases. The gain difference seen in Figure 2.16 between the copolarized and the orthogonally polarized cases comes from this initial difference. In the case of backward pumping, the PMD diffusion length becomes so small (about 0.2 m) that this difference completely disappears. The level of signal fluctuations depends strongly on the relative directions of pump and signal propagation. In the case of forward pumping, $\sigma_s$ grows monotonically with the distance, reaching 24% at the end of the 10-km fiber. In contrast, $\sigma_s$ is only 0.8% even for a 10-km-long amplifier in the case of backward pumping, a value 30 times smaller than that occurring in the forward-pumping case.

## 2.3.3  Probability Distribution of Amplified Signal

The moment method used to obtain the average and variance of the amplified signal becomes increasingly complex for higher order moments. It is much more useful if one can determine the probability distribution of the amplified signal because it contains, by definition, all the statistical information. It turns out that it is possible to find this probability distribution in an analytic form [86].

The fluctuating amplifier gain $G(L)$ for a Raman amplifier of length $L$ can be found from Eqs. (2.3.26)–(2.3.27) as (in decibel units)

$$G_{dB} = a \left[ \frac{g_R}{2} (1 + 3\mu) P_{in} L_{eff} - \alpha_s L \right]$$
$$+ \frac{a}{2} g_R (1 + \mu/3) \int_0^L P_p(z) \left[ \hat{p}(z) \cdot \hat{s}(z) \right] dz, \qquad (2.3.35)$$

where $a = 10/\ln(10) = 4.343$ and $\hat{s}$ is the unit vector in the direction of $s$. Using $s = s_0 \hat{s}$ in Eq. (2.3.27), $\hat{s}$ is found to satisfy

$$\frac{d\hat{s}}{dz} = \frac{g_R}{2} P_p(z) \left[ \hat{p} - (\hat{p} \cdot \hat{s}) \hat{s} \right] - \Omega_R b \times \hat{s}. \qquad (2.3.36)$$

In this equation, $\hat{p}$ represents the pump polarization at the input end. When polarization scrambling is used to randomize $\hat{p}$, both $\hat{p}$ and $\hat{s}$ become random. However, it is only the scalar product $\hat{p} \cdot \hat{s} \equiv \cos\theta$ that determines $G_{dB}$.

To find the probability distribution of $G_{dB}$, we note that the integral in Eq. (2.3.35) can be written as a sum if we divide the fiber multiple segments. Clearly, the random variable $G_{dB}$ is formed from a sum of a large number of random variables with identical statistics. According to the central limit theorem [88], the probability density of $G(L)$ should be Gaussian, no matter what the statistics of $\hat{p}(z)$ and $\hat{s}(z)$ are, as long as the correlation between $\cos\theta(z)$ and $\cos\theta(z')$ goes to zero sufficiently rapidly as $|z - z'|$ increases. In practice, this correlation decays exponentially to zero over the PMD diffusion length $L_d$. For fiber lengths $L \gg L_d$, we thus expect $G_{dB}$ to follow a Gaussian distribution. Indeed, the Gaussian nature of $G_{dB}$ has been observed experimentally [89].

It is clear from Eq. (2.3.35) that $\ln [P_s(L)]$ will also follow a Gaussian distribution. As a result, the probability distribution of the amplified signal power, $P_s(L)$, at the amplifier output corresponds to a *lognormal*

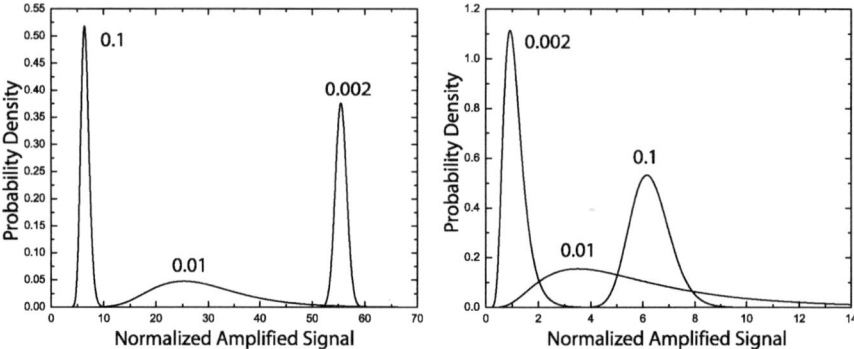

**Figure 2.18:** Probability density of amplified signal for three values of $D_p$ (in ps/$\sqrt{\text{km}}$) for copolarized (left) and orthogonally polarized (right) signals. The amplified signal is normalized to the input signal power. (After Ref. [86]; © 2003 OSA.)

*distribution* and can be written as

$$
p\left[P_s\left(L\right)\right] = \frac{\left[\ln\left(\sigma_s^2+1\right)\right]^{-1/2}}{P_s\left(L\right)\sqrt{2\pi}} \exp\left[-\frac{1}{2\ln\left(\sigma_s^2+1\right)}\right.
$$
$$
\left. \ln^2\left(\frac{P_s\left(L\right)\sqrt{\sigma_s^2+1}}{\langle P_s\left(L\right)\rangle}\right)\right], \qquad (2.3.37)
$$

where $\sigma_s^2$ is defined in Eq. (2.3.28). Figure 2.18 shows how the probability density changes with the PMD parameter in the cases of copolarized and orthogonally polarized input signals, respectively, under the conditions of Figure 2.16. When the PMD effects are relatively small, the two distributions are relatively narrow and are centered at quite different locations in the copolarized and orthogonally polarized cases. When $D_p = 0.01$ ps/$\sqrt{\text{km}}$, the two distributions broaden considerably and begin to approach each other. For larger values of the PMD parameter $D_p = 0.1$ ps/$\sqrt{\text{km}}$, they become narrow again and their peaks almost overlap because the amplified signal becomes independent of the input SOP.

## 2.3.4  Polarization-Dependent Gain

Similar to the concept of differential group delay used for describing the PMD effects on pulse propagation, the PMD effects in Raman amplifiers can be quantified using the concept of polarization-dependent gain (PDG), a quantity defined as the difference between the maximum and the minimum values of $G$ realized while varying the SOP of the input signal. The gain difference $\Delta = G_{max} - G_{min}$ is itself random because both $G_{max}$ and $G_{min}$ are random. It is useful to know the statistics of $\Delta$ and its relationship to the operating parameters of a Raman amplifier because they can identify the conditions under which PDG can be reduced to acceptable levels.

The polarization-dependent loss (PDL) is often described by introducing a PDL vector [90–92]. The same technique can be used for the PDG in Raman amplifiers. The PDG vector $\boldsymbol{\Delta}$ is introduced in such a way that its magnitude gives the PDG value $\Delta$ (in dB) but its direction coincides with the direction of $\boldsymbol{S}(L)$ for which the gain is maximum. Using Eq. (2.3.27), $\boldsymbol{\Delta}$ is found to satisfy the following linear Langevin equation (see Appendix C of Ref. [86] for details):

$$\frac{d\boldsymbol{\Delta}}{dz} = ag_R \boldsymbol{P} - \Omega \boldsymbol{b} \times \boldsymbol{\Delta}. \tag{2.3.38}$$

This equation can be readily solved because of its linear nature. The solution is given by

$$\boldsymbol{\Delta}(L) = ag_R \overset{\leftrightarrow}{R}(L) \int_0^L \overset{\leftrightarrow}{R}^{-1}(z) \boldsymbol{P}(z) dz, \tag{2.3.39}$$

where the rotation matrix $\overset{\leftrightarrow}{R}(z)$ is obtained from $d\overset{\leftrightarrow}{R}/dz = -\Omega_R \boldsymbol{b} \times \overset{\leftrightarrow}{R}$. For fibers much longer than the correlation length, $\boldsymbol{\Delta}$ follows a three-dimensional Gaussian distribution [75].

The moments of $\boldsymbol{\Delta}$ can be obtained from Eq. (2.3.38) following the same procedure used for calculating the average gain and signal fluctuations [88]. The first two moments satisfy [85]

$$\frac{d\langle \boldsymbol{\Delta} \rangle}{dz} = -\eta \langle \boldsymbol{\Delta} \rangle + ag_R P_p(z) \hat{\boldsymbol{p}}, \qquad \frac{d\langle \Delta^2 \rangle}{dz} = 2ag_R P_p(z) \hat{\boldsymbol{p}} \cdot \langle \boldsymbol{\Delta} \rangle. \tag{2.3.40}$$

These equations can be easily integrated over the fiber length $L$. In the case of forward pumping, the solution is given by

$$\langle \mathbf{\Delta} \rangle = \frac{a g_R P_{in} \hat{\mathbf{p}}}{\eta - \alpha_p} \left[1 - \alpha_p L_{eff} - \exp(-\eta L)\right], \qquad (2.3.41)$$

$$\langle \mathbf{\Delta}^2 \rangle = \frac{2 (a g_R P_{in})^2}{\eta^2 - \alpha_p^2} \left[\left(1 - \alpha_p L_{eff}\right) \exp(-\eta L) - 1 \right.$$
$$\left. + \left(\alpha_p + \eta\right) L_{eff} \left(1 - \alpha_p L_{eff}/2\right)\right]. \qquad (2.3.42)$$

In the case of backward pumping, we can use these equations provided $\alpha_p$ is replaced with $-\alpha_p$, $P_{in}$ is replaced with $P_{in} \exp(-\alpha_p L)$, $L_{eff}$ is redefined as $\left[\exp\left(\alpha_p L\right) - 1\right]/\alpha_p$, and $\Omega = -\left(\omega_p + \omega_s\right)$.

The probability density function of $\mathbf{\Delta} \equiv \Delta_1 \hat{\mathbf{e}}_1 + \Delta_2 \hat{\mathbf{e}}_2 + \Delta_3 \hat{\mathbf{e}}_3$ can be written as

$$p(\mathbf{\Delta}) = \frac{(2\pi)^{-3/2}}{\sigma_\| \sigma_\perp^2} \exp \left[ -\frac{(\Delta_1 - \Delta_0)^2}{2\sigma_\|^2} - \frac{\Delta_2^2 + \Delta_3^2}{2\sigma_\perp^2} \right], \qquad (2.3.43)$$

where $\Delta_0 = |\langle \mathbf{\Delta} \rangle|$ while $\sigma_\|^2$ and $\sigma_\perp^2$ are variances of the PDG vector in the direction parallel and perpendicular to $\hat{\mathbf{p}}$, respectively, and are given by [86]

$$\sigma_\|^2 = \eta \int_0^L \left[\langle \mathbf{\Delta}^2 \rangle - \langle \mathbf{\Delta} \rangle^2\right] \exp\left[-3\eta (L - z)\right] dz, \qquad (2.3.44)$$

$$\sigma_\perp^2 = \eta \int_0^L \langle \mathbf{\Delta}^2 \rangle \exp\left[-3\eta (L - z)\right] dz. \qquad (2.3.45)$$

These equations show that $\sigma_\| < \sigma_\perp$ when PMD is small because the pump mostly amplifies the copolarized signal.

In practice, one is often interested in the statistics of the PDG magnitude $\Delta$. Its probability density can be found from Eq. (2.3.43) after writing $\mathbf{\Delta}$ in spherical coordinates and integrating over the two angles. The result is found to be [86]

$$p\left(\Delta\right) = \frac{\Delta}{2\sigma_{\|}\sigma} \exp\left[-\frac{\Delta^2\left(r-1\right)-r\Delta_0^2}{2\sigma^2}\right]$$
$$\times \left[\mathrm{erf}\left(\frac{\Delta\left(r-1\right)+r\Delta_0}{\sqrt{2}\sigma}\right) + \mathrm{erf}\left(\frac{\Delta\left(r-1\right)-r\Delta_0}{\sqrt{2}\sigma}\right)\right],$$

$$(2.3.46)$$

where $\sigma^2 = \sigma_\perp^2\left(r-1\right), r = \sigma_\perp^2/\sigma_\|^2$, and $\mathrm{erf}\left(x\right)$ is the error function.

Figure 2.19 shows how $p(\Delta)$ changes with $D_p$ in the case of forward pumping. All parameters are the same as in Figure 2.16. The PDG values are normalized to the average gain $G_{\mathrm{av}} = ag_R(1 + 3\mu)P_{\mathrm{in}}L_{\mathrm{eff}}/2$ (in dB) so that the curves are pump-power independent. In the limit $D_p \to 0$, $p(\Delta)$ becomes a delta function located at the maximum gain difference (almost $2G_{\mathrm{av}}$) as little gain exists for the orthogonally polarized signal. As $D_p$ increases, $p(\Delta)$ broadens quickly because PMD changes the signal SOP

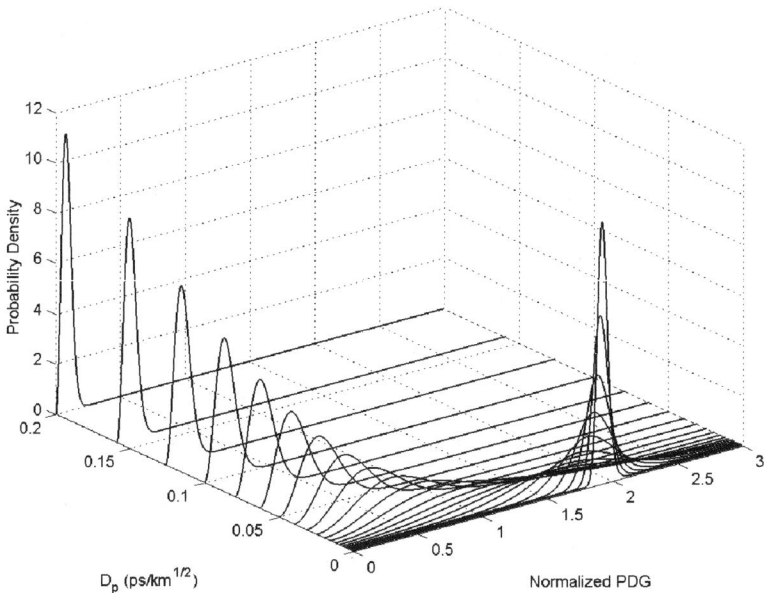

**Figure 2.19:** The probability distribution of PDG as a function of $D_p$ under conditions of Figure 2.16. The PDG value is normalized to the average gain $G_{\mathrm{av}}$. (After Ref. [86]; © 2003 OSA.)

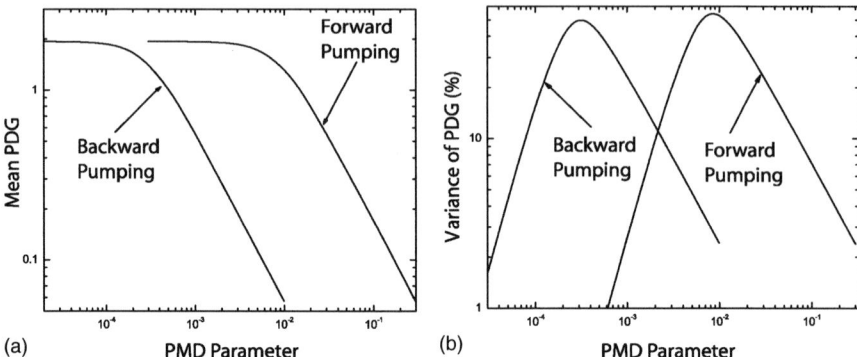

**Figure 2.20:** (a) Mean PDG and (b) variance $\sigma_\Delta$ (both normalized to the average gain $G_{av}$) as a function of PMD parameter under forward- and backward-pumping conditions. (After Ref. [86]; © 2003 OSA.)

randomly. If $D_p$ is relatively small, the diffusion length $L_d$ is larger than or comparable to the effective fiber length $L_{eff}$, and $p(\Delta)$ remains centered at almost the same location but broadens because of large fluctuations. Its shape mimics a Gaussian distribution. When $D_p$ is large enough that $L_{eff} \gg L_d$, $p(\Delta)$ becomes Maxwellian and its peaks shifts to smaller values. This behavior has been observed experimentally [82].

The mean value of PDG $\langle \Delta \rangle$, and the variance of PDG fluctuations, $\sigma_\Delta^2 = \langle \Delta^2 \rangle - \langle \Delta \rangle^2$, can be calculated using the PDG distribution and are given in Eqs. (2.3.41) and (2.3.42). Figure 2.20 shows how these two quantities vary with the PMD parameter for the same Raman amplifier used for Figure 2.16. Both $\langle \Delta \rangle$ and $\sigma_\Delta$ are normalized to the average gain $G_{av}$. As expected, the mean PDG decreases monotonically as $D_p$ increases. The mean PDG $\langle \Delta \rangle$ is not exactly $2G_{av}$ when $D_p = 0$ because the gain is not zero when pump and signal are orthogonally polarized. Note however that $\langle \Delta \rangle$ can be as large as 30% of the average gain for $D_p > 0.05$ ps/$\sqrt{km}$ in the case of forward pumping and it decreases slowly with $D_p$ after that, reaching a value of 8% for $D_p = 0.2$ ps/$\sqrt{km}$. This is precisely what was observed in Ref. [82] through experiments and numerical simulations. In the case of backward pumping, the behavior is nearly identical to that in the case of forward pumping except that the curve shifts to a value of $D_p$ smaller by about a factor of 30.

As seen in Figure 2.20b, the Root Mean Square (RMS) value of PDG fluctuations increases rapidly as $D_p$ becomes nonzero, peaks to a value close to 56% of $G_{av}$ for $D_p$ near 0.01 ps/$\sqrt{km}$ in the case of forward pumping, and then begins to decrease. Again, fluctuations can exceed 7% of the average gain level even for $D_p = 0.1$ ps/$\sqrt{km}$. A similar behavior holds for backward pumping, as seen in Figure 2.20. Both the mean and the RMS values of PDG fluctuations decrease with $D_p$ inversely for $D_p > 0.03$ ps/$\sqrt{km}$ ($L_d < 0.5$ km for $\Omega/2\pi = 13.2$ THz). This $D_p^{-1}$ dependence agrees very well with the experimental results in Ref. [82].

The $D_p^{-1}$ dependence of $\langle \Delta \rangle$ and $\sigma_\Delta$ can be deduced analytically from Eqs. (2.3.41) and (2.3.42) in the limit $L_{eff} \gg L_d$. In this limit, the PDG distribution $p(\Delta)$ becomes approximately Maxwellian and the average and RMS values of PDG in the case of forward pumping are given by

$$\langle \Delta \rangle \approx \frac{4ag_R P_{in}}{\sqrt{\pi} D_p |\Omega|} \left[ L_{eff} \left( 1 - \alpha_p L_{eff}/2 \right) \right]^{1/2}. \tag{2.3.47}$$

$$\sigma_\Delta \approx (3\pi/8 - 1)^{1/2} \langle \Delta \rangle. \tag{2.3.48}$$

The same equations hold in the case of backward pumping except that $|\Omega| = \omega_p - \omega_s$ should be replaced with $\omega_p + \omega_s$.

## 2.4 Ultrafast Raman Amplification

The CW regime considered so far assumes that both the pump and the signal fields are in the form of CW beams. This is rarely the case in practice. Although pumping is often continuous in lightwave systems, the signal is invariably in the form of a pulse train consisting of a random sequence of 0 and 1 bits. Fortunately, the CW theory can be applied to such systems, if the signal power $P_s$ is interpreted as the average channel power, because the Raman response is fast enough that the entire pulse train can be amplified without any distortion. However, for pulses shorter than 10 ps or so, one must include the dispersive and nonlinear effects that are likely to affect the amplification of such short pulses in a Raman amplifier.

The situation is quite different when a Raman amplifier is pumped with optical pulses and is used to amplify a pulsed signal. Because of the dispersive nature of silica fibers, the pump and signal pulses travel with different group velocities, $v_{gp}$ and $v_{gs}$, respectively, because of their different wavelengths. Thus, even if the two pulses were overlapping initially, they separate after a distance known as the walk-off length. The CW theory can still be applied for relatively wide pump pulses (width $T_0 > 1$ ns) if the walk-off length $L_W$, defined as

$$L_W = T_0/|v_{gp}^{-1} - v_{gs}^{-1}|, \tag{2.4.1}$$

exceeds the fiber length $L$. However, for shorter pump pulses for which $L_W < L$, Raman amplification is limited by the group-velocity mismatch and occurs only over distances $z \sim L_W$ even if the actual fiber length $L$ is considerably larger than $L_W$. At the same time, the nonlinear effects such as SPM and XPM become important and affect considerably the evolution of the pump and signal pulses [9]. This section discusses the pulsed case assuming that pulse widths remain larger than the Raman response time ($\sim$50 fs) so that transient effects are negligible. The PMD effects are ignored to simplify the following discussion.

### 2.4.1  Pulse-Propagation Equations

To study the propagation of a short pulse in a dispersive nonlinear medium, we use Eq. (2.3.1) for the induced third-order polarization but simplify it by assuming that the optical field $E$ maintains its SOP and can be written as $\boldsymbol{E} = \hat{\boldsymbol{e}}E$, where $\hat{\boldsymbol{e}}$ is a unit vector that does not change with propagation. In this scalar approximation, Eq. (2.3.1) reduces to

$$P^{(3)}(\boldsymbol{r}, t) = \frac{\varepsilon_0}{2}\sigma E^3(\boldsymbol{r}, t) + \varepsilon_0 E(\boldsymbol{r}, t)\int_{-\infty}^{t} h_R(t-\tau) E^2(\boldsymbol{r}, \tau)d\tau, \tag{2.4.2}$$

where $h_R(t) = a(t) + b(t)$ represents the Raman response function. Writing $E$ and $P^{(3)}$ as in Eqs. (2.3.2) and (2.3.3), neglecting fiber birefringence but accounting for the frequency dependence of the dielectric constant, we arrive at the following set of two coupled propagation equations [93]:

$$
\frac{\partial A_s}{\partial z} + \frac{\alpha_s}{2} A_s + \beta_{1s} \frac{\partial A_s}{\partial z} + \frac{i\beta_{2s}}{2} \frac{\partial^2 A_s}{\partial t^2} - \frac{\beta_{3s}}{6} \frac{\partial^3 A_s}{\partial t^3}
$$

$$
= i\gamma_s \left( 1 + \frac{i}{\omega_s} \frac{\partial}{\partial t} \right) \left[ (1 - f_R) \left( |A_s(t)|^2 + 2 |A_p(t)|^2 \right) A_s(t) \right.
$$

$$
+ f_R A_s(t) \int_{-\infty}^{t} h_R \left( t - t' \right) \left( |A_s(t')|^2 + |A_p(t')|^2 \right.
$$

$$
\left. + A_s(t') A_p(t') e^{-i\Omega(t-t')} \right) dt' \right], \tag{2.4.3}
$$

$$
\frac{\partial A_p}{\partial z} + \frac{\alpha_p}{2} A_p + \beta_{1p} \frac{\partial A_p}{\partial z} + \frac{i\beta_{2p}}{2} \frac{\partial^2 A_p}{\partial t^2} - \frac{\beta_{3p}}{6} \frac{\partial^3 A_p}{\partial t^3}
$$

$$
= i\gamma_p \left( 1 + \frac{i}{\omega_p} \frac{\partial}{\partial t} \right) \left[ (1 - f_R) \left( |A_p(t)|^2 + 2 |A_s(t)|^2 \right) A_p(t) \right.
$$

$$
+ f_R A_p(t) \int_{-\infty}^{t} h_R \left( t - t' \right) \left( |A_p(t')|^2 + |A_s(t')|^2 \right.
$$

$$
\left. + A_p(t') A_s(t') e^{-i\Omega(t-t')} \right) dt' \right], \tag{2.4.4}
$$

where $f_R$ represents the fractional Raman contribution ($f_R \approx 0.18$), $h_R(t)$ has been normalized such that $\int_0^\infty h_R(t)dt = 1$, $\gamma_j = n_2\omega_j/(cA_{\text{eff}})$ is the nonlinear parameter at the frequency $\omega_j$, and $n_2$ is related to $\sigma$ appearing in Eq. (2.4.2).

The three dispersion parameters, $\beta_1$, $\beta_2$, and $\beta_3$, include the effects of fiber dispersion to the first three orders. They become increasingly important for shorter pulses with a wider spectrum. The first-order dispersion parameter is related to the group velocity $v_{gj}$ of a pulse as $\beta_{1j} = 1/v_{gj}$, where $j = p$ or $s$. The second-order parameter $\beta_{2j}$ represents dispersion of group velocity and is known as the group-velocity dispersion (GVD) parameter [9]. For a similar reason, $\beta_{3j}$ is called the third-order dispersion parameter.

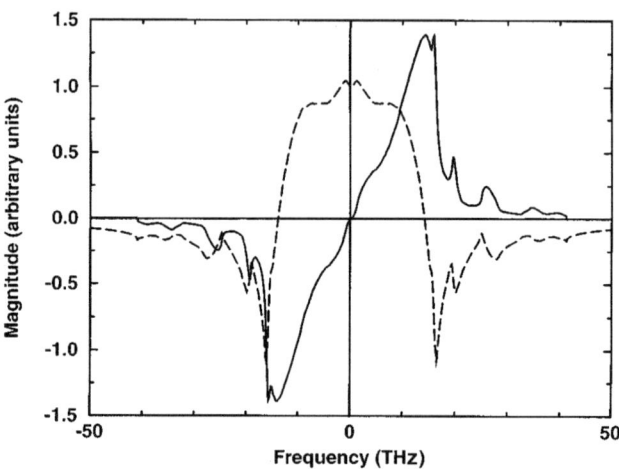

**Figure 2.21:** Imaginary (solid curve) and real (dashed curve) parts of the Raman response as a function of frequency shift. (After Ref. [93]; © 1996 OSA.)

The Raman response function $h_R(t)$ plays an important role for short optical pulses [94–98]. The Raman-gain spectrum is related to the imaginary part of its Fourier transform $\tilde{h}_R(\omega)$. The real part of $\tilde{h}_R(\omega)$ can be obtained from the imaginary part by using the Kramers–Kronig relations. Figure 2.21 shows the real and imaginary parts of $\tilde{h}_R(\omega)$ obtained by using the experimentally measured Raman gain spectrum [16]. Physically, the real part $\tilde{h}_R(w)$ leads to Raman-induced index changes.

Although Eqs. (2.4.3) and (2.4.4) should be solved for femtosecond pulses, they can be simplified considerably for picosecond pulses [93]. For relatively broad pulse widths (>1 ps), $A_s$ and $A_p$ can be treated as constants compared with the time scale over which $h_R(t)$ changes (<0.1 ps). The integrals in Eqs. (2.4.3) and (2.4.4) can then be performed analytically. Neglecting the index changes associated with the real part of $\tilde{h}_R(\omega)$ and introducing a reduced time $T = t - z/v_{gp}$ in a frame of reference moving with the pump pulse, these equations reduce to [9]

$$\frac{\partial A_s}{\partial z} - d\frac{\partial A_s}{\partial T} + \frac{i\beta_{2s}}{2}\frac{\partial^2 A_s}{\partial T^2}$$
$$= i\gamma_s\left[|A_s|^2 + 2|A_p|^2\right]A_p + \frac{g_s}{2}|A_p|^2 A_s - \frac{\alpha_s}{2}A_s, \qquad (2.4.5)$$

$$\frac{\partial A_p}{\partial z} + \frac{i\beta_{2p}}{2} \frac{\partial^2 A_p}{\partial T^2}$$

$$= i\gamma_p \left[ |A_p|^2 + 2|A_s|^2 \right] A_p - \frac{g_p}{2} |A_s|^2 A_p - \frac{\alpha_p}{2} A_p, \qquad (2.4.6)$$

where the Raman gain coefficients are defined as

$$g_j = 2 f_R \gamma_j \left| \mathrm{Im} \left[ \tilde{h}_R \left( \Omega \right) \right] \right| \qquad (j = p, s), \qquad (2.4.7)$$

and the walk-off parameter $d$ accounts for the group-velocity mismatch:

$$d = \beta_{1p} - \beta_{1s} = v_{gp}^{-1} - v_{gs}^{-1}. \qquad (2.4.8)$$

Typical values of $d$ are in the range 2–6 ps/m. The GVD parameter $\beta_{2j}$, the nonlinearity parameter $\gamma_j$, and the Raman gain coefficient $g_j$ are slightly different for the pump and signal pulses because of their different wavelengths. In terms of the wavelength ratio $\lambda_p/\lambda_s$, these parameters are related as

$$\beta_{2s} = \frac{\lambda_p}{\lambda_s} \beta_{2p}, \qquad \gamma_s = \frac{\lambda_p}{\lambda_s} \gamma_p, \qquad g_s = \frac{\lambda_p}{\lambda_s} g_p. \qquad (2.4.9)$$

Equations (2.4.5) and (2.4.6) reduce to Eqs. (2.1.6) and (2.1.7) when the dispersive and nonlinear terms are neglected and pump and signal powers are introduced as $P_j = |A_j|^2 (j = p, s)$.

Four length scales can be introduced to determine the relative importance of various terms in Eqs. (2.4.5) and (2.4.6). For pump pulses of duration $T_0$ and peak power $P_0$, these are defined as

$$L_D = \frac{T_0^2}{|\beta_{2p}|}, \qquad L_W = \frac{T_0}{|d|}, \qquad L_{NL} = \frac{1}{\gamma_p P_0}, \qquad L_G = \frac{1}{g_p P_0}. \qquad (2.4.10)$$

The dispersion length $L_D$, the walk-off length $L_W$, the nonlinear length $L_{NL}$, and the Raman gain length $L_G$ provide, respectively, the length scales over which the effects of GVD, walk-off, nonlinearity (both SPM and XPM), and Raman gain become important. The shortest length among them plays the dominant role. Typically, $L_W \sim 1$ m for $T_0 < 10$ ps whereas $L_{NL}$ and $L_G$ become smaller or comparable to it for $P_0 > 100$ W.

In contrast, $L_D \sim 1$ km for $T_0 = 10$ ps. Thus, the GVD effects are generally negligible for pulses as short as 10 ps. The situation changes for pulse widths $\sim 1$ ps or less because $L_D$ decreases faster than $L_W$ with a decrease in the pulse width. The GVD effects can then affect the amplification process significantly, especially in the anomalous-dispersion regime.

## 2.4.2   Effects of Group-Velocity Mismatch

When the second-derivative terms in Eqs. (2.4.5) and (2.4.6) are neglected by setting $\beta_{2p} = \beta_{2s} = 0$, these equations can be solved analytically [99–103]. The analytic solution takes a simple form if the signal pulse is assumed to be relatively weak and pump depletion is neglected by setting $g_p = 0$. We also neglect fiber losses for simplicity. Equation (2.4.5) for the pump pulse then yields the solution

$$A_p(z, T) = A_p(0, T) \exp\left[i\gamma_p \left|A_p(0, T)\right|^2 z\right],  \qquad (2.4.11)$$

where the XPM term has been neglected assuming $|A_s|^2 \ll |A_p|^2$. For the same reason, the SPM term in Eq. (2.4.6) can be neglected. The solution of Eq. (2.4.6) is then given by [99]

$$A_s(z, T) = A_s(0, T + zd) \exp\left[(g_s/2 + 2i\gamma_s)\,\psi(z, T)\right],  \qquad (2.4.12)$$

where

$$\psi(z, T) = \int_0^z \left|A_p\left(0, T + zd - z'd\right)\right|^2 dz'.  \qquad (2.4.13)$$

Equation (2.4.11) shows that the pump pulse of initial amplitude $A_p(0, T)$ propagates without change in its shape. However, the SPM-induced phase shift imposes a frequency chirp on the pump pulse that broadens its spectrum. The signal pulse, in contrast, changes both its shape and spectrum as it propagates through the fiber; temporal changes occur owing to Raman gain whereas spectral changes have their origin in XPM. Because of pulse walk-off, both kinds of changes are governed by an overlap factor $\psi(z, T)$ that takes into account the relative separation between the two pulses along the fiber. This factor depends on the pulse shape. For a Gaussian pump pulse with the input amplitude

$$A_p(0, T) = \sqrt{P_0} \exp\left(-T^2/2T_0^2\right),$$ (2.4.14)

the integral in Eq. (2.4.13) can be performed in terms of error functions with the result

$$\psi(z, \tau) = [\text{erf}(\tau + \delta) - \text{erf}(\tau)]\left(\sqrt{\pi}P_0 z/\delta\right),$$ (2.4.15)

where $\tau = T/T_0$ and $\delta$ is the propagation distance in units of the walk-off length; that is, $\delta = zd/T_0 = z/L_W$. An analytic expression for $\psi(z, \tau)$ can also be obtained for pump pulses having "sech" shape [100]. In both cases, the signal pulse compresses initially, reaches a minimum width, and then begins to rebroaden as it is amplified through SRS. It also acquires a frequency chirp through XPM. This qualitative behavior persists even when pump depletion is included [100–102].

The analytic solution (2.4.12) can be used to obtain both the shape and the spectrum of the signal pulse during the initial stages of Raman amplification. The spectral evolution is governed by the XPM-induced frequency chirp. Note, however, that the signal pulse travels faster than the pump pulse in the normal-GVD regime. As a result, chirp is induced mainly near the trailing edge in the case of normal dispersion and near the leading edge in the case of anomalous dispersion. It should be stressed that both pulse shapes and spectra are considerably modified when pump depletion is included because the growing signal pulse affects itself through SPM and the pump pulse through XPM.

When fiber length is comparable to the dispersion length $L_D$, it is important to include the GVD effects. Such effects cannot be described analytically, and a numerical solution of Eqs. (2.4.5) and (2.4.6) is necessary to understand the Raman amplification process. A generalization of the split-step Fourier method can be used for this purpose [9]. The method requires specification of both the pump and the signal pulse shapes at the input end of the fiber.

For numerical purposes, it is useful to introduce the normalized variables. A relevant length scale along the fiber length is provided by the walk-off length $L_W$. By defining

$$z' = \frac{z}{L_W}, \quad \tau = \frac{T}{T_0}, \quad U_j = \frac{A_j}{\sqrt{P_0}}, \tag{2.4.16}$$

and using Eq. (2.4.9), Eqs. (2.4.5) and (2.4.6) become

$$\frac{\partial U_p}{\partial z'} \pm \frac{i}{2} \frac{L_W}{L_D} \frac{\partial^2 U_p}{\partial \tau^2} = \frac{i L_W}{L_{NL}} \left[ |U_p|^2 + 2|U_s|^2 \right] U_p - \frac{L_W}{2L_G} |U_s|^2 U_p, \tag{2.4.17}$$

$$\frac{\partial U_s}{\partial z'} - \frac{\partial U_s}{\partial \tau} \pm \frac{ir}{2} \frac{L_W}{L_D} \frac{\partial^2 U_s}{\partial \tau^2} = \frac{ir L_W}{L_{NL}} \left[ |U_s|^2 + 2|U_p|^2 \right] U_s + \frac{rL_W}{2L_G} |U_p|^2 U_s, \tag{2.4.18}$$

where the choice of sign depends on the sign of the GVD parameter (negative for anomalous dispersion). The parameter $r \equiv \lambda_p/\lambda_s$ is about 0.94. Figure 2.22 shows the evolution of the pump and signal pulses in the normal-GVD regime over three walk-off lengths for $L_D/L_W = 1000$, $L_W/L_{NL} = 24$, and $L_W/L_G = 12$. The pump pulse is assumed to have a Gaussian shape initially. The input signal pulse is also Gaussian but is assumed to have a peak power of only $2 \times 10^{-7}$ W initially.

Several features of Figure 2.22 are noteworthy. The signal pulse starts to grow exponentially initially but the growth slows down because of the walk-off effects. In fact, energy transfer from the pump pulse stops after $z = 3L_W$ as the two pulses are then physically separated because of the group-velocity mismatch. Since the signal pulse moves faster than the pump pulse in the normal-GVD regime, the energy for Raman amplification comes from the leading edge of the pump pulse. This is clearly apparent at $z = 2L_W$ where energy transfer has led to a two-peak structure in the pump pulse as a result of pump depletion—the hole near the leading edge corresponds exactly to the location of the signal pulse. The small peak near the leading edge disappears with further propagation as the signal pulse walks through it. The pump pulse at $z = 3L_W$ is asymmetric in shape and appears narrower than the input pulse as it consists of the trailing portion of the input pulse. The signal pulse is also narrower than the input pulse and is asymmetric with a sharp leading edge.

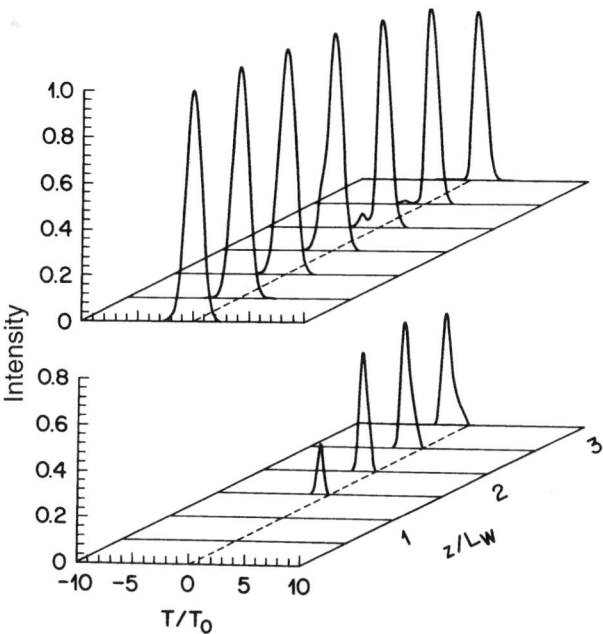

**Figure 2.22:** Evolution of pump (top) and signal (bottom) pulses over three walk-off lengths when $L_D/L_W = 1000$, $L_W/L_{NL} = 24$, and $L_W/L_G = 12$. (After Ref. [9]; © 2001 Elsevier.)

The spectra of pump and signal pulses display many interesting features resulting from the combination of SPM, XPM, group-velocity mismatch, and pump depletion. Figure 2.23 shows the pump and Raman spectra at $z/L_W = 2$ (left column) and $z/L_W = 3$ (right column). The asymmetric nature of these spectra is due to XPM. The high-frequency side of the pump spectra exhibits an oscillatory structure that is characteristic of SPM. In the absence of SRS, the spectrum would be symmetric with the same structure appearing on the low-frequency side. As the low-frequency components occur near the leading edge of the pump pulse and because the pump is depleted on the leading side, the energy is transferred mainly from the low-frequency components of the pump pulse. This is clearly seen in the pump spectra of Figure 2.23. The long tail on the low-frequency side of the Raman spectra is also partly for the same reason. The Raman spectrum is nearly featureless at $z = 2L_W$ but develops considerable internal structure

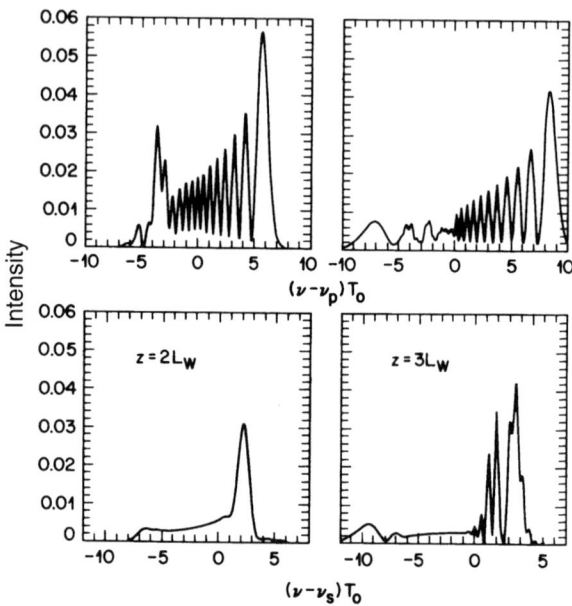

**Figure 2.23:** Spectra of pump (upper row) and signal (lower row) pulses at twice (left column) and thrice (right column) the walk-off length for parameter values of Figure 2.22. (After Ref. [9]; © 2001 Elsevier.)

at $z = 3L_W$. This is due to the combination of XPM and pump depletion; the frequency chirp across the signal pulse induced by these effects can vary rapidly both in magnitude and in sign and leads to a complicated spectral shape.

The temporal and spectral features seen in Figures 2.22 and 2.23 depend on the peak power of input pulses through the lengths $L_G$ and $L_{NL}$ in Eqs. (2.4.17) and (2.4.18). When peak power is increased, both $L_G$ and $L_{NL}$ decrease by the same factor. Numerical results show that because of a larger Raman gain, the signal pulse grows faster and carries more energy than that shown in Figure 2.22. More importantly, because of a decrease in $L_{NL}$, both the SPM and the XPM contributions to the frequency chirp are enhanced, and the pulse spectra are wider than those shown in Figure 2.23. An interesting feature is that the signal-pulse spectrum becomes considerably wider than the pump spectrum. This is due to the stronger effect of

XPM on the signal pulse compared with that of the pump pulse. Numerical results that include all of these effects show an enhanced broadening by up to a factor of three. Direct measurements of the frequency chirp also show an enhanced chirp for the signal pulse compared with that of the pump pulse.

### 2.4.3 Anomalous-Dispersion Regime

When the wavelengths of the pump and signal pulses lie inside the anomalous-dispersion regime of the optical fiber used for Raman amplification, both of them experience the soliton-related effects [104–114]. Under suitable conditions, almost all of the pump-pulse energy can be transferred to the signal pulse, which propagates undistorted as a fundamental soliton. Numerical results show that this is possible if most of the energy transfer occurs at a distance at which the pump pulse, propagating as a higher order soliton, achieves its minimum width [104]. In contrast, if energy transfer is delayed and occurs at a distance where the pump pulse has split into its components, the signal pulse does not form a fundamental soliton, and its energy rapidly disperses.

Equations (2.4.17) and (2.4.18) can be used in the anomalous-GVD regime by simply choosing the negative sign for the second-derivative terms. As in the case of normal GVD shown in Figure 2.22, energy transfer to the signal pulse occurs near $z \approx L_W$. This condition implies that $L_W$ should not be too small in comparison with the dispersion length $L_D$. Typically, $L_W$ and $L_D$ become comparable for femtosecond pulses of widths $T_0 \sim 100$ fs. For such ultrashort pump pulses, the distinction between pump and signal pulses gets blurred as their spectra begin to overlap considerably. This can be seen by noting that the Raman gain peak in Figure 2.3 corresponds to a spectral separation of about 13 THz while the spectral width of a 100-fs pulse is $\sim$10 THz. Equations (2.4.17) and (2.4.18) do not provide a realistic description for femtosecond pump pulses, and one should solve Eqs. (2.4.3) and (2.4.4) in their place.

An interesting situation occurs when a single ultrashort pulse propagates inside the fiber without a pump pulse. Figure 2.24 shows the evolution of an ultrashort pulse in the anomalous-GVD regime of the fiber by solving Eq. (2.4.3) numerically with an input of the form $A(0, t) = \sqrt{P_0}\, \text{sech}\,(t/T_0)$ using $T_0 = 50$ fs. Input peak power $P_0$ of

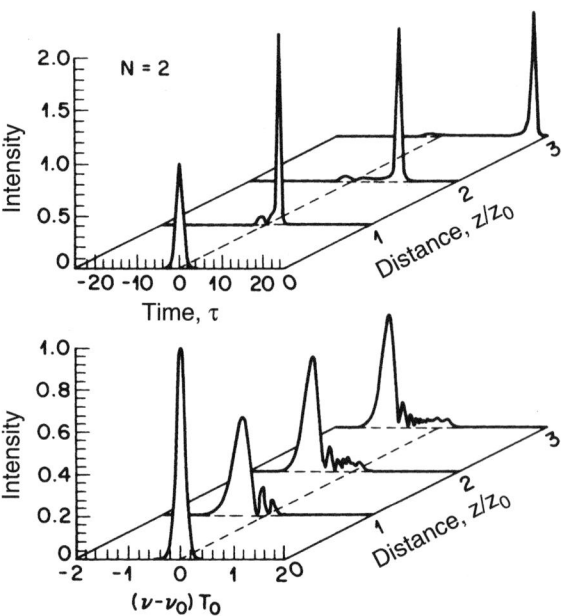

**Figure 2.24:** Pulse shapes and spectra of a 50-fs pulse in the anomalous dispersion regime showing Raman-induced spectral shift and temporal delay. The soliton period $z_0$ is related to the dispersion length as $z_0 = (\pi/2)L_D$. (After Ref. [9]; © 2001 Elsevier.)

the pulse was chosen such that the pulse will form a second-order soliton in the absence of higher order effects. As seen in Figure 2.24, pulse is initially distorted but most of its energy is carried by a pulse whose center shifts to the right. This temporal shift indicates a reduction in the group velocity of the pulse.

To understand why group velocity of the pulse changes, we need to consider modifications in the pulse spectrum. As seen in Figure 2.24, pulse spectrum shifts toward the red side, and this red shift continues to increase as pulse propagates further down the fiber. This shift has its origin in the integral appearing in Eq. (2.4.3) and containing the Raman response function. Physically speaking, high-frequency components of an optical pulse pump the low-frequency components of the same pulse through SRS, a phenomenon referred to as *intrapulse* Raman scattering. Since energy in the blue part of the original pulse spectrum is used to amplify the red-shifted components, the pulse spectrum shifts toward lower frequencies, as the pulse propagates down the fiber. This spectral shift is known as the soliton self-frequency shift [106] and was first observed in a 1986 experiment [107]. As the pulse spectrum shifts toward the red side, speed of pulse changes because group velocity in a dispersive medium depends on the central frequency at which pulse spectrum is located. More specifically, the pulse is delayed because the red-shifted pulse travels slower in the anomalous-GVD regime of an optical fiber. However, it should be stressed that anomalous dispersion is not a prerequisite for spectrum to shift. In fact, the Raman-induced spectral shift can occur even in the normal-dispersion regime although its magnitude is relatively small because of a rapid broadening of the pulse in the case of normal dispersion [115]. We turn to this case next.

## 2.4.4 Normal-Dispersion Regime

In this section we discuss what happens to a short pulse being amplified inside a Raman amplifier while experiencing normal dispersion so that it cannot form a soliton. We expect the pulse to undergo some reshaping through the amplification process. It turns out that the pulse acquires a parabolic shape and its width is always the same irrespective of the shape and energy of input pulses. Parabolic pulses have attracted considerable attention in recent years since it was observed experimentally and discovered theoretically that such a shape exists for pulses being amplified inside an optical fiber while experiencing normal dispersion [116–123]. Parabolic pulses were first observed in a 1996 experiment using a fiber amplifier in which gain was provided by erbium ions [116]. They were later observed in ytterbium-doped fiber amplifiers [117–119].

There is no fundamental reason why parabolic pulses should not form in a distributed Raman amplifier provided the dispersion is normal for the signal pulse. Indeed, such pulses were observed in a Raman amplifier in a 2003 experiment [122]. The amplifier was pumped using a CW laser operating at 1455 nm and capable of delivering up to 2 W of power. Signal pulses of 10-ps width were obtained from a mode-locked fiber laser. The pump and signal were copropagated inside a 5.3-km-long fiber whose GVD parameter was about 5 $ps^2$/km (normal GVD) at the signal wavelength.

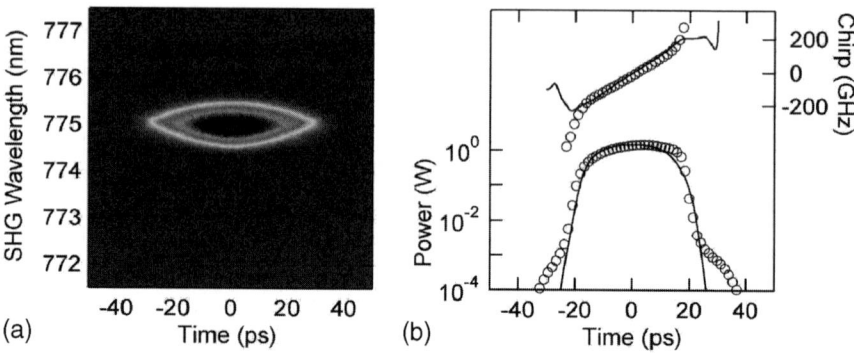

**Figure 2.25:** (a) FROG trace of a 0.75-pJ signal pulse amplified by 17 dB using Raman gain in a 5.3-km-long fiber. (b) Intensity and chirp profiles (circles) retrieved from the FROG trace. Solid lines show the results of numerical simulations. (After Ref. [122]; © 2003 OSA.)

The amplified pulses were characterized using the frequency-resolved optical gating (FROG) technique. Figure 2.25 shows the FROG trace and the pulse shape and frequency chirp deduced from it when a pulse with 0.75-pJ energy was amplified by 17 dB. The chirp is nearly linear and the pulse shape is approximately parabolic.

One can predict the pulse shape and chirp profiles by solving Eqs. (2.4.3) and (2.4.4) numerically. For a CW pump, the dispersive effects can be neglected but the nonlinear effects should be included. Figure 2.26 shows the evolution toward a parabolic shape when a "sech" input pulse is amplified over the 5.3-km length of the Raman amplifier. The shape of the output pulse can be fitted quite well with a parabola but the fit is relatively poor when the shape is assumed to be either Gaussian (dashed curve) or hyperbolic secant (dotted curve). The chirp is approximately linear. Both the predicted pulse shape and the chirp profile agree well with the experimental data, as shown in Figure 2.25. It was also confirmed experimentally that the final pulse characteristics were independent of input pulse parameters and were determined solely by the Raman amplifier. Both the width and the chirp changed when the gain of the amplifier was changed, but the pulse shape remained parabolic, and the frequency chirped across the pulse remained nearly linear. These properties are in agreement with the theoretical solution of the underlying equation.

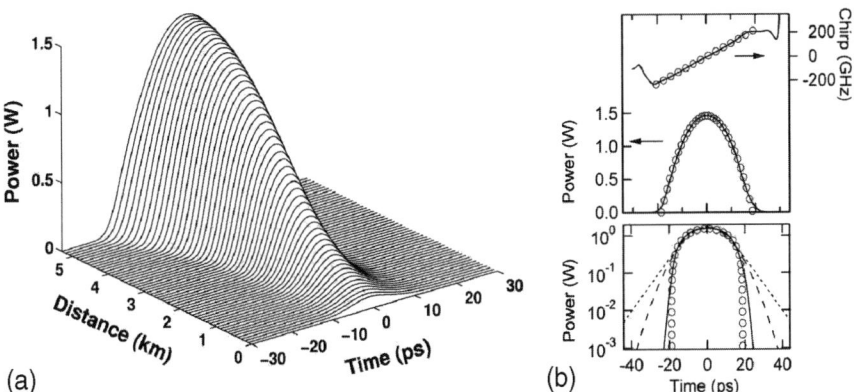

**Figure 2.26:** (a) Evolution toward a parabolic shape of a "sech" input pulse over the 5.3-km length of the Raman amplifier. (b) Intensity and chirp (solid lines) profiles of the output pulse together with parabolic and linear fits, respectively (circles). The bottom part shows the same results on a logarithmic scale together with the Gaussian (dashed curve) and "sech" (dotted curve) fits. (After Ref. [122]; © 2003 OSA.)

Such solutions are referred to as being self-similar and are of considerable interest both from the fundamental and the applied points of view [117–121].

# References

[1]  C. V. Raman, *Indian J. Phys.* **2,** 387 (1928).

[2]  E. J. Woodbury and W. K. Ng, *Proc. IRE* **50,** 2347 (1962).

[3]  R. W. Hellwarth, *Phys. Rev.* **130,** 1850 (1963); *Appl. Opt.* **2,** 847 (1963).

[4]  E. Garmire, E. Pandarese, and C. H. Townes, *Phys. Rev. Lett.* **11,** 160 (1963).

[5]  Y. R. Shen and N. Bloembergen, *Phys. Rev.* **137,** A1786 (1965).

[6]  W. Kaiser and M. Maier, in *Laser Handbook*, Vol. 2, F. T. Arecchi and E. O. Schulz-Dubois, Eds. (North-Holland, Amsterdam, 1972), Chap. E2.

[7]  Y. R. Shen, *The Principles of Nonlinear Optics* (Wiley, New York, 1984), Chap. 10.

[8]   R. W. Boyd, *Nonlinear Optics*, 2nd ed. (Academic Press, San Diego, 2003), Chap. 9

[9]   G. P. Agrawal, *Nonlinear Fiber Optics*, 3rd ed. (Academic Press, San Diego, 2001), Chap. 8.

[10]  R. H. Stolen, E. P. Ippen, and A. R. Tynes, *Appl. Phys. Lett.* **20,** 62 (1972).

[11]  R. H. Stolen and E. P. Ippen, *Appl. Phys. Lett.* **22,** 276 (1973).

[12]  R. H. Stolen, *IEEE J. Quantum Electron.* **15,** 1157 (1979).

[13]  F. L. Galeener, *Phys. Rev. B.* **19,** 4292 (1979).

[14]  R. H. Stolen, *Proc. IEEE* **68,** 1232 (1980).

[15]  N. Shibata, M. Horigudhi, and T. Edahiro, *J. Noncrys. Solids* **45,** 115 (1981).

[16]  R. H. Stolen, J. P. Gordon, W. J. Tomlinson, and H. A. Haus, *J. Opt. Soc. Am. B.* **6,** 1159 (1989).

[17]  D. J. Dougherty, F. X. Kartner, H. A. Haus, and E. P. Ippen, *Opt. Lett.* **20,** 31 (1995).

[18]  D. Mahgerefteh, D. L. Butler, J. Goldhar, B. Rosenberg, and G. L. Burdge, *Opt. Lett.* **21,** 2026 (1996).

[19]  M. Ikeda, *Opt. Commun.* **39,** 148 (1981).

[20]  G. A. Koepf, D. M. Kalen, and K. H. Greene, *Electron. Lett.* **18,** 942 (1982).

[21]  Y. Aoki, S. Kishida, H. Honmou, K. Washio, and M. Sugimoto, *Electron. Lett.* **19,** 620 (1983).

[22]  E. Desurvire, M. Papuchon, J. P. Pocholle, J. Raffy, and D. B. Ostrowsky, *Electron. Lett.* **19,** 751 (1983).

[23]  A. R. Chraplyvy, J. Stone, and C. A. Burrus, *Opt. Lett.* **8,** 415 (1983).

[24]  M. Nakazawa, M. Tokuda, Y. Negishi, and N. Uchida, *J. Opt. Soc. Am. B.* **1,** 80 (1984).

[25]  K. Nakamura, M. Kimura, S. Yoshida, T. Hikada, and Y. Mitsuhashi, *J. Lightwave Technol.* **2,** 379 (1984).

[26]  M. Nakazawa, T. Nakashima, and S. Seikai, *J. Opt. Soc. Am. B.* **2,** 215 (1985).

[27]  L. F. Mollenauer, J. P. Gordon, and M. N. Islam, *IEEE J. Quantum Electron.* **22,** 157 (1986).

[28]  N. A. Olsson and J. Hegarty, *J. Lightwave Technol.* **4,** 391 (1986).

[29]  K. Vilhelmsson, *J. Lightwave Technol.* **4,** 400 (1986).

[30]  E. Desurvire, M. J. F. Digonnet, and H. J. Shaw, *J. Lightwave Technol.* **4,** 426 (1986).

[31]  Y. Aoki, S. Kishida, and K. Washio, *Appl. Opt.* **25,** 1056 (1986).

[32]  S. Seikai, T. Nakashima, and N. Shibata, *J. Lightwave Technol.* **4,** 583 (1986).

[33] K. Mochizuki, N. Edagawa, and Y. Iwamoto, *J. Lightwave Technol.* **4,** 1328 (1986).

[34] N. Edagawa, K. Mochizuki, and Y. Iwamoto, *Electron. Lett.* **23,** 196 (1987).

[35] R. W. Davies, P. Melman, W. H. Nelson, M. L. Dakss, and B. M. Foley, *J. Lightwave Technol.* **5,** 1068 (1987).

[36] L. F. Mollenauer and K. Smith, *Opt. Lett.* **13,** 675 (1988).

[37] M. J. O'Mahony, *J. Lightwave Technol.* **6,** 531 (1988).

[38] S. V. Chernikov, Y. Zhu, R. Kashyap, and J. R. Taylor, *Electron. Lett.* **31,** 472 (1995).

[39] S. G. Grubb, *Proc. Conf. on Optical Amplifiers and Applications* (Optical Society of America, Washington, DC), 1995.

[40] E. M. Dianov, *Laser Phys.* **6,** 579 (1996).

[41] D. I. Chang, S. V. Chernikov, M. J. Guy, J. R. Taylor, and H. J. Kong, *Opt. Commun.* **142,** 289 (1997).

[42] P. B. Hansen, L. Eskilden, S. G. Grubb, A. J. Stentz, T. A. Strasser, J. Judkins, J. J. DeMarco, J. R. Pedrazzani, and D. J. DiGiovanni, *IEEE Photon. Technol. Lett.* **9,** 262 (1997).

[43] A. Bertoni, *Opt. Quantum Electron.* **29,** 1047 (1997).

[44] A. Bertoni and G. C. Reali, *Appl. Phys. B.* **67,** 5 (1998).

[45] E. M. Dianov, M. V. Grekov, I. A. Bufetov, V. M. Mashinsky, O. D. Sazhin, A. M. Prokhorov, G. G. Devyatykh, A. N. Guryanov, and V. F. Khopin, *Electron. Lett.* **34,** 669 (1998).

[46] D. V. Gapontsev, S. V. Chernikov, and J. R. Taylor, *Opt. Commun.* **166,** 85 (1999).

[47] H. Masuda, S. Kawai, K. Suzuki, and K. Aida, *IEEE Photon. Technol. Lett.* **10,** 516 (1998).

[48] J. Kani and M. Jinno, *Electron. Lett.* **35,** 1004 (1999).

[49] H. Masuda and S. Kawai, *IEEE Photon. Technol. Lett.* **11,** 647 (1999).

[50] Y. Emori, K. Tanaka, and S. Namiki, *Electron. Lett.* **35,** 1355 (1999).

[51] S. A. E. Lewis, S. V. Chernikov, and J. R. Taylor, *Electron. Lett.* **35,** 1761 (1999).

[52] H. D. Kidorf, K. Rottwitt, M. Nissov, M. X. Ma, and E. Rabarijaona, *IEEE Photon. Technol. Lett.* **12,** 530 (1999).

[53] H. Suzuki, J. Kani, H. Masuda, N. Takachio, K. Iwatsuki, Y. Tada, and M. Sumida, *IEEE Photon. Technol. Lett.* **12,** 903 (2000).

[54] S. Namiki and Y. Emori, *IEEE J. Sel. Topics Quantum Electron.* **7,** 3 (2001).

[55] B. Zhu, L. Leng, L. E. Nelson, Y. Qian, L. Cowsar, S. Stulz, C. Doerr, L. Stulz, S. Chandrasekhar, et al., *Electron. Lett.* **37,** 844 (2001).

[56] K. Rottwitt and A. J. Stentz, in *Optical Fiber Telecommunications*, Vol. **4A,** I. Kaminow and T. Li, Eds. (Academic Press, San Diego, CA, 2002), Chap. 5.

[57] J. Bromage, P. J. Winzer, and R. J. Essiambre, in *Raman Amplifiers for Telecommunications*, M. N. Islam, Ed. (Springer, New York, 2003), Chap. 15.

[58] H, Suzuki, N. Takachio, H. Masuda, and K. Iwatsuki, *J. Lightwave Technol.* **21,** 973 (2003).

[59] M. Morisaki, H. Sugahara, T. Ito, and T. Ono, *IEEE Photon. Technol. Lett.* **15,** 1615 (2003).

[60] B. Zhu, L. E. Nelson, S. Stulz, A. H. Gnauck, C. Doerr, J. Leuthold, L. Grner-Nielsen, M. O. Pedersen, J. Kim, and R. L. Lingle, Jr., *J. Lightwave Technol.* **22,** 208 (2004).

[61] D. F. Grosz, A. Agarwal, S. Banerjee, D. N. Maywar, and A. P. Küng, *J. Lightwave Technol.* **22,** 423 (2004).

[62] J. Bromage, *J. Lightwave Technol.* **22,** 79 (2004).

[63] G. P. Agrawal, *Applications of Nonlinear Fiber Optics* (Academic Press, San Diego, 2001), Chap. 7.

[64] E. Desuvire, *Erbium-Doped Fiber Amplifiers: Principles and Applications* (Wiley, New York, 1994).

[65] P. C. Becker, N. A. Olsson, and J. R. Simpson, *Erbium-Doped Fiber Amplifiers: Fundamentals and Technology* (Academic Press, San Diego, CA), 1999.

[66] E. Desuvire, D. Bayart, B. Desthieux, and S. Bigo, *Erbium-Doped Fiber Amplifiers: Device and System Development* (Wiley, New York, 2002).

[67] G. P. Agrawal, *Fiber-Optic Communication Systems*, 3rd ed. (Wiley, New York, 2002).

[68] P. Wan and J. Conradi, *J. Lightwave Technol.* **14,** 288 (1996).

[69] P. B. Hansen, L. Eskilden, A. J. Stentz, T. A. Strasser, J. Judkins, J. J. DeMarco, R. Pedrazzani, and D. J. DiGiovanni, *IEEE Photon. Technol. Lett.* **10,** 159 (1998).

[70] M. Nissov, K. Rottwitt, H. D. Kidorf, and M. X. Ma, *Electron. Lett.* **35,** 997 (1999).

[71] S. R. Chinn, *IEEE Photon. Technol. Lett.* **11,** 1632 (1999).

[72] S. A. E. Lewis, S. V. Chernikov, and J. R. Taylor, *IEEE Photon. Technol. Lett.* **12,** 528 (2000).

[73] C. H. Kim, J. Bromage, and R. M. Jopson, *IEEE Photon. Technol. Lett.* **14,** 573 (2002).

[74] C. R. S. Fludger, V. Handerek, and R. J. Mears, *J. Lightwave Technol.* **19,** 1140 (2001).

[75] G. J. Foschini and C. D. Poole, *J. Lightwave Technol.* **9,** 1439 (1991).

[76] P. K. A. Wai and C. R. Menyuk, *J. Lightwave Technol.* **14,** 148 (1996).

[77] J. P. Gordon and H. Kogelnik, *Proc. Natl. Acad. Sci. USA* **97,** 4541 (2000).

[78] D. Wang and C. R. Menyuk, *J. Lightwave Technol.* **19,** p. 487 (2001).

[79] H. Kogelnik, R. M. Jopson, and L. E. Nelson, in *Optical Fiber Telecommunications IV B*, I. P. Kaminow and T. Li, Eds. (Academic Press, San Diego, 2002), Chap. 15.

[80] D. Mahgerefteh, H. Yu, D. L. Butler, J. Goldhar, D.Wang, E. Golovchenko, A. N. Phlipetskii, C. R. Menyuk, L. Joneckis, Proc. Conf. on Lasers and Electro-optics (Optical Society of America, Washington, DC, 1997), p. 447.

[81] A. Berntson, S. Popov, E. Vanin, G. Jacobsen, and J. Karlsson, Proc. Opt. Fiber Commun. Conf. (Optical Society of America, Washington, DC, 2001), Paper MI2-1.

[82] P. Ebrahimi, M. C. Hauer, Q. Yu, R. Khosravani, D. Gurkan, D. W. Kim, D. W. Lee, and A. E. Willner, Proc. Conf. on Lasers and Electro-optics (Optical Society of America, Washington, DC, 2001), p. 143.

[83] S. Popov, E. Vanin, and G. Jacobsen, *Opt. Lett.* **27,** 848 (2002).

[84] Q. Lin and G. P. Agrawal, *Opt. Lett.* **27,** 2194 (2002).

[85] Q. Lin and G. P. Agrawal, *Opt. Lett.* **28,** 227 (2003).

[86] Q. Lin and G. P. Agrawal, *J. Opt. Soc. Am. B.* **20,** 1616 (2003).

[87] R. Hellwarth, J. Cherlow, and T. Yang, *Phys. Rev. B.* **11,** 964 (1975).

[88] C. W. Gardiner, *Handbook of Stochastic Methods*, 2nd ed. (Springer, New York, 1985).

[89] H. H. Kee, C. R. S. Fludger, and V. Handerek, Proc. Opt. Fiber Commun. (Optical Society of America, Washington, DC, 2002), p. 180.

[90] B. Huttner, C. Geiser, and N. Gisin, *IEEE J. Sel. Topics Quantum Electron.* **6,** 317 (2000).

[91] A. Mecozzi and M. Shtaif, *IEEE Photon. Technol. Lett.* **14,** 313 (2002).

[92] A. Galtarossa and L. Palmieri, *IEEE Photon. Technol. Lett.* **15,** 57 (2003).

[93] C. Headley III, and G. P. Agrawal, *J. Opt. Soc. Am. B.* **13,** 2170 (1996).

[94] K. J. Blow and D. Wood, *IEEE J. Quantum Electron.* **25,** 2665 (1989).

[95] P. V. Mamyshev and S. V. Chernikov, *Opt. Lett.* **15,** 1076 (1990).

[96] S. V. Chernikov and P. V. Mamyshev, *J. Opt. Soc. Am. B.* **8,** 1633 (1991).

[97] P. V. Mamyshev and S. V. Chernikov, *Sov. Lightwave Commun.* **2,** 97 (1992).

[98] R. H. Stolen and W. J. Tomlinson, *J. Opt. Soc. Am. B.* **9,** 565 (1992).

[99] J. T. Manassah, *Appl. Opt.* **26,** 3747 (1987); J. T. Manassah and O. Cockings, *Appl. Opt.* **26,** 3749 (1987).

[100] J. Hermann and J. Mondry, *J. Mod. Opt.* **35,** 1919 (1988).

[101] R. Osborne, *J. Opt. Soc. Am. B.* **6,** 1726 (1989).

[102] D. N. Cristodoulides and R. I. Joseph, *IEEE J. Quantum Electron.* **25,** 273 (1989).

[103] Y. B. Band, J. R. Ackerhalt, and D. F. Heller, *IEEE J. Quantum Electron.* **26,** 1259 (1990).

[104] V. A. Vysloukh and V. N. Serkin, *JETP Lett.* **38,** 199 (1983).

[105] E. M. Dianov, A. M. Prokhorov, and V. N. Serkin, *Opt. Lett.* **11,** 168 (1986).

[106] J. P. Gordon, *Opt. Lett.* **11,** 662 (1986).

[107] F. M. Mitschke, and L. F. Mollenauer, *Opt. Lett.* **11,** 659 (1986).

[108] P. Beaud, W. Hodel, B. Zysset, and H. P. Weber, *IEEE J. Quantum Electron.* **23,** 1938 (1987).

[109] A. S. Gouveia-Neto, A. S. L. Gomes, J. R. Taylor, and K. J. Blow, *J. Opt. Soc. Am. B.* **5,** 799 (1988).

[110] A. Höök, D. Anderson, and M. Lisak, *J. Opt. Soc. Am. B.* **6,** 1851 (1989).

[111] E. A. Golovchenko, E. M. Dianov, P. V. Mamyshev, A. M. Prokhorov, and D. G. Fursa, *J. Opt. Soc. Am. B.* **7,** 172 (1990).

[112] C. Headley III, and G. P. Agrawal, *J. Opt. Soc. Am. B* **10,** 2383 (1993).

[113] R. F. de Souza , E. J. S. Fonseca, M. J. Hickmann, and A. S. Gouveia-Neto, *Opt. Commun.* **124,** 79 (1996).

[114] K. Chan and W. Cao, *Opt. Commun.* **158,** 159 (1998).

[115] J. Santhanam and G. P. Agrawal, *Opt. Commun.* **222,** 413 (2003).

[116] K. Tamura and M. Nakazawa, *Opt. Lett.* **21,** 68 (1996).

[117] M. E. Fermann, V. I. Kruglov, B. C. Thomsen, J. M. Dudley, and J. D. Harvey, *Phys. Rev. Lett.* **26,** 6010 (2000).

[118] V. I. Kruglov, A. C. Peacock, J. D. Harvey, and J. M. Dudley, *J. Opt. Soc. Am. B* **19,** 461 (2002).

[119] J. Limpert, T. Schreiber, T. Clausnitzer, K. Zollner, H. J. Fuchs, E. B. Kley, H. Zellmer, and A. Tunnermann, *Opt. Exp.* **10,** 628 (2002).

[120] A. C. Peacock, N. G. R. Broderick, and T. M. Monro, *Opt. Commun.* **218,** 167 (2003).

[121] V. I. Kruglov, A. C. Peacock, J. D. Harvey, *Phys. Rev. Lett.* **90,** 113902 (2003).

[122] C. Finot, G. Millot, C. Billet, and J. M. Dudley, *Opt. Exp.* **11,** 1547 (2003).

[123] T. Hirooka and M. Nakazawa, *Opt. Lett.* **29,** 498 (2004).

# Chapter 3

# Distributed Raman Amplifiers

**Karsten Rottwitt**

There are mainly three reasons for the recent renewed interest in Raman amplification. One is the capability to provide distributed amplification, another is the possibility to provide gain at any wavelength by selecting appropriate pump wavelengths, and the third is the fact that the amplification bandwidth may be broadened simply by adding more pump wavelengths.

In this chapter the application, as a means to achieve distributed amplification, will be described in detail. The term *distributed amplification* refers to the method of cancellation of the intrinsic fiber loss. As opposed to discrete amplification, which will be discussed in Chapter 4, the loss in distributed amplifiers is counterbalanced at every point along the transmission fiber in an ideal distributed amplifier. As a consequence, the signal-to-noise ratio is improved when compared to using discrete amplification, which is sometimes also referred to as lumped amplification where gain is provided at a discrete location. This holds true even when the discrete amplifier is assumed ideal regarding its noise properties.

In the late eighties, Raman amplification was perceived as the way to overcome attenuation in optical fibers and research on long haul transmission was carried out demonstrating transmission over several thousand kilometers using distributed Raman amplification [1]. However, with the development and commercialization of erbium-doped fiber amplifiers through the early nineties, work on distributed Raman amplifiers was abandoned because of its poor pump power efficiency when compared to erbium-doped fiber amplifiers (EDFAs). In the mid-nineties, high-power pump lasers became available and in the years following, several system experiments demonstrated the benefits of distributed Raman amplification including repeater-less undersea experiments [2], high-capacity terrestrial [3, 4] as well as submarine systems transmission experiments [5], shorter span single-channel systems including 320 Gbit/s pseudolinear transmission [6], and in soliton systems [7].

The capability to improve noise performance by using distributed amplification was demonstrated in distributed erbium-doped amplifiers in the early nineties [8, 9] and more recently in the above-mentioned system demonstrations using distributed Raman amplification. In both of these distributed amplification schemes, the distributed erbium-doped fiber amplifiers or the distributed Raman fiber amplifiers, the transmission fiber is, in itself, turned into an amplifier.

This makes it more challenging to optimize the fiber design with respect to amplifier performance because the fiber at the same time has to be optimized for signal transmission, that is, with constraints to the group velocity dispersion and the nonlinearities at the signal wavelengths.

In addition, when evaluating the performance of a distributed amplifier, further noise sources need attention compared to the noise sources known from conventional lumped amplified transmission systems. These new noise sources include effects due to Rayleigh scattering, nonlinear interactions between pump and signal channels, and in the case of Raman amplification effects such as pump and signal crosstalk. These noise sources become relevant because gain is accumulated over tens of kilometers in distributed amplifiers, and especially the Raman process has a very fast response time on the order of femtoseconds [10], enhancing the pump and signal crosstalk.

In this chapter the benefits, challenges, and fundamental designs of distributed Raman amplifiers will be discussed. In Section 3.1, the benefits of distributed Raman amplification will be highlighted with emphasis on some of the first experimental demonstrations of the advantages including *improved noise performance* as compared to discrete amplification, *upgradeability* of existing discretely amplified systems, *amplification at any wavelength* controlled simply by selecting the appropriate pump wavelength, *extended bandwidth* achieved by using multiple pumps when compared to amplification using EDFAs, and finally control of the *spectral shape of the gain and the noise figure*, which may be adjusted by combining and controlling the wavelength and power among multiple pumps.

The above benefits are all based on the characteristics of the Raman amplification process. In Section 3.2, these will be described with a focus on those especially relevant for the benefits of distributed Raman amplification. Emphasis will be given to the benefits from the Raman scattering process including the gain spectrum achieved when combining multiple pump wavelengths, and to the benefits from the distributed Raman gain process that is the noise properties including accumulation of spontaneous emission.

In Section 3.3, the discussion of characteristics will continue, however with emphasis on major challenges in using distributed Raman amplifiers. These discussions include Rayleigh scattering, nonlinear impairments on the signal, and the time response enabling pump–signal cross talk, and signal–pump–signal cross talk.

Because of the fast response time from the Raman process, the conventional pumping configuration is the counter- or backward-pumped amplifier in which the pump and signal propagate in opposite directions; Section 3.4 highlights the counterpumped Raman amplifier configurations including pumping using multiple wavelengths, broadened pumps, and frequency swept pumps as well as time division multiplexed pumps.

In Section 3.5, we describe more advanced distributed Raman amplifiers that represent solutions to some of the challenges in using distributed Raman amplifiers. Most important is the co- or forward-pumped amplifier configuration, in which the pump and signal propagate in the same direction. The implementation of this configuration offers an improved noise performance over a counter-pumped distributed Raman amplifier. The desire for such implementation has led to the development of new pumping configurations of distributed Raman amplifiers. These include second-order or cascaded pumping and the use of quiet pumps.

# 3.1 Benefits of Distributed Raman Amplification

In the mid-nineties, research led to the development and commercialization of powerful high-power lasers. Examples of such include a laser consisting of several broad area lasers combined into a double-clad ytterbium fiber followed by a cascaded Raman resonator [11]. This seeded further research related to the application of such high-power lasers and enabled especially research on Raman amplifiers.

The research on distributed Raman amplification has demonstrated several benefits of this amplification method that will be highlighted in this section. In Section 3.1.1, the possibility for upgrading an existing transmission system will be described. In Section 3.1.2, we describe the possibility for improving the capacity and reach in repeatered transmission systems. Both of the benefits in Section 3.1.1 and Section 3.1.2 are based on the improved noise performance due to the distributed amplification. In Section 3.1.3 and Section 3.1.4, we focus on benefits that are specific to the Raman amplification. Section 3.1.3 discusses the ability to extend the useable bandwidth by using multiple pump wavelength and in Section 3.1.4 the ability to obtain gain at any wavelength is discussed.

The description of various benefits is not a complete or thorough description but serves to motivate the significant research interest in the topic of distributed Raman amplification that will be described in the remainder of this chapter.

## 3.1.1   Upgradability

One of the first experiments that demonstrated a benefit of distributed amplification was in an undersea system experiment in which remotely pumped inline EDFAs were pumped at 1480 nm [12]. In this experiment, 145 mW of pump power was launched into a 66-km-long dispersion shifted fiber at the receiver side. This gave a Raman gain of 5.3 dB for a signal at 1558 nm. The remaining pump power launched into the remote EDFA was 3.8 mW, giving rise to a 15.4-dB gain. The power budget was improved by 11.0 dB due to the remote EDFA and the Raman gain.

This demonstrated the potential application of distributed Raman amplification as a means to upgrade existing systems. A more elaborate investigation was reported in 1997 using the above-mentioned high-power lasers [13]. Without the use of any remotely pumped EDFA a power budget improvement of 7.5 dB was demonstrated by launching 1.0 W of 1453-nm pump light from the receiver end. This improvement was sufficient either to increase the bitrate in a single channel from 2.5 Gb/s to 10 Gb/s or to increase the channel count from one 10 Gb/s channel to four 10 Gb/s channels centered around 1550 nm.

By propagating pump power through a significant part of the signal path, Raman gain is obtained along the signal path in the above experiments. As a result, the signal is prevented from decaying as much as it otherwise would have in the absence of pump power, leading to a signal-to-noise ratio at the output that is much better when compared to the same system without the Raman gain.

Since these initial experiments, numerous experiments have demonstrated the improved noise performance when comparing distributed against lumped amplification. To facilitate such a comparison between lumped and distributed amplification, the effective noise figure, as defined in Chapter 1, was introduced in [14] and measured versus pump power in a dispersion-shifted fiber and a silica core fiber. It was found that there was an optimum pump power level of 600 mW resulting in an effective noise figure of −3 dB in the investigated dispersion-shifted fiber. In a silica core fiber, the corresponding optimum pump power was found to be 1300 mW, resulting in an effective noise figure of −4 dB. Below this optimum pump power the noise performance improves with increasing pump power because the minimum signal power within the span increases with increasing pump power. For higher pump-power levels, the optical signal-to-noise ratio decreases,which is explained by Rayleigh scattering. These results spurred renewed interest in distributed Raman amplification since they clearly demonstrated that the method could be used as a means to upgrade existing unrepeatered systems simply by applying a pump module at the output end of the transmission allowing for distributed Raman amplification.

## 3.1.2 Noise Improvements in Repeatered Systems

One of the major benefits, only obtainable through the use of distributed amplification, is that signal gain may be pushed into a transmission span preventing the signal from decaying as much as it otherwise would have if no amplification was provided within the span. As a consequence, the signal-to-noise ratio does not drop as much as it would have in a system based on transmission through a passive fiber followed by a discrete amplifier.

This improved noise performance may be used in different ways. One way is to extend the reach between repeaters [15], another is to extend the total reach of a transmission system, and finally a third is to improve the transmission capacity in a long-haul transmission system. The latter has been demonstrated in numerous experiments. In 1997, an all-Raman transmission experiment made it possible to transmit 10 channels each carrying 10 Gbit/s through 7200 km of fiber [5]. Another demonstration illustrating the distributed Raman amplification as being an enabling technology was presented in 1999. In this experiment, 40 channels each carrying 40 Gbit/s, were transmitted through four 100-km-long fiber spans [15a].

In relation to these experiments it is important to note that the Raman amplification process is inherent to germano-silicate fibers. This is a great advantage because, as no special fiber is needed to obtain Raman amplification, the transmission fiber acts as an amplifier itself. However, any steps taken to alternate the Raman gain coefficient by changing the fiber design may also impact the transmission of the signals.

### 3.1.3   Bandwidth/Flatness

Another major benefit of Raman amplification, though not specific to distributed amplification, is that the amplification bandwidth may be extended simply by combining power at multiple pump wavelengths. In an experiment presented in 1998, two different pump wavelengths, one at 1453 nm and one at 1493 nm, were combined, which resulted in amplification over more than 92 nm [16]. In the experiment, the 92-nm bandwidth was defined by the wavelength range over which there was sufficient amplification to counterbalance the intrinsic fiber loss occurring through a 45-km-long span. The measured gain and noise figure is illustrated in Figure 3.1. In the experiment, 206 mW was used at 1453 nm whereas 256 mW of pump power was used in the 1493-nm pump beam. The effective noise figure, as defined in Chapter 1, is the noise figure read from Figure 3.1 minus the intrinsic loss (~9 dB).

After the result was described, several publications demonstrated that the bandwidth may be expanded to nearly 100 nm with a flatness that improves as the number of pump wavelengths increases. In one experiment [17], an average gain of 6.5 dB with a flatness of ±0.5 dB was achieved over 100 nm by using 12 semiconductor pump lasers.

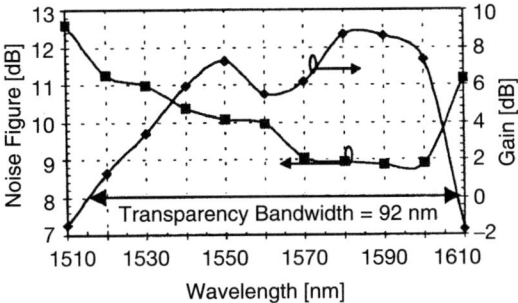

**Figure 3.1:** Measured gain and noise figure of a 45-km-long distributed amplifier. The fiber was a dispersion-shifted fiber and 206 mW at 1453 nm was launched together with 256 mW at 1495 nm. (Figure from [16].)

### 3.1.4    Gain at Any Wavelength

Another important feature of the Raman amplification process is that amplification is achievable at any wavelength by choosing the pump wavelength in accordance with the signal wavelength; that is, in germano-silicate fibers the maximum gain is achieved when the pump is located at a frequency 13 THz above the frequency of the signal. This was used in 1998 when an experiment was performed with signals located both at 1400 nm and at 1550 nm [18]. The first wavelength was chosen primarily to demonstrate transmission through a so-called all-wave fiber in which the usual OH losses at 1385 nm are removed. Amplification at 1400 nm was provided through the use of a Raman amplifier. In the experiment, four channels at 10 Gbit/s around the 1400-nm wavelength were transmitted simultaneously with 16 channels at 2.5 Gbit/s centered at 1550 nm. The transmission span was 80 km long.

## 3.2    Beneficial Characteristics of Distributed Raman Amplifiers

Raman amplification is a result of stimulated Raman scattering of light on optical phonons in glass. This process has some unique characteristics, some of which are advantageous in the use as an amplifier and some that are challenging and need to be given more attention when designing an amplifier. The first include distributed amplification, gain at any wavelength, and intrinsic to germano-silicate fibers. The latter include response time, Rayleigh scattering, nonlinearities, and polarization dependence. We return to these challenges in Section 3.3.

In some of the early work in the eighties on application of Raman scattering for counterbalancing the intrinsic fiber loss, the pump-power efficiency was a significant problem and powerful color-center lasers were necessary to provide sufficient power to counteract the intrinsic fiber loss. The pump-power efficiency, measured in Raman gain per unit pump power, approximate 60 dB/W in a dispersion shifted fiber that is much longer than the effective length as defined in Chapter 1 ($\sim$20 km). Thus, when the erbium-doped fiber amplifier with an efficiency in gain per unit pump power of several dB/mW was matured, much work on Raman amplifiers was halted. However, with the recent development of high-power lasers, especially broad area semiconductor lasers, the basis for renewed interest in Raman amplifiers was created.

In this section we focus on advantages of the distributed Raman amplifiers and more specifically in Section 3.2.1 on advantages that are not specific to distributed amplification but are specific to the Raman amplification process, whereas in Section 3.2.2 we focus on the main advantages for distributed amplification, that is, the improved noise performance.

## 3.2.1   General Benefits from the Raman Scattering Process

The application of stimulated Raman scattering in optical amplifiers has many general advantages to both discrete and distributed Raman amplifiers. In the following we briefly highlight the general advantages that have been exploited in Raman amplifiers before we give a detailed description of the specific advantages of distributed Raman amplifiers.

### 3.2.1.1   Simple—Intrinsic to the Fiber

Raman scattering is a process that occurs in germano-silicate fibers. Thus, it is present in all glass-based transmission fibers. As a consequence, all that is needed to turn a transmission span into an amplifier is one or more pump lasers and couplers to get the pump power into the fiber.

This is a great advantage and also makes the amplifier configuration very simple. This has some extra benefits simply because the fewer components that go into an amplifier, the more reliable the amplifier is.

### 3.2.1.2   Gain at Any Wavelength

Since the Raman process originates from the interaction of a pump and a signal with phonons, as discussed in Chapter 2, gain may be achieved if the difference between the pump and signal energy matches the phonon energy. As a consequence, gain may be achieved at any signal wavelength simply by choosing an appropriate pump wavelength. In typical germano-silica-based transmission fibers, the phonon energy, or more specifically the phonon frequency, has a large distribution ranging from zero to more than 26 THz as represented in the Raman gain coefficient spectrum; see Figure 3.2.

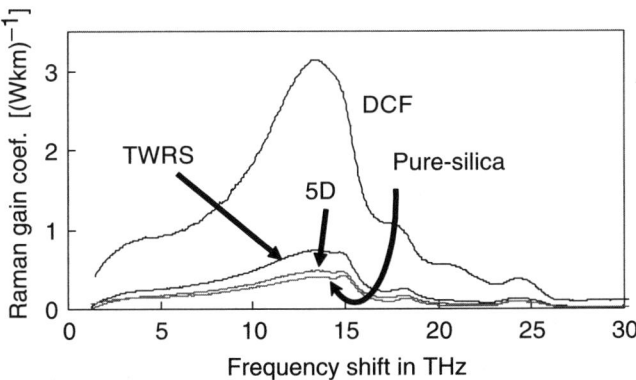

**Figure 3.2:** Wavelength spectrum of the Raman gain coefficient measured on four different fiber types as marked in the figure (DCF, dispersion compensation fiber with a high germanium content; TWRS, OFS TrueWave RS fiber, dispersion shifted high capacity transmission fiber with reduced dispersion slope; 5D, nondispersion shifted fiber; and Pure-silica, a fiber without germanium). The frequency shift between pump and signal is in THz ($= c/\lambda_p - c/\lambda_s$). The pump beam is 1425 nm and unpolarized.

### 3.2.1.3    Predictable Gain Coefficient

Figure 3.2 illustrates that maximum Raman gain is achieved when the frequency difference between the pump and the signal equals 13 THz. The spectrum in the figure is recorded using 1425-nm light. Scaling the spectrum to other pump wavelengths, may be done by multiplying with the ratio of 1425 nm relative to the new pump wavelength. In addition, the Raman gain coefficient also needs to be scaled in accordance with the wavelength dependence of the spatial overlap between the pump and the signal modes, in the following referred to as the Raman effective area.

The Raman gain coefficient scales inversely with the Raman effective area. This latter scaling may be done in different ways. The simplest approximation is achieved by replacing the Raman effective area with the effective area known from self-phase modulation. To perform this scaling one may use experimental data on the wavelength dependence of the self-phase modulation effective area. Theoretically, the effective area and its wavelength dependence may be evaluated by assuming a Gaussian mode profile or calculated from the refractive index profile of the particular fiber [19].

Using the theory described in [19] it is possible to predict the Raman gain coefficient in an arbitrary germano-silicate fiber. In this method the Raman gain coefficient is calculated on the basis of the refractive index profile of the fiber and the corresponding transverse mode profiles of the pump and signal. In addition, the radial distribution of germano-silicate and silica is evaluated from the refractive index profile. From these data the Raman gain coefficient is then determined by the use of material spectra of the Raman scattering, overlap integrals of pump and signal, and the germano-silicate cross-sectional distribution.

Figure 3.3a illustrates the predicted and measured spectrum of the Raman gain coefficient in a OFS TrueWave RS fiber in accordance with [19]. The figure illustrates that it is possible to predict the Raman gain coefficient to a high degree of accuracy. Figure 3.3b shows predicted Raman gain coefficients from a so-called graded refractive index fiber; that is, the refractive index difference $\Delta$ between the core and the cladding is expressed as $\Delta(r) = \Delta(0)[1 - (r/a)^{\alpha}]$, $(r < a)$, where $a$ is the core radius. The curves in the figure are evaluated for the case in which the refractive index $\Delta(0)$ corresponds to 5 mol% germanium and a core radius $a$ of 3 $\mu$m. The spectrum of the Raman gain coefficient is calculated for various values of

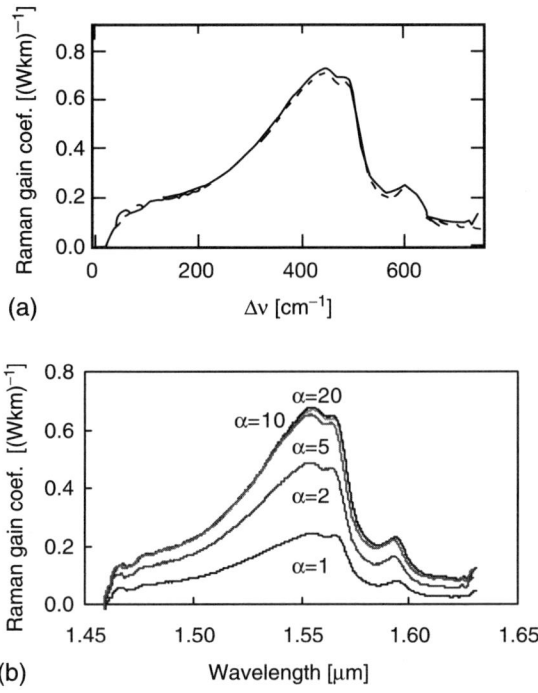

**Figure 3.3:** (a) Raman gain coefficient of a True Wave RS fiber predicted (solid) and measured (dashed). From [19], (b) predicted Raman gain coefficients for a graded refractive index profile fiber.

$\alpha$; $\alpha = 1$ corresponds to a triangular profile and $\alpha \to \infty$ corresponds to a rectangular index profile. As seen in the figure, 70% of the asymptotic spectrum, achieved for the step index profile $\alpha \to \infty$, is achieved for alpha larger than 2.

### 3.2.1.4 Combining Pump Wavelengths

As opposed to some other optical amplification methods, for example, erbium-doped fiber amplifiers, where gain is associated with a material specific emission spectrum, the wavelength range over which amplification is obtained may be extended simply by using a multiple of pump wavelengths. The composite amplification spectrum is determined from the mutual interactions among the pump wavelengths.

This mutual interaction among pump and signal beams may be described through coupled equations and the composite spectrum predicted [20]. The composite amplification spectrum may thus be modified by selecting appropriate pump wavelengths as well as the appropriate power in each wavelength. Examples will be given in Section 3.4.

## 3.2.2   Noise Properties of Distributed Raman Amplifiers

Even though the intrinsic loss of a transmission fiber is extremely low, about 0.2 dB/km at 1550 nm, the loss is responsible for one of the major limitations when transmitting light through an optical fiber. To overcome the loss, amplification is needed. This may be achieved all optically, for example, by passing the signal through an erbium-doped fiber amplifier or a Raman amplifier. As opposed to the erbium-doped fiber amplifier, Raman amplification is obtained without codoping the fiber; that is, the process is intrinsic to the fiber, as described in Section 3.2.1. Because the pump power propagates through many kilometers of the fiber, the signal attenuation is counterbalanced over a distance of tens of kilometers, and the signal may even be amplified.

One of the consequences of distributing the gain over many kilometers is that the signal is prevented from decaying as much as it otherwise would have if the amplification were discrete and located at the end of the fiber span. The result of this is that the signal-to-noise ratio is improved when applying distributed amplification rather than lumped amplification. This is illustrated in Figure 2.12.

The above-mentioned benefits of distributed amplification may also be explained by using an example. Consider the electrical signal-to-noise ratio at the output of a 100-km-long fiber span, and more specifically one may compare three cases. In the first case, not illustrated in Figure 2.12, no amplification is provided to counterbalance the loss. In the second case the loss in the 100-km-long fiber is counterbalanced by an ideal erbium-doped fiber amplifier located at the output end of the fiber. This is case b in Figure 2.12. Finally, in the third case Raman amplification is used to counterbalance the attenuation in the fiber. This is case Raman in Figure 2.12.

In the first case, the passive fiber, the signal-to-noise ratio decays exponentially throughout the fiber. If a loss of 0.2 dB/km is assumed, the signal-to-noise ratio has dropped by 20 dB after 100 km. In the second case, an amplifier is now located after the 100-km passive fiber. Assuming that the amplifier is ideal, for example, an ideal erbium-doped fiber amplifier with a noise figure $F = 2 - 1/G$, that is, $F = 3$ dB for high gain, $G$, the signal-to-noise ratio will drop an additional 3 dB at the amplifier. In this case the signal-to-noise ratio has dropped a total 23 dB; see case b in Figure 2.12.

In the third case, the distributed Raman amplifier, the signal-to-noise ratio also decays exponentially as long as there is no pump power. However, when the pump begins to provide gain, the signal-to-noise ratio drops at a slower rate, resulting in a signal-to-noise ratio that is significantly better than in the two other cases. Assuming a pump loss of 0.25 dB/km and a Raman gain coefficient of 0.7 $(Wkm)^{-1}$, the 100-km fiber is transparent when 305 mW of pump power is launched into the fiber. In this case, the signal-to-noise ratio has dropped by 17 dB. Since the noise figure is defined as the decay in signal-to-noise ratio from input to output, the noise figure in this example equals 17 dB. The noise figures in these three examples are summarized in Table 3.1.

The results in Table 3.1 depend on the length of the transmission span; the Raman gain, that is, the pump power; the signal and pump wavelengths, and so forth. It is noted that the table demonstrates the benefits of distributed Raman amplification by comparing the Raman amplifier against the passive fiber and the amplifier, where the amplifier is located at the output end of the fiber. If, on the other hand, the discrete amplifier is located at the beginning of the span, the noise figure would be lower when compared directly to that of the backward pumped Raman amplifier in Table 3.1; see also Figure 2.12.

The comparisons in Table 3.1 illustrate the reason for introducing the so-called effective noise figure or the equivalent noise figure as discussed in the text following Equation (2.2.11) in Chapter 2. The effective noise

**Table 3.1 Noise figure in three examples of 100-km spans, based on electrical signal-to-noise ratio degradation**

| Case | Noise Figure, $F$ (dB) |
|---|---|
| Passive fiber | 20 |
| Passive fiber and amplifier | 23 |
| Raman amplifier | 17 |

figure is simply the noise figure that one would require from a discrete amplifier placed after a passive fiber of the same length as the Raman amplifier assuming that the Raman amplifier and the imaginary transmission span should perform equally well. So, for example, the effective noise figure of the Raman amplifier in Table 3.1 would be $-3$ dB.

In the above example, the Raman gain exactly counterbalances the intrinsic fiber losses. In other words, since a loss of 0.2 dB/km is assumed in Table 3.1, that is, a total loss of 20 dB over the 100 km, the Raman gain equals 20 dB. This is often referred to as the on–off Raman gain, because it is simply the gain one gets by turning the pump laser on. Note that the gain of the discrete amplifier used in Table 3.1 equals the on–off Raman gain. However, often it makes more sense just to consider the gain or loss of a fiber span, that is, the output power relative to the input power. In the example just referred to, the gain equals 0 dB.

The noise figure is evaluated as described in Chapter 2, from the spontaneously emitted power, or even simpler from the number of spontaneous photons in the signal mode $n_{\text{ASE}}$, that is, $F = (2n_{\text{ASE}} + 1)/G$, where ASE refers to amplified spontaneous emission. The number of spontaneously emitted photons may be evaluated directly from the noise power $P_{\text{ASE}}$ given by [21]:

$$P_{\text{ASE}} = h\nu B_o \eta_T g_R \int_{z=0}^{L} P_p(z)/G(z)\, dz, \qquad (3.2.1)$$

where $h\nu$ is the photon energy, $B_0$ the considered bandwidth, $\eta_T$ the thermal equilibrium phonon number [22], $P_p(z)$ the pump power, $G(z)$ the gain of the fiber span, $g_R$ the Raman gain coefficient, and $L$ the length of the fiber span.

In the general case in which the loss rate at the pump wavelength differs from the loss rate at the signal wavelength, a simple numerical integration is required to evaluate the noise figure. However, when the loss rate at the pump $\alpha_p$ and signal $\alpha_s$ are equal, $\alpha = \alpha_s = \alpha_p$ the noise figure may be evaluated analytically, as will be demonstrated in the following.

Using the approximation of equal loss, the spontaneous emission power at the output end of the fiber equals

$$P_{\text{ASE}} = h\nu B_0 \eta_T \left\{ G - 1 + \frac{G\alpha}{g_r P_p^L} \left\{ \exp(\alpha L) - \frac{1}{G} \right\} \right\}, \qquad (3.2.2)$$

where $G$ is the net gain $G = \exp(g_r P_p^L L_{\text{eff}}) \exp(-\alpha L)$, and $L_{\text{eff}}$ is the effective length $L_{\text{eff}} = (1 - \exp(-\alpha_p L))/\alpha_p$ and $P_p^L$ the launched pump power. Introducing the on–off Raman gain $G_R$ as $G_R = \exp(g_r P_p^L L_{\text{eff}})$, and considering the case of a transparent amplifier, that is, $G = 1$, this corresponds to a noise figure of

$$F = 1 + 2 \frac{\eta T \alpha}{g_r P_p^L} (G_R - 1). \tag{3.2.3}$$

At the limit when the amplifier length approaches infinity $(L \to \infty)$, the noise figure in decibels approximates

$$F|_{\text{dB}} \to G_R|_{\text{dB}} + 10 \log \left( 2 \frac{\eta T \alpha}{g_r P_p^L} \right). \tag{3.2.4}$$

This equation shows that for large fiber lengths the noise figure grows nearly decibel for decibel in the on–off Raman gain, whereas when the length approaches zero $(L \to 0)$ the noise figure approaches 0 dB, as would be expected according to Eq. (3.2.3).

The noise figure as a function of the on–off Raman gain for a transparent backward pumped Raman amplifier is displayed in Figure 3.4. The figure displays the noise figure both when assuming that the loss rate at the pump wavelength equals the loss at the signal wavelength, $\alpha_p = \alpha_s = 0.20$ dB/km, and more realistically when the loss rate at the pump is $\alpha_p = 0.25$ dB/km and the loss rate at the signal wavelength is $\alpha_s = 0.20$ dB/km. The pump and signal wavelengths are chosen to match the gain peak in the Raman gain spectrum; that is, in Figure 3.4, the pump wavelength $\lambda_p = 1455$ nm and the signal wavelength $\lambda_s = 1555$ nm.

Finally, Figure 3.4 also illustrates the ideal noise figure achieved in the limit where the loss is counterbalanced along every point of the span. This limit is [23]

$$F|_{\text{dB}} \approx 10 \log \{2\alpha_s L + 1\}, \tag{3.2.5}$$

where $\alpha_s$ is the loss rate and $L$ is the span length; that is, $\exp(\alpha_s L)$ is the gain necessary to counterbalance the span loss. If, for example, a transmission fiber is 100 km long with a loss rate of 0.2 dB/km, the span loss equals 20 dB, resulting in an asymptotic value in the noise figure of 10 dB.

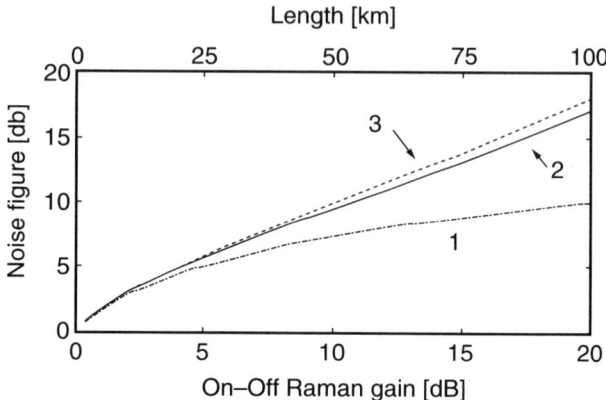

**Figure 3.4:** The noise figure versus the on–off Raman gain or similarly the length of a transparent span. The lower trace (1) represents the ideal lower limit, whereas the middle trace (2) represents the case in which the loss rate at the pump equals the loss rate at the signal wavelength 0.2 dB/km. Finally the upper trace (3) is the realistic case in which the loss at the pump is $\alpha_p = 0.25$ dB/km and the loss at the signal is $\alpha_s = 0.2$ dB/km. All traces assume room temperature.

The numbers displayed in Table 3.1 correspond to 20 dB in on–off Raman gain in Figure 3.4.

From the two curves in Figure 3.4 illustrating the noise figure for realistic loss rates and equal loss rates, it is obvious that assuming the same loss rate may be acceptable to obtain simple insight into amplifier behavior. However, accurate predictions clearly require the use of accurate values of loss rates.

Comparing the three graphs on Figure 3.4, the effect of excursions in the signal power becomes clear. As the on–off Raman gain, or in analogy the length of the fiber span, increases the signal excursions along the span deviate more and more from the ideal distributed amplification. This causes an 8-dB penalty for the realistic case relative to the ideal distributed amplifier when the span length equals 100 km or 20 dB on–off Raman gain.

In Table 3.1 the improvement in noise figure, when comparing a Raman amplifier against a passive fiber followed by a discrete amplifier, was illustrated in a specific case in which the amplifier length was 100 km. However, this improvement strongly depends on the amplifier length. Assuming as

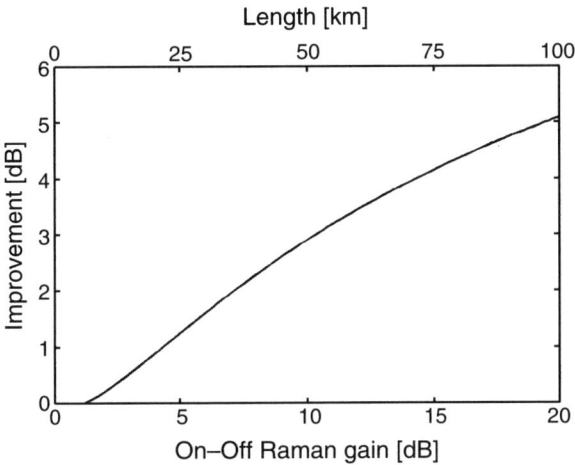

**Figure 3.5:** The improvement in noise figure versus fiber length on–off Raman gain. The improvement is relative to a passive span succeeded by an ideal discrete erbium-doped fiber amplifier. The on–off gain may be directly translated to fiber length by dividing the values on the *x*-axis with the signal loss rate.

in Table 3.1 that the amplifier is transparent and the noise figure of the discrete amplifier is ideal, the expected improvement as a function of the on–off Raman gain is displayed in Figure 3.5.

As discussed earlier, the benefits from the distributed Raman amplification relate to the fact that signal gain is pushed into the transmission span, preventing the signal from decaying as much as it otherwise would have. It is obvious that by pumping in the forward direction, the accumulated spontaneous emission will be reduced. This is illustrated in Figure 3.6, which displays the noise figure of a transparent 100-km-long Raman amplifier; that is, the gain is adjusted exactly to counterbalance the intrinsic fiber loss, where the ratio of backward to forward pump power is varied. More specifically, the ratio $x$ is defined as the backward pump power $P_b$ relative to the total pump power $P_{tot}$. The forward pump power $P_f$ is then related to the total pump power as $P_f = P_{tot}(1 - x)$. The value 0 on the *x*-axis corresponds to a purely forward-pumped amplifier whereas the value 1 corresponds to a conventional purely backward-pumped amplifier.

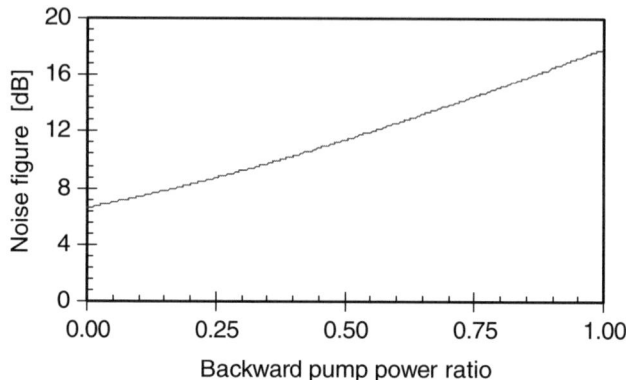

**Figure 3.6:** Noise figure versus ratio of backward to forward pump power. The data are calculated using the simplified model [24], and $\alpha_p = 0.25$ dB/km, $\alpha_s = 0.20$ dB/km, and $g_r = 0.7$ (Wkm)$^{-1}$. Denoting the backward pump power ratio $x$, the forward pump power $P_f$ equals $P_f = P_{tot}(1 - x)$, whereas the backward pump power $P_b$ equals $P_b = P_{tot}x$, room temperature.

Figure 3.6 illustrates an improvement of approximately 10 dB when using forward pumping as opposed to backward pumping in the example of a 100-km-long amplifier. However, the benefits in signal-to-noise ratio do not take into account the possible drawback due to Rayleigh scattering, nonlinear effects, or possible cross-coupling between the signal and the pump. The latter is most detrimental when the pump and signal propagate in the same direction, that is, in the forward-pumped Raman amplifier configuration. Further details regarding these issues will be discussed in the following section.

## 3.3   Challenging Characteristics of Distributed Raman Amplifiers

Although the distributed Raman amplification process offers several advantages, it is also accompanied by challenges. Most of these can be overcome by choosing an amplifier design that mitigates the challenges.

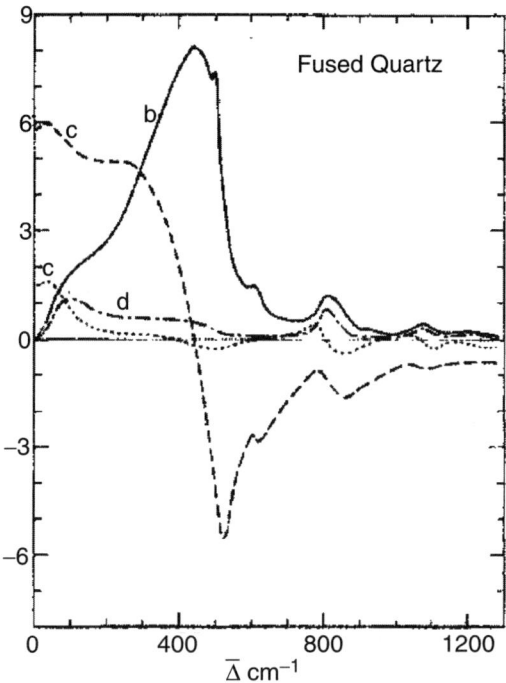

**Figure 3.7:** The Raman susceptibility ($10^{15}$ times the values in e.s.u. units) versus the frequency shift between input and scattered light measured in wavenumbers ($cm^{-1}$). (From [25].)

One of the challenges is the polarization dependence of the Raman gain. The Raman scattering is strongly polarization dependent. Consider an experiment in which plane polarized light is launched into a silica sample and detected at the output of the sample through a polarizer. If the output polarizer is oriented first in the same direction as at the input and then turned 90° the two spectra labeled b and d in Figure 3.7 appear.

Figure 3.7 displays the Raman susceptibility. Curves a and c are the real part of the susceptibility whereas curves b and d are the imaginary part, which is proportional to the Raman gain coefficients. The curves are measured by launching plane polarized light at a sample and in cases a and b the measured output polarization is parallel to the input whereas for cases c and d the output polarization is orthogonal.

A standard single-mode fiber, in which both the pump and the signal propagate in the fundamental mode, supports two orthogonal polarization states in each mode. Thus, even though the fiber is single mode, the pump and signal may propagate in orthogonal polarizations. In this case the gain coefficient will vary as shown in Figure 3.7 between curve b if the pump and signal have the same polarizations and curve d if they are orthogonal. To overcome this polarization dependence, Raman amplifiers are typically pumped using unpolarized pump beams. As a result the Raman gain coefficient is the average of the curves b and d [9]. Un-polarized pump beams may easily be obtained using the types of high power pump lasers described in Section 3.1 or by polarization multiplexing two uncorrelated semiconductor lasers. Application of such unpolarized pump lasers is assumed in the following. A more careful analysis of the polarization-dependent gain is shown in Chapter 2.

In addition to the polarization dependence, the Raman gain also depends on the spatial overlap of the pump and signal mode as mentioned in Section 3.2.1. On one hand, this allows for a fiber designed for optimum spatial overlap between the pump and signal modes and the core. On the other hand, the spatial overlap restricts the design space because single-mode propagation of both pump and signal is desired.

To illustrate this, Figure 3.8 displays the Raman gain coefficient as a function of the radius of the core, $a$, in a step index fiber with an extremely high germanium concentration $\sim$20 mol%. A change in the core radius is analogous to a change in the normalized frequency $V$ defined as $V = (2\pi/\lambda_s)a\sqrt{n_1^2 - n_2^2}$, where $\lambda_s$ is the wavelength of the signal $n_1$, and $n_2$ is the refractive index of the core and cladding, respectively. A pump at 1450 nm and a signal at 1555 nm is assumed. When $V$ exceeds 2.405 the fiber becomes multimode. For a core diameter below 3.6 $\mu$m the fiber is single mode for both pump and signal. As explained above, even in this case the fiber supports two modes of orthogonal polarizations, leading to the curves b and d in Figure 3.7.

In Figure 3.8 the Raman gain coefficient for 13 THz between pump and signal is illustrated for the case in which the pump and signal are in the same polarization. In the figure, the diameter of the core is varied to describe the single-mode and multimode operation. In the single-mode region for both pump and signal the maximum Raman gain coefficient is found for a core diameter close to 2.8 $\mu$m. This is the geometry that provides the maximum overlap of both the pump and the signal mode and their overlap to the core.

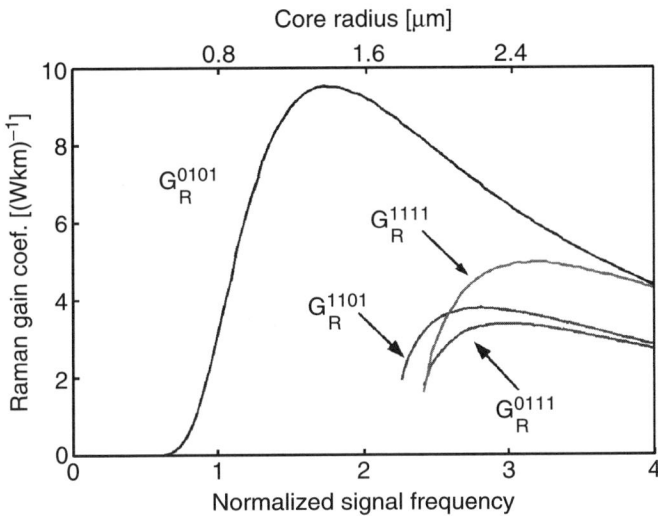

**Figure 3.8:** Calculated Raman gain coefficients versus the normalized frequency at the signal wavelength. Each curve is labeled according to the mode of the pump and signal. For example, $G_R^{1101}$ is the gain coefficient corresponding to the pump being in the LP$_{11}$ mode and the signal in the LP$_{01}$ mode. The relatively large gain coefficients are due to the high index contrast between core and cladding, that is, a large content of germanium in the center.

When the diameter of the core is between 3.6 and 3.8 μm the pump becomes multimode whereas the signal is still single mode. As a consequence the Raman gain coefficient may now assume two different values depending on whether the pump is in the fundamental mode or in the higher order mode. This is the case even when the pump and signal are in the same polarization state. In addition to the reduced Raman efficiency of the amplifier (gain per unit pump power), the variation in the Raman gain coefficient may translate into a noise contribution, if for example the power in one mode at some random distance within the span transfers to the other mode.

Finally, when the diameter of the core exceeds 3.8 μm both the pump and the signal become multimode. Thus the Raman gain coefficient may now assume several values because of the degeneracy of pump and signal modes, including also degeneracy in the angular coordinate of a given cross-section of the fiber [26]. Assuming that the pump and signal are in the same polarization state and the same trigonometric mode, the data in Figure 3.8. are found for the Raman gain coefficients as a function of the core radius of the fiber. If the core radius exceeds 5.7 μm, even more modes are guided, and the description gets further complicated.

Figure 3.8 shows that there is a strong dependence in the Raman gain coefficient due to the mode profile of the pump and signal. In addition, the figure illustrates the complexity with respect to the Raman gain in dealing with more than the fundamental mode of pump and signal. However, in the following we assume that both the pump and the signal are in the fundamental mode. A careful description of the reduced efficiency and the noise properties is outside the scope of this chapter.

Other challenges are more difficult to overcome. Two of the most important are the impact of Rayleigh reflections and nonlinearities. These are discussed in Section 3.3.1 and 3.3.2, respectively. Finally, since the Raman scattering is interactions between light and noninstantaneous vibrational quanta, which are much faster than for example the metastable lifetime of an erbium ion ($\sim$10 ms), the Raman amplifier is sensitive to cross-coupling between the pump and the signal and also sensitive to coupling between two signal channels mediated through the pump; that is, a signal transfers a footprint to the pump which transfers to another signal. These two cross-coupling effects are most critical in forward-pumped Raman amplifiers. In Section 3.3.3 these will be described.

### 3.3.1   Nonlinearities

The refractive index of germanosilicate fibers changes with the intensity of the light propagating through the fiber. The rate of change is characterized by the so-called intensity-dependent refractive index, which for a typical transmission fiber (silica) is $3.2 \cdot 10^{-20}$ m$^2$/W [27]. This intensity-dependent refractive index is extremely important.

The nonlinear propagation effects relevant to the evolution of the electric field amplitude involve a cubic term in the electric field; see Eq. (2.4.3). In a communication system with multiple signal channels, the nonlinear effects are typically classified by the number of field amplitudes involved in the process, each associated with its own frequency.

The term *self-phase modulation* (SPM) is used if all three fields belong to the same frequency as the channel of interest. As a result of SPM the phase of a given signal channel is modulated by the signal itself. In the absence of group velocity dispersion, SPM causes broadening of the pulse in the frequency domain. In a typical transmission system, the interplay of group velocity dispersion and SPM is complicated. Under certain conditions the SPM may be used to counteract the group velocity dispersion in a soliton. In other conditions, it causes unwanted pulse distortion.

A phase change appearing on a signal channel may also be caused by signals, a tone, or two other channels. In this case this is referred to as cross-phase modulation (XPM). The phase modulation caused by other signal channels causes a signal distortion and hence a degradation of the system performance. The XPM is strongly dependent on walk-off between the signal channel and the channels causing the XPM. Thus it depends on the channel's spacing, the group-velocity dispersion, and the polarization of the channels.

Finally, if all three amplitudes in the cubic nonlinear term belong to channels other than the considered, one refers to four-wave mixing (FWM). In this process power is transferred to new frequencies from the signal channels. The appearance of additional waves and the depletion of the signal channels will degrade the system performance through both cross talk and depletion. The efficiency of the FWM depends on channel dispersion and channel spacing.

The above classification is based on the assumption that each channel may be treated as quasi-monochromatic waves. However, in more recent experiments, where the bitrate per channel is pushed above 100 Gbit/s, the assumption of quasi-monochromatic waves is not valid, and a more careful description needs to be carried out, which also includes Intra XPM and FWM [28]. In the following we restrict ourselves to conventional XPM and FWM.

The impact of SPM and XPM is typically evaluated through the accumulated nonlinear phase shift

$$\varphi_{NL}^{SPM} = \int_0^L \gamma(z) P_s(z) dz \qquad \text{for SPM, and} \qquad (3.3.1a)$$

$$\varphi_{NL}^{XPM} = \int_0^L \sum_{i,j} \gamma(z) \psi_i \psi_j dz \quad \text{for XPM,} \qquad (3.3.1b)$$

where $\gamma$ is the nonlinear strength that more specifically is related to the nonlinear refractive index $n_2$ divided by the wavelength $\lambda$ and the effective area $A_{\text{eff}}$ through $\gamma = 2\pi n_2/(\lambda A_{\text{eff}})$ [27]. The sum in Eq. (3.3.1b) is carried out over all channels. Around a signal wavelength of 1.55 μm, a typical transmission fiber has an effective area $A_{\text{eff}}$ around 75 μm². Using these values together with $3.2 \cdot 10^{-20}$ m²/W for the nonlinear refractive index $n_2$, the nonlinear strength, $\gamma$, approximates $1.7 \cdot 10^{-3}$ (mW)$^{-1}$. It should be noted that different fibers have different nonlinear strength. This is mainly due to different content of germanium and different effective areas, that is, spot size of the transverse electrical field in the fiber. This has led to an effort within optimization of so-called engineered fiber spans, where the task is to include both the effect of nonlinear phase shift and spontaneous emission and fiber dispersion in an evaluation of the performance of a transmission span consisting of several types of fibers including large effective area fibers, inverse dispersion fibers, and so forth [29].

In the following we assume that the nonlinear strength is independent of the position coordinate, that is, constant throughout an entire fiber span. From this, the nonlinear impairments due to SPM and XPM scale linearly with the effective length, $L_{\text{eff}}^s$, of the signal, $L_{\text{eff}}^s = (1 - \exp(-\alpha_s L))/\alpha_s$, whereas FWM scales quadratic with the effective length of the signal [30]. This is true in the case of short interaction lengths in which the signals approximately propagate together. For long lengths, the fiber dispersion, or more specifically the walk-off, is important.

For a passive fiber the signal effective length approximates the physical length $L$ for short physical lengths, whereas for long lengths the effective length approximates the fixed value $1/\alpha_s$. This is illustrated in Figure 3.9, which displays the effective length versus the physical length. The figure illustrates both the effective length for a passive fiber, that is, a lumped amplified system, and a distributed Raman amplifier. For the Raman amplifier, the signal effective length is [24]

$$L_{\text{eff}}^s = \int_{z=0}^{L} P(z)dz/P_0, \qquad (3.3.2)$$

where $P_0$ is the input signal power and $P(z)$ is the signal power versus distance. For a transmission span pumped to transparency, the required on–off Raman gain increases as the span length increases. Starting at very short span lengths, the required Raman gain is close to zero and the effective length approximates the value of a passive fiber. As the span length increases, the required Raman gain also increases and the effective

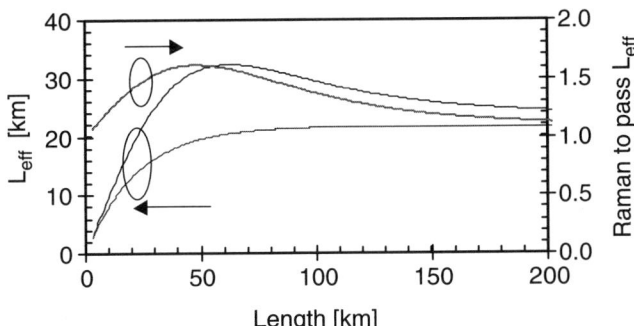

**Figure 3.9:** The effective length of the signal versus the physical length. The lower trace is for a passive fiber whereas the middle trace is for a transparent Raman amplifier. The top trace is the ratio of the effective length for the Raman amplifier to the effective length of a passive fiber as read on the right axis. The data are calculated using $\alpha_p = 0.25$ dB/km, $\alpha_s = 0.20$ dB/km, and $g_r = 0.7\,(\text{Wkm})^{-1}$.

length increases. In the limit when the length of the fiber span goes to infinity, the Raman amplification, even though it is distributed, occurs over a short length (tens of kilometers) relative to the span length and hence appears discrete. Thus, in this limit, the effective length again approximates the effective length of a passive fiber span of the same length; see Figure 3.9.

From Figure 3.9 it is clear that there is always an additional enhancement of the nonlinearities in a Raman amplifier as compared against a passive fiber span, followed by a discrete amplifier, when launching the same input power. The figure also shows that the penalty in the effective length is at its maximum when the span length is about 50 km, where the effective length is 1.6 times that of a passive fiber span of the same length. Increasing the span length above this value brings the performance, with respect to nonlinear effects, of the Raman amplifier span closer to the passive fiber span.

Figure 3.9 is calculated for a purely backward-pumped Raman amplifier. In the case of a purely forward-pumped amplifier, the effective length exceeds the span length as opposed to a backward-pumped amplifier in which the effective length always is shorter than the span length. Figure 3.10 illustrates the effective length for a transparent 100-km-long

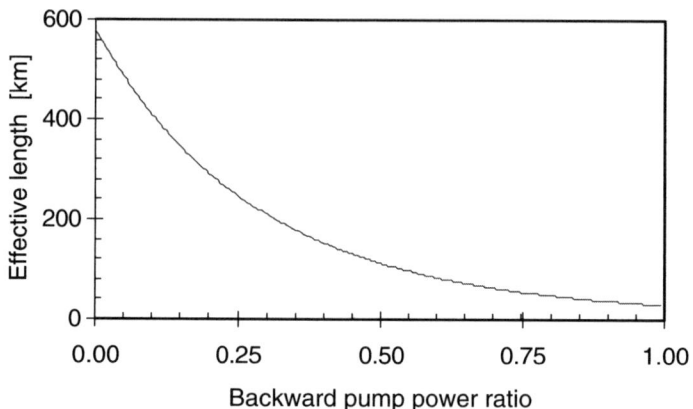

**Figure 3.10:** The effective length of a 100-km-long transparent Raman amplifier versus the ratio of backward pump power to forward pump power. $\alpha_p = 0.25$ dB/km, $\alpha_s = 0.20$ dB/km, and $g_r = 0.7$ (Wkm)$^{-1}$.

Raman amplifier versus the fraction of backward to forward pump power, the latter defined as described in the figure caption of Figure 3.6.

For the example in Figure 3.10, the passive fiber is 100 km long, that is, an effective length of 21.5 km for an equivalent length of unpumped fiber. The effective length of the forward-pumped amplifier is close to 600 km, whereas the effective length of the backward-pumped Raman amplifier is close to 30 km. The effective length equals the physical length when the backward pump power ratio equals 0.54 corresponding to having 0.54 times the total pump power in the backward direction and 0.46 times the total pump power in the forward direction.

From Figure 3.10 it is clear the effective length of a forward-pumped Raman amplifier is much longer than that of a backward-pumped Raman amplifier. In Figure 3.10, the effective length for the forward-pumped amplifier is approximately six times the physical length, whereas it is about three times shorter than the physical length when the amplifier is solely backward pumped. From Figure 3.10 it is clear that there will be a further enhancement of the nonlinear effects when pumping in a forward configuration as opposed to a backward configuration, based on the same input power.

As mentioned in the beginning, the system capacity depends on a balance between noise accumulation and pulse distortion typically from nonlinearities, that is, XPM, SPM, and FWM. In a simplified approximation, considering noise accumulation as outlined in Section 3.2 and including XPM between many channels in a wavelength-division-multiplexed (WDM) system, the transmission capacity may be evaluated analytically.

Until this point, evaluation of different system spans, with respect to their nonlinearities, has been based on the effective signal length. However, this comparison may be too simple, mainly because the system impact from nonlinear cross talk also depends on the walk-off between individual channels, the channel count, the channel spacing, and the channel power. To arrive at a more accurate model, one needs to solve the propagation equation for a WDM system with the right input conditions and the correct signal power evolution.

The propagation equation for a channel in a communication system with many signal channels may be solved when the spontaneous emission is decoupled from the nonlinearities and by using a multiplicative Gaussian noise model [31]. In this model each span is treated separately, and noise is simply added span per span in a long-haul system. Applying this model, and assuming furthermore that the walk-off length, $L_w = (DB\Delta\lambda)^{-1}$, where $D$ is the fiber group velocity dispersion, $B$ is the channel frequency bandwidth, and $\Delta\lambda$ is the channel wavelength separation, is shorter than the effective length, then the maximum transmission capacity $\eta$ expressed as the maximum bits per second per unit bandwidth, denoted the spectral efficiency [32], may be predicted as $\eta = \log_2(1 + S/N)$, where S is the signal power and N the average noise power, which consists of accumulated ASE and XPM. It should be noted that a spectral efficiency of 1 bit per Hertz per second is achieved for an S/N of 1. In a real communication system the success criterion for a communication system is whether an information carrying signal can be transmitted with a bit error rate much less than a predefined value, for example, $10^{-9}$. For a 10-GB/s channel this requires an S/N in excess of 20 dB [33].

Consider a rather optimistic long-haul transmission system consisting of 40 spans, each being 100 km long. In one case the intrinsic loss is counterbalanced with EDFAs with a 3-dB noise figure, whereas in another case the loss is cancelled by a backward-pumped Raman amplifier with a noise figure of 17 dB, that is, an effective noise figure of –3 dB. In the first case the effective length equals 21 km, whereas in the Raman case the effective length equals 30 km. Assuming furthermore a dispersion of 1.5 ps/nm/km, a modulation bandwidth of 40 GHz, and a channel spacing

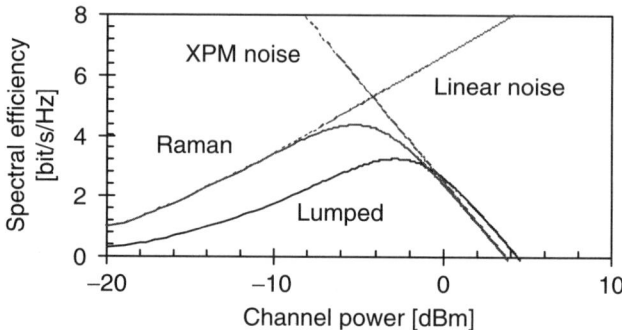

**Figure 3.11:** The spectral efficiency versus the channel power in a system with 100 channels occupying 16 THz of bandwidth. A nonlinearity of $1.5 \times 10^{-3}$ $(Wm)^{-1}$ is used at the center wavelength 1.555 $\mu$m.

of 160 GHz, that is, the walk-off length equals 13 km, which is shorter than the effective length, the result in Figure 3.11 is predicted.

Figure 3.11 shows the system performance evaluated as the spectral efficiency for the two cases described, the Raman case and the lumped case. Both examples are shown as a function of the channel power; that is, the power in each channel is assumed to be identical. In each of the two cases, the spectral efficiency is limited by the linear signal-to-noise ratio, that is, accumulation of ASE from amplifiers for low channel power, whereas the spectral efficiency is limited by XPM as the channel power increases. The optimum system performance in each individual case is achieved when the two types of penalties, linear noise and XPM, are equal in size. For the discretely amplified system this happens for a channel power slightly below −2 dBm, whereas for the all-Raman system the optimum channel power is close to −5 dBm.

On the one hand, the figure shows that the accumulation of ASE significantly improves when Raman gain is used as opposed to discrete amplification. On the other hand, the figure also shows that the XPM is worse in the Raman amplified system as compared to the discretely amplified system. The important result is that because of the reduction in accumulated ASE the channel power may be lowered in the all-Raman system as compared to the discretely amplified system. This leads to reduced penalties from XPM in the Raman system and as a result an improvement of 40% in the spectral efficiency when comparing the Raman system against the discretely amplified system.

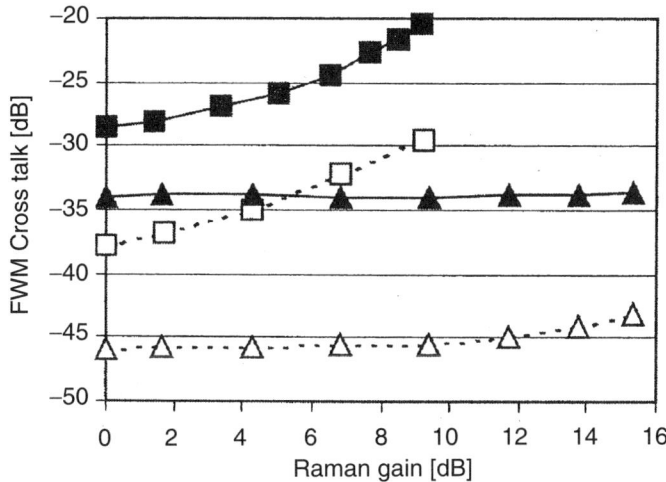

**Figure 3.12:** Measured four-wave mixing cross talk as a function of the on–off Raman gain. Open marks represent 2 channels whereas filled marks represent 39 channels placed with 100 GHz channel spacing. The square marks represent the short 22-km-long amplifier whereas the triangles represent the 100-km-long amplifier. (From [34].)

In the above it was assumed that the walk-off length is much shorter than the effective length in order to apply the model. If this is not the case, the cross talk will depend strongly on the effective length. In [34] this is pointed out and illustrated by comparing the four-wave mixing cross talk in two Raman amplifiers of different lengths, one being 22 km and another being 100 km. The results are obtained using non-zero dispersion shifted high capacity transmission fiber, corning SMF-LS. The results are shown in Figure 3.12, which displays the FWM cross talk as a function of the on–off Raman gain. In addition to the length dependence, the figure also illustrates the increase in cross talk as the number of channels increases from 2 to 39.

In the case of the 22-km-long distributed Raman amplifier, the four-wave mixing cross talk is directly related to the effective length; that is, it increases as the Raman gain, and hence the effective length, increases. However, this dependence disappears when the length of the Raman amplifier is increased to 100 km, that is, when the length is much longer than the length where the signals remain phase matched.

In the above we have focused on the impact of the nonlinear refractive index. For completeness it should be noted that there are other nonlinear

effects that may impact the system performance including Raman scattering among channels and intrachannel effects in ultra-high bitrate single
channel systems [28].

In the process of Raman scattering among channels, signal channels will
experience gain from channels at lower wavelengths whereas other channels will be depleted by channels at higher wavelengths. This will cause a
gain tilt that to first order may be counterbalanced by the proper design of
the pump configuration of the Raman amplifier. However, since the channels are copropagating with each other, the Raman scattering among the
channels will cause extra noise since the channel-by-channel Raman gain
will fluctuate according to the bit pattern in the channels. This penalty has
been evaluated for discretely amplified systems and it has been found that
the statistics of the signal modulation and the number of channels are to
important numbers [35].

To explain this, it is easy to imagine the situation with only two channels.
In this case the lowest wavelength channel will act as a pump to the highest
wavelength channel. However, the Raman process only happens when the
two channels have a mark within a defined timeslot. If more channels
are added and only the interaction with the lowest wavelength channel
is considered for simplicity without losing any general argument, then if
all other channels have a mark at the same bit slot this gives the highest
depletion on the lowest wavelength channel, whereas there is no depletion if
all channels, except the lowest, have spaces at a given bit slot. However, the
more channels there are, the lower the probability that all other channels
contain a mark or space within a given bit slot. In fact it only takes a
few channels before the Raman interaction among channels appears more
deterministic [35].

It is worth noting that the above process is not specific to the use of
distributed Raman amplification and would show up with the same strength
in a transmission system based on lumped amplification if the channel path
average power was the same as in the all-Raman amplified system. For
further details we refer to Section 2.2.3.

### 3.3.2  Rayleigh Reflections

A fraction of the intrinsic loss in an optical fiber is due to Rayleigh
scattering. In this process light is elastically scattered; that is, the frequency of the scattered light is identical to that of the incoming light.

However, a small part of the scattered light is recaptured, half of which propagates in the direction of the launched light and the other half propagates in the opposite direction. In a typical transmission fiber of ideally infinite length, in practice about 40 km or longer, the back-reflected power measured at the input end of a fiber is approximately 30 dB lower than the launched power.

Consider a Raman amplifier in which the signal travels in one direction, the forward direction. Because of Rayleigh reflections, a fraction of the signal is scattered and after this, half of the recaptured Rayleigh scattered signal-light propagate in the same direction as the signal (undistinguishable from the signal) and the other half in the opposite direction. The part that propagates in the opposite direction may again undergo a reflection and now after two reflections it propagates in the same direction as the signal. This causes a severe penalty, as this double-reflected signal is directly at the same frequency as the original signal but appears like an echo of the signal and interferes with the original signal. Naturally, the penalty contains multiple reflections rather than two, but is typically referred to as the double-Rayleigh-reflected penalty or alternatively multipath interference (MPI). This penalty will be the main focus of this section.

Thus far, only reflections of the signal were discussed. However, in a real communication system, the Rayleigh process cannot distinguish signal from noise. The implication of this is that the spontaneous emission contains a contribution caused by Rayleigh reflections of the noise propagating in the opposite direction relative to the signal. This is referred to as the single-reflected backward-propagating ASE. We also return briefly to this in the following.

Before we discuss the impact of multiple reflections in distributed amplification, it is useful to describe the impact of reflections in a system with in-line discrete amplifiers. Thus in the following we describe three situations: (1) a discrete amplifier surrounded by two discrete reflections, (2) a discrete amplifier with distributed reflections, and (3) a distributed Raman amplifier.

### 3.3.2.1 A Discrete Amplifier and Two Discrete Reflections

It is well established that discrete reflections may introduce an additional penalty to the performance of an optical communication system [36]. A simple example of this is found when an in-line amplifier is surrounded by discrete reflections. Assuming a realistic gain and two low-value power reflections, $R_1$ and $R_2$, the double-reflected signal power, $P_{dbr}$, relative to the transmitted signal power, $P_{through}$, approximates

$$P_{dbr}/P_{through} = R_1 G^2 R_2. \qquad (3.3.3)$$

This ratio is also referred to as the Rayleigh cross talk and denoted $R_c$.

In a digital communication system, marks and spaces, that is, light pulses or no pulses, define the transmitted information. Because of noise on the signal, even in the transmitter, the marks and spaces each have a probability distribution, that is, a mean value and a variance [33]. During transmission, the mean and variance of both the marks and the spaces change. In the receiver the received optical power is transferred to an electrical current with associated mean and variance. Based on the actual current in a bit slot, it is then decided whether a received bit has sufficient current to represent a mark or if it represents a space. This decision is made in an electric circuit based on a threshold current level. In a testing laboratory, where it is well known what information was sent from the transmitter, the number of errors made by the detection circuitry can be evaluated. This number is the bit error rate (BER).

Theoretically the BER can be evaluated by assuming well-defined probability distributions for the marks and spaces in addition to a well-defined decision threshold current level. Assuming that the probability distributions for the marks and spaces are Gaussian, the BER is then found by integrating the tails of the probability distributions from the threshold current. More specifically, the BER is the sum of the probability of receiving a mark given a space has been sent plus the probability of detecting a space given a mark has been sent. These integrals may be determined under the Gaussian assumption and the BER approximated through BER $= (\exp(-Q^2/2))/(Q\sqrt{2\pi})$, where the argument $Q$ may be expressed as the mean current for the marks relative to the sum of the standard deviation for marks and spaces, that is, $Q = I_1/(\sigma_1 + \sigma_0)$. It is noted that $Q^2$ is the optical signal-to-noise ratio, illustrating that since a BER of $10^{-9}$ is achieved for $Q = 6$, this corresponds to an optical signal-to-noise ratio of 20 dB.

To describe the system impact of multiple path interference, it is beneficial first to imagine the communication system without MPI. In the simplest case the probability distributions of the marks and spaces, that is, their mean and variance, originate from absorption within the fiber and spontaneous emission in amplifiers. As a result of this a decision threshold current is defined and the system may be characterized with respect to the BER. If now we expand the considerations to also include MPI, the probability distributions of the marks and spaces change due to an increased variance of the signal. As a consequence the BER increases. However, to keep a fixed BER the transmitted signal power may be increased, corresponding to an increased mean value of the marks. The increase in power defines the so-called power penalty.

The threshold current may be either fixed or adjusted for minimum BER. Assuming a fixed threshold current level, the power penalty, that is, the power penalty due to multiple path interference in the signal path, is generally expressed through [37–39].

$$\text{Penalty} = -5 \log(1 - 4\sigma_{\text{RIN}}^2 Q^2), \tag{3.3.4}$$

where $\sigma_{\text{RIN}}^2$ is the variance of the signal and $Q$ is the argument to the BER function as previously described.

Assuming a Gaussian probability distribution as described above, the power penalty due to the double-reflected signal power may be evaluated. If the variance of the signal has two main contributions, one from spontaneous emission and one from the double-reflected signal, $\sigma_{\text{RIN}}^2 = \sigma_{\text{ASE}}^2 + \sigma_{\text{dbr}}^2$, the power penalty is [36]

$$\text{Penalty} = -5 \log(1 - 144(N_{\text{sp}} h\nu B / P_{\text{in}} + G^2 R^2)), \tag{3.3.5}$$

where the term $(N_{\text{sp}} h\nu B / P_{\text{in}})$ accounts for the contribution from ASE whereas the term $G^2 R^2$ accounts for discrete reflections, and where $N_{\text{sp}}$ is the spontaneous emission noise factor, $h\nu$ is the photon energy, $B$ is the optical bandwidth of the signal, $P_{\text{in}}$ is the signal power, $G$ is the gain of the discrete amplifier, and finally $R^2 = R_1 R_2$ is the power reflection of two discrete reflections $R_1$ placed before and $R_2$ after the amplifier.

### 3.3.2.2 A Discrete Amplifier and Distributed Reflections (Rayleigh Reflections)

In the above, the reflections were imagined to originate from discrete points; however, because of Rayleigh scattering in fibers there will always be an unavoidable contribution to the power penalty due to distributed Rayleigh reflections. If the in-line amplifier is surrounded by fiber rather than discrete reflections, the variance is then modified. Assuming $\tilde{n}$ numbers of periodic spans, each of length $L$, the total power penalty, neglecting other contributions, equals [39, 40]

$$\text{Penalty} = -10\log\left\{1 - \frac{Q^2}{2}\tilde{n}\left(\frac{B_R^2 G^2}{4\gamma^2} + \frac{B_R^2}{4\alpha^2}\left(e^{-2\alpha L} + 2\alpha L - 1\right)\right)\right\},$$
(3.3.6)

where $B_R$ is the Rayleigh scattering coefficient times the recapture factor, $\gamma$ is the average gain coefficient of the fiber within the lumped amplifier, and $\alpha$ is the attenuation coefficient.

### 3.3.2.3   Distributed Raman Amplifier

To this point only discrete amplification has been discussed. In distributed amplifiers the effect due to Rayleigh reflections may be even more severe because of the long lengths of the amplifier over which gain is accumulated. However, the reflected power depends on the signal gain distribution, which may allow for reduction of the double-reflected power by optimized design. Consider, for example, a fixed gain that is localized in one part of the fiber, that is, the beginning or the end corresponding to a forward- or a backward-pumped Raman amplifier respectively. In both cases, the double-reflected power scales in a manner similar to Eq. (3.2.5), that is, like $G^2$. If, however, the gain is split with a loss element between the reflections, the double-reflected power would drop as in the case of having two discrete amplifiers each with gain $\sqrt{G}$ with an isolator in between.

Analogous to the discrete in-line amplifier, the distributed Raman amplifier also adds spontaneous emission to the signal. Thus, in addition to the Rayleigh reflected signal the Rayleigh reflection process also causes an extra contribution to the spontaneous emission. The most severe reflected noise contribution originates from the backward-propagating noise, which only needs an odd number of reflections, in the simplest picture only one, to mix with the forward-propagating noise. However, this contribution is smaller than the conventional forward-propagating ASE for on–off Raman gain values below 20 dB [24]. Thus in the following we focus only on the double Rayleigh reflected signal.

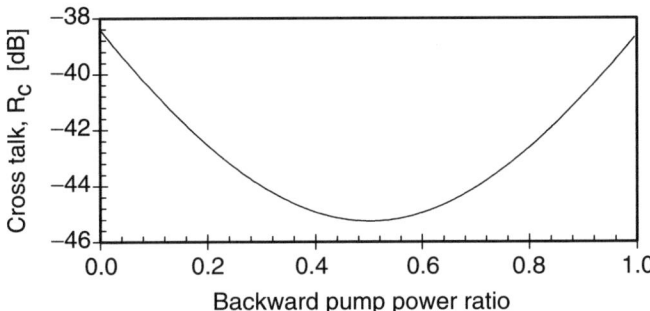

**Figure 3.13:** The ratio of double-reflected signal power to transmitted power. Calculated as in Refs. [24, 41] using $\alpha_p = 0.25$ dB/km, $\alpha_s = 0.20$ dB/km, $g_r = 0.7$ (Wkm)$^{-1}$, and a Rayleigh reflection coefficient $B_R = 100 \cdot 10^{-9}$ m$^{-1}$ corresponding to $-30$ dB back-reflected power.

In a general case the double-reflected signal power relative to the transmitted signal power $R_c$ is evaluated from [24, 41]

$$R_c = P_{\text{drb}}/P_{\text{through}} = (B_R)^2 \int_0^L \frac{1}{G^2(\tilde{z})} \int_{\tilde{z}}^L G^2(x) dx \, d\tilde{z}. \qquad (3.3.7)$$

where $B_R$ is the Rayleigh-backscatter coefficient.

In the above expression all considerations regarding polarization have been neglected. However, assuming that the distance between two successive reflections is long, the reflected power that beats with the signal needs to be multiplied by 5/9 due to the polarization rotation of the double-reflected signal relative to the original signal [42].

Figure 3.13 illustrates the double-reflected signal power relative to the signal output power ($= R_c$) in a 100-km-long fiber Raman amplifier in which the ratio of forward to backward pump power is varied, similar to Figure 3.6 and 3.10. In the curve, no assumptions to polarization are made, and the intrinsic loss over the 100 km equals 20 dB.

The figure clearly illustrates the impact of having the gain divided into two blocks. If the gain is either at the input or at the output end, $R_c$ is 7 dB higher than if the gain is equally distributed between the input and the output end with a transmission loss between.

The cross talk displayed in Figure 3.13 is expected to depend on the depletion of the amplifier since the depletion affects the gain distribution within the fiber and hence the cross talk in accordance with Eq. (3.3.7). For a backward-pumped Raman amplifier this leads to a slightly reduced cross talk if the amplifier is pushed into depletion [43]. In a specific example of a 100-km-long TrueWave RS reduced dispersion slope fiber and a Raman gain of 20.7 dB, the degree of pump depletion, defined through a reduction in residual pump power, was varied from 0 dB to 5.5 dB. As a result of this the Rayleigh cross talk was only reduced by 0.3 dB [43].

Finally, the power penalty may be evaluated as explained above using Eq. (3.3.4), resulting in a penalty

$$\text{Penalty} = -10 * \log(1 - \frac{5}{9} Q^2 \tilde{n} R_C), \tag{3.3.8}$$

where $Q^2$ is the BER threshold value, $\tilde{n}$ the number of spans, and $R_c$ the Rayleigh cross talk. Figure 3.14 illustrates the penalty when comparing lumped and distributed amplification using Eq. (3.3.6) and Eq. (3.3.8),

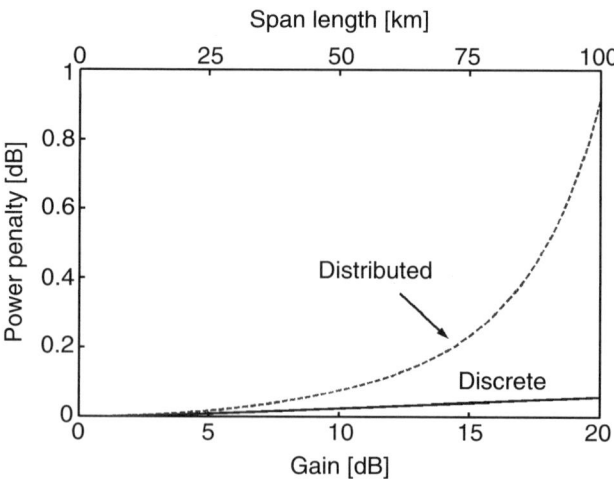

**Figure 3.14:** Power penalty versus gain (or span length) in a transmission consisting of 75 amplifiers. For the discrete amplifier, an amplifier fiber length of 25 m and a fiber loss of 0.2 dB/km is used, whereas the data for the distributed Raman amplifier are $\alpha_p = 0.25$ dB/km, $\alpha_s = 0.20$ dB/km, $g_r = 0.7$ (Wkm)$^{-1}$, and a Rayleigh reflection coefficient $B_R = 100\ 10^{-9}\ \text{m}^{-1}$.

respectively. In the figure the span length is varied and the penalty calculated assuming that the gain of the discrete amplifier and the distributed
amplifier is adjusted to make the span transparent. Compared to Eq. (3.3.6)
and Eq. (3.3.8), only half the cross-talk power is included to take into
account that the noise power is reduced in half when mark density is
0.5 [38].

Figure 3.14 clearly demonstrates the increased penalty due to the distributed amplification process as the length of the fiber span increases. For
a 100-km-long fiber span, the power penalty is more than 10 times higher in
the distributed amplification case as compared to the lumped amplifier case.

The power penalty discussed previously may alternatively be given as
a penalty in the threshold value $Q$. Such penalty is measured using a multipath interferometer and verified against theoretical predictions according
to the above theoretical considerations. The results from measurements on
a 40-Gbit/s carrier suppressed return to zero signal is shown in Figure 3.15
together with predictions [44].

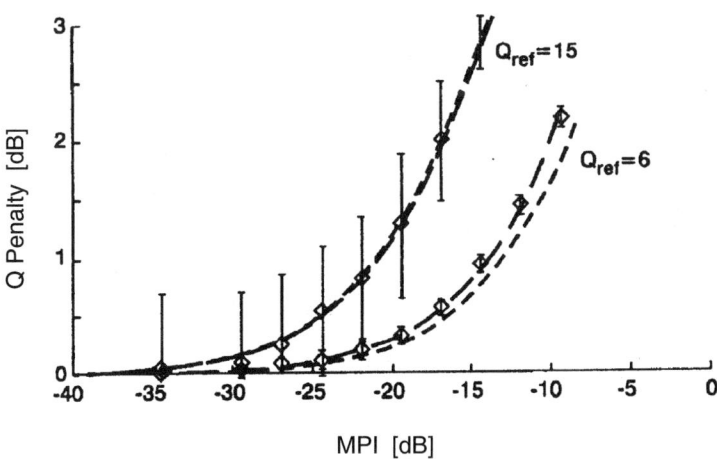

**Figure 3.15:** The penalty in $Q$ versus the multiple path interference as defined
in the beginning of Section 3.3.2. The transmitted signal was a 40-Gbit/s carrier
suppressed return-to-zero. $Q_{ref}$ is the initial $Q$ value. The short dash is analytical, whereas the long dash is numerical predictions. Marks are measured data.
(From [44].)

The results in this section demonstrate that there is a drawback to Raman amplifiers for large on–off Raman gains because of Rayleigh reflections when comparing against systems using discrete amplification. The penalty from Rayleigh reflections may be reduced by applying bidirectional pumping.

For completeness, it should be noted that the challenge of translating the amount of Rayleigh reflected power into a system penalty might be approached in several ways. Here we have focused on the Raman cross-coupling, $R_c$, which may be translated into a relative intensity noise [43]. In addition to calculating a power penalty, one may also evaluate a noise figure [45] or alternatively a signal-to-noise ratio [46].

### 3.3.3   Time Response

The Raman scattering may be described through a third-order susceptibility, of which the imaginary part is proportional to the Raman gain coefficient spectrum and the real part proportional to a change in the refractive index (for frequency shift approaching zero) as described in Chapter 2 and in Figure 3.7. From the third-order susceptibility, the Raman response function in the time domain is then evaluated through a Fourier transformation. The results from such analysis demonstrate a Raman response time in the order of tens of femtoseconds [10].

To explain the response time, Hollenbeck and Cantrell [47] propose a model based on an ensemble of damped harmonic oscillators with Gaussian distributed resonant frequencies. The response function is in this case a convolution of a Gaussian and a Lorentzian response or more specifically a series of such convolution terms. This so-called intermediate broadening model gives an excellent fit to both the gain spectrum and the time response function when using 13 terms in the series.

The very fast response time enables mutual interactions between the pump and the signal in addition to cross-coupling between two signal channels mediated by the pump. Both of these interactions are crucial when considering forward pumping in which pump and signal propagate in the same direction. In addition, they are especially critical in distributed Raman amplifiers because the amplifying fiber is the transmission fiber and thus a modification of the walk-off between pump and signal achieved through a modified dispersion will not only mitigate the penalties but also impact the dispersion of the signal.

### 3.3.3.1 Pump-Signal Cross Talk

Because of the very short response time, any fluctuations in pump power slower than the response time can cause gain fluctuations if the pump and signal are propagating in the same direction. This results in additional noise on the signal. If, on the other hand, the pump and signal are counter-propagating, the signal passes the pump that propagates in the opposite direction. This introduces a strong averaging of the accumulated gain and hence reduces the effect of pump–signal cross talk. In the copumped Raman amplifier the signal and pump propagate through the fiber in the same direction and only the effect of walk-off between pump and signal causes averaging. The direct coupling of pump fluctuations to signal fluctuations is demonstrated in Figure 3.16.

The efficiency of the averaging effect is strongly correlated to the frequency of the fluctuations in the pump. In the limit in which the frequency is very low, all variations in the pump power are directly transferred, whereas in the other limit in which the frequencies are much higher than the time of a bit slot, no fluctuations will be transferred to the signal.

To further analyze this, a simple mathematical model that allows evaluation of relevant parameters may be used. In the model, the pump power is written as a constant power with a sinusoidal variation on top:

$$P_p = P_p^0 \exp(-\alpha_p z) \left\{ 1 + m \sin((\omega_p t - \beta_p z) + \theta) \right\}, \qquad (3.3.9)$$

where $m$ is the modulation depth, $\beta_p$ is the propagation constant at the angular pump frequency $\omega_p$, and $\theta$ is a phase constant. By putting this into the propagation equation for the amplifier both the forward- and the backward-pumped amplifier may be described [49–51].

### *Forward-Pumped Case*

For the copumped amplifier the pump and signal travel in the same direction, and if the group velocity at the pump and signal wavelength is identical the pump and signal travel at the same speed. In this situation in which there is no walk-off between pump and signal, the fluctuations in the pump power will be transferred directly to the signal as each bit may experience a different pump power.

In the more realistic situation, the pump and signal propagate at different speeds because of group velocity dispersion. This walk-off velocity,

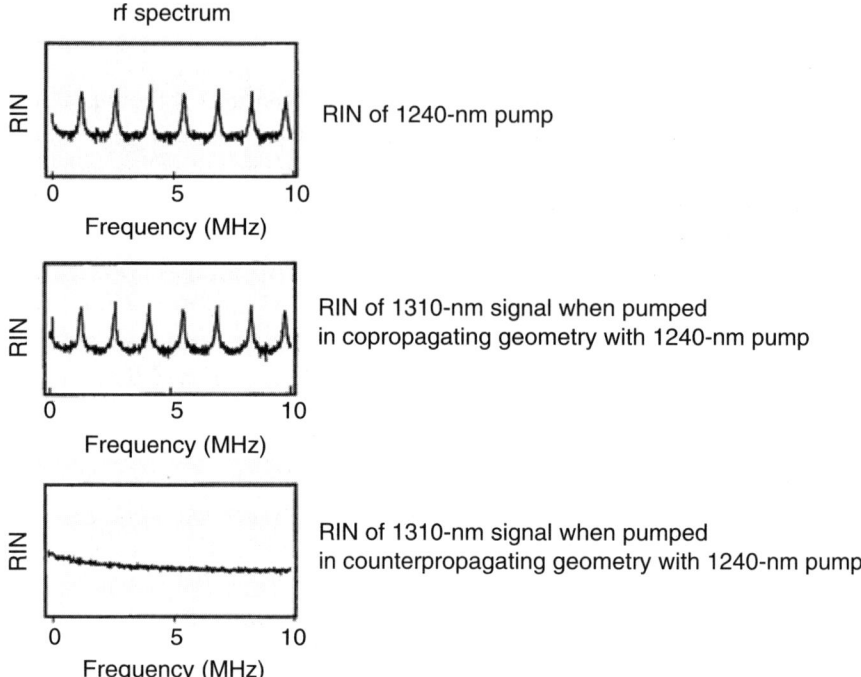

**Figure 3.16:** Signal fluctuations caused by signal pump signal cross talk may be characterized using relative intensity noise (RIN), defined in terms of detected density of the photocurrent in a 1-kHz bandwidth at a specified frequency divided by the power of the photocurrent. (Top) Spectrum of pump, (middle) transfer of pump RIN to signal in copropagating geometry, and (bottom) transfer of RIN in counterpropagating geometry. (From [48].)

$v_w$, when measured in meters per second, is inversely proportional to the group velocity dispersion, $D$ (for simplicity, $D$ is the simple average of the dispersion at the signal and pump wavelength), times the wavelength separation, $\Delta\lambda$, between the signal and pump; that is, $v_w^{-1} = D\Delta\lambda$ (see also the beginning of Section 3.3.1).

Inserting the modulation of the pump beam from Eq. (3.3.9) in the propagation equation for the signal, the relative intensity noise of the signal $R_s$ is calculated directly from the fluctuations in the signal gain [50]. For an infinitely long fiber, in practice much longer than the

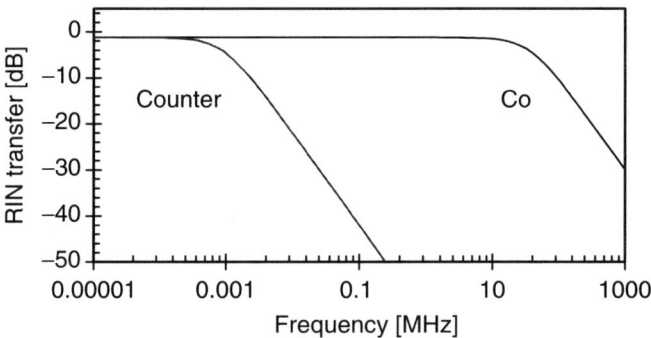

**Figure 3.17:** The relative RIN transferred to the signal due to the pump to signal cross talk versus the frequency of the oscillations of the pump. The curve labeled Counter is for a counterpumped amplifier whereas the curve labeled Co is for a copumped amplifier. There is an average dispersion of 2.5 ps/nm/km, a loss rate at the pump of 0.25 dB/km, and a group velocity of $2.0 \times 10^8$ m/s. A wavelength separation between the pump and the signal of $\Delta\lambda = 100$ nm is used.

effective length $L \gg 1/\alpha_p$, the relative intensity noise of the signal in decibels is

$$R_s \approx R_p + 20\log(\ln G_R) + 10\log\left(\frac{(V/L_{\text{eff}})^2}{(\alpha_p V)^2 + (2\pi D\Delta\lambda V f)^2}\right),$$

$$(3.3.10)$$

where $R_p$ is the relative intensity of the pump in decibels, $f$ is the frequency of the modulation of the pump, $(\omega_P = 2\pi f)$ is the group velocity of the signal, and $G_R$ is the on–off Raman gain. Equation (3.3.10) shows that the relative intensity (RIN) of the signal in general is larger than the RIN in the pump, $R_p$, except for very low gain values. Figure 3.17 displays the relative intensity noise of the signal according to Eq. (3.3.10) when the RIN of the pump and the contribution from the on–off gain is omitted.

Equation (3.3.10) shows that the RIN transfer, that is, the third-order term in Eq. (3.3.10), reduces as the frequency exceeds the so-called corner frequency $f_c$ defined as

$$f_c = \frac{\alpha_p}{2\pi D\Delta\lambda}.$$

$$(3.3.11)$$

As discussed in the beginning of this section, the corner frequency is related to the walk-off length divided by the effective length of the pump in the amplifier, $L_{\text{eff}} = 1/\alpha_p$.

### Backward-Pumped Case

For the backward-pumped amplifier the fluctuations in the pump power will be averaged simply because a single bit in the information carrying signal will get a gain contribution from a relatively large segment of the pump. This may also be understood by considering the time it takes a slot of the pump to pass the information carrying bit sequence. Assuming that the pump and signal have the same group velocity, $V$, just with opposite sign, then the time, it takes two segments of length $L_{\text{eff}}$ to pass each other; that is, the transit time is $L_{\text{eff}}/V$. Thus, any variations slower than this transit time will be transferred to the signal whereas fluctuations faster than this will not be transferred. The relative intensity noise of the signal in decibels is

$$R_s \approx R_p + 20\log(\ln G_R) + 10\log\left(\frac{(V/L_{\text{eff}})^2}{\left(\alpha_p V\right)^2 + (4\pi f)^2}\right). \qquad (3.3.12)$$

Equation (3.3.12) for the counterpumped amplifier has a similar form as Eq. (3.3.10) for the copumped amplifier. However, the corner frequency no longer depends on the walk-off, but is

$$f_c = \frac{\alpha_p V}{4\pi}. \qquad (3.3.13)$$

Figure 3.17 displays the RIN of the signal for both the co- and the counter-pumped amplifier. The corner frequency for the counterpumped amplifier is close to 1 kHz, whereas the corner frequency for the copumped amplifier is close to 10 MHz.

Figure 3.17 clearly illustrates that the corner frequency for the counter-pumped amplifier is much lower than the corner frequency of the copumped amplifier. Fluctuations in the pump power at frequencies above the corner frequency are only weakly transferred. This is explained by the averaging effect. From this it is clear that the corner frequency is an important parameter. According to Eq. (3.3.11) and Eq. (3.3.13) the corner frequency is mainly determined by the fiber loss and the walk-off between the pump

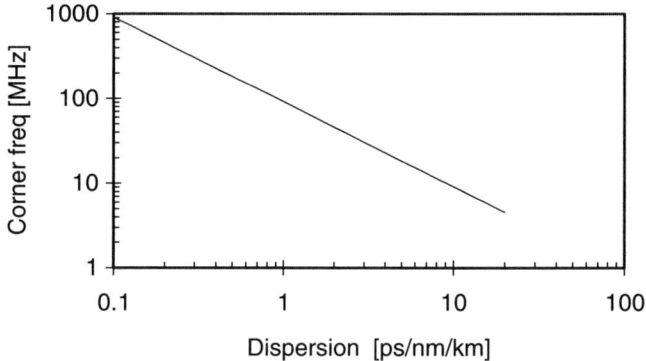

**Figure 3.18:** The corner frequency of a forward-pumped Raman amplifier as a function of the group velocity dispersion. The trace is calculated for a dispersion of 2.5 ps/nm/km, a loss rate at the pump of 0.25 dB/km, and wavelength separation between pump and signal of $\Delta\lambda = 100$ nm.

and the signal; that is, the dispersion and the wavelength separation between the pump and the signal.

Figure 3.18 illustrates the corner frequency versus the group velocity dispersion for a copumped amplifier. Since the walk-off is inversely proportional to the group velocity dispersion, the figure illustrates the effect of walk-off as the corner frequency reduces when the walk-off length reduces. For comparison, one finds from Eq. (3.3.11) that the corner frequency in the counterpumped amplifier is around 1 kHz.

In addition to the direct coupling of fluctuations from the pump power to the signal gain, noise on the signal may also occur during propagation of the pump through the fiber. These noise contributions include mode-partitioning noise, mode beating between lasers that, for example, are polarization multiplexed together.

At the output of a laser, optical power is distributed among many different modes, and the power in each individual mode fluctuates, but because of the many modes, the total output power of the laser is relatively stable. However, when the different modes propagate, they all propagate at different velocities due to the dispersion. Thus, the power fluctuations in one mode are not necessarily counterbalanced by fluctuations in other modes along every position of the fiber. As a consequence, gain fluctuations are different at various positions along the fiber. Choosing a fiber with a low dispersion at the pump wavelength minimizes the mode partitioning noise since modes stay together and counterbalance each other at all positions along the fiber.

Another type of noise that also becomes apparent only under propagation is mode beating, which turns into a noise source when the mode beating falls in the electrical bandwidth of the signal. An example of this is polarization multiplexing of semiconductor lasers. Each laser has many modes, and when these propagate they couple through cross-coupling of polarizations, causing an additional noise penalty. By tuning the modes of the lasers in the two polarizations, the beating may be minimized.

Finally, noise contributions may exist because of nonlinear interactions among individual pump beams mixing up at the signal wavelengths or between the pump and the signal. The first contribution is caused by four-wave mixing among the pumps in an amplifier pumped using multiple pump wavelengths. This may produce strong four-wave mixing components in the signal band for certain fiber types. We return to this in Section 3.4. The other noise contribution originating from nonlinear interactions among pump and signal beams originates from individual pump modes that interact directly with the signal through four-wave mixing. To avoid this, the dispersion zero is critical and should be chosen to be shorter than the shortest pump wavelength (or longer than the longest wavelength signal channel).

### 3.3.3.2   Signal–Pump–Signal Cross Talk

In the Raman process, the amplitude modulation of the encoded signals is impressed upon the pump. Thus even a perfectly fictive noise free pump, that is, no fluctuations in wavelength or amplitude, sometimes referred to as a quiet pump with reference to the previous discussion on pump–signal cross talk, will become noisy during the Raman amplification process. Thereby induced noise on the pump may be transferred to other signals through the Raman process, in the following denoted as signal–pump–signal (SPS) cross talk.

The amount of SPS cross talk depends on (a) the relative directions of the pump and signal beams, (b) the walk-off between pump and signal, (c) the amount of pump depletion, and (d) the statistics of signals including the number of channels. In the following we discuss these dependencies with emphasis on the dependency on pump depletion and statistics.

### Directional Dependence

In analogy to the pump–signal cross talk, the pump mediated signal cross talk also depends strongly on the relative directions of the pump and signal beams. This may simply be understood by dividing the signal–pump–signal cross talk process into two parts as outlined above. In the first part a footprint of a given signal channel is transferred to the pump in the second part; this is then transferred to another signal channel. The second part corresponds to the situation of simple pump–signal cross talk. Following this argument, it is clear that the cross talk is most severe for copumped amplifiers as compared to counterpumped amplifiers.

### Dependence on Walk-off

The relative propagation speed difference between the pump and the signal as well as between the signals themselves causes averaging of signal–pump–signal cross talk, similar to the discussion of pump–signal cross talk, and results in a limited cross-talk bandwidth over which the signal–pump–signal cross talk occurs.

### Dependence on Pump Depletion

In addition to the directional and walk-off dependence, the signal–pump–signal cross talk depends on the pump depletion and the size of the applied Raman gain. In general, the smaller the pump depletion and the smaller the Raman gain, the smaller the signal–pump–signal cross talk [51]. Thus, conventionally copumped Raman amplifiers are limited to applications with small levels of Raman gain and small degrees of pump depletion.

In the forward-pumped Raman amplifier, the pump power is gradually depleted. At any given point, however, the pump depletion varies due to the different power requirements of different signal patterns. Theoretical calculations show that unacceptable large cross talk may occur even though the amplifier works in the linear region, that is, when the pump decays exponentially $P(z) = P_0 \exp(-\alpha_p z)$ and the on–off gain in decibels is expressed simply as $G|_{dB} = 4.34 g_r P_0 L_{eff}$, where $P_0$ is the launched pump power. To ensure a high ratio of output power in a given channel relative to the power in the same channel originating from other channels, the injected pump power must be restricted well below a factor of 2 from the pump power where the gain is 3 dB in depletion. The latter is predicted under the assumption that the pump and signals propagate at the same speed, that is, identical group velocity of pump and signal [51].

### Dependence on Signal Statistics

It is the collective modulation of the $N_{ch}$ signal channels that serves as the noise source for SPS cross talk. Thus it is important to consider the statistics of the $N_{ch}$ transmitted channels as for example the probability that a mark is simultaneously transmitted on all $N_{ch}$ channels. This probability decreases as the number of channels increases. Assuming that all the signal channels interact equally strong with the pump, the RIN of the collective signal may be evaluated theoretically as the variance of the collective signal power relative to the squared mean value of the power. Such calculations show that the RIN decays quadratically; that is, it is beneficial to have many channels.

It is rather complicated to make accurate theoretical predictions on the coupling between signal pump and signal, since many parameters need to be included. First is the propagation of the $N_{ch}$ signal channels and the set of pump wavelengths. This propagation needs to include walk-off. In addition, the propagation also needs to include many bits within the bit pattern and finally all the signal channels. Second, the Raman amplification needs to be treated as accurately as possible; that is, the complete gain profile as well as depletion need to be accounted for. This complete problem is rather complicated to solve and often simplifications are done.

Neglecting the effect of dispersion, or in effect walk-off, the system impact is strongly affected by the degree of amplifier depletion [51]. This demonstrates the complexity of the SPS cross talk, since on one hand the signal–pump–signal interaction is expected to increase with pump depletion, whereas on the other hand, the strong depletion is often due to many channels and hence since the RIN reduces strongly with the number of channels, the SPS cross talk may indirectly decrease with depletion.

To minimize the signal–pump–signal cross talk it has been proposed to constrain the Raman gain to values well below the saturation point [51].

Alternatively, a signal modulation may be chosen that minimizes the power within the cross-talk bandwidth [49, 52].


## 3.4 Backward Pumping

The strong interest in using Raman amplification is due to the fact that the system performance may be improved, that gain at any signal wavelength may be obtained simply by choosing the proper pump wavelength, and, just as important, that the gain spectrum may be broadened by combining multiple pump sources at different wavelengths.

Among the first type of lasers to be used in transmission experiments using distributed Raman amplification was high-power pump lasers based on a cascaded fiber Raman resonator [11] pumped either by a double-clad ytterbium fiber laser at around 1100 nm or a neodymium laser at 1064 nm [53]. These high-power pump sources are well suited for backward pumping of Raman amplifiers because they are unpolarized and are emitting light at many longitudinal modes within a few nanometers.

Since the Raman process is polarization dependent, the preferred pump needs to be unpolarized as mentioned above. In addition, the spectrum of a particular pump should be relatively broad, a few nonometers, to avoid the onset of Brillauin scattering. Finally, because of the relative low gain coefficient of the Raman gain, pump powers on the order of a few hundreds of milliwatts are required.

The backward-pumping configuration is preferred because of the fast response time of the Raman process (below 100 fs) [10, 47], which as described enables amplitude fluctuations in the pump power to transfer directly to the signal gain in a forward-pumped configuration. In the backward-pumped configuration amplitude fluctuations in the pump power are averaged out when the signal and pump pass quickly by each other in opposite directions.

Even with the constraints described above regarding the pumping configuration and the pump laser, there are several free parameters including the number of pump wavelengths, the actual pump wavelengths, and the power in each pump wavelength. In this section some of the progress and different approaches to do backward pumping will be described. The majority of research within selecting pumping configurations and pump lasers has been concentrating on achieving as wide and as flat a gain spectrum as possible. Only little effort has focused on creating both flat gain and flat noise spectrum. We return briefly to this in Section 3.5.2.

In addition to the high-power fiber-based pump lasers, there are several alternative pumping configurations including multiple discrete semiconductor pump lasers discussed in Section 3.4.1, broadened discrete pumps and continuous spectrum pumps discussed in Section 3.4.2, and time-division multiplexing (TDM) of pumps or frequency sweeping of a single pump laser. The latter has been proposed to minimize the interaction among pumps as, for example, four-wave mixing among pumps where the four-wave signal ends up within the signal band. This will be addressed in Section 3.4.3.

## 3.4.1  Wavelength Multiplexing of Pumps

When combining multiple discrete pumps, the task is to adjust the wavelength and amplitude of each pump laser in order to minimize the peak-to-peak variations in the gain spectrum and to broaden the gain spectrum or to optimize the noise spectrum. Finally, in the design of an amplifier with a fixed total bandwidth there is a trade-off between using many pumps to minimize the peak-to-peak variations and using as few pumps as possible for simplicity and cost.

An example of using two high-power pump sources, one at 1453 nm and one at 1495 nm, has already been mentioned in Section 3.1. This setup resulted in a 45-km-long dispersion shifted fiber being transparent over a bandwidth of 92 nm. More specifically an on–off Raman gain of approximately 13 dB ± 4 dB was demonstrated.

In one of the first demonstrations of combining several relatively low power semiconductor diodes, 12 wavelengths were combined using a WDM planar light-wave circuitry [17]. Unpolarized light at each wavelength is obtained by combining two independent laser diodes with orthogonal polarization states. The constructed amplifier had a 100-nm bandwidth with a gain of 6.5 dB ± 0.5 dB in a dispersion shifted fiber.

In the intermediate a gain of 10.9 dB ± 0.5 dB over the C band from, for example, 1530 to 1565 nm is demonstrated in a 2100-m-long dispersion-compensating fiber. This is obtained by using 5 pump wavelengths [53a].

In the above three examples the profile of the gain spectrum is adjusted for maximum flatness by carefully selecting the pump wavelengths and adjusting the power in the individual pump wavelengths. It is obvious that this is not a trivial procedure and a practical approach to optimizing the pump wavelengths and their power settings is desired both in the process of designing the amplifier and also as a feature in an already realized amplifier configuration.

The design process typically involves a theoretical optimization with the goal being to return pump wavelengths and pump power levels for a given specified amplifier performance. At first this task sounds rather simple; however, many factors must be considered in the design of the amplifier and the systems that use them. A thorough understanding of some of the key factors is required including pump-to-pump power transfer, signal-to-signal power transfer, pump depletion, Rayleigh scattering, and spontaneous emission. There are several approaches regarding numerical solutions to this task [20, 54]; however, they are considered outside the scope of this chapter.

Experimentally, the problem of arriving at an expected amplifier performance is not easy. When, for example, the pump power at one wavelength is adjusted it automatically affects the pump power at other wavelengths because of the Raman coupling between pumps. One solution to this problem is to feed the measured gain spectrum into a computer and do the gain flattening through the use of a software code created to find the optimum gain flatness via computer controlled pump power setting.

To maximize the chance for success in flattening the gain shape, knowledge of the Raman gain spectrum and the fact that power is transferred from the shortest pump wavelength to the longest pump wavelength [55] is essential. In addition, it has been proposed to reduce the number of adjustable parameters by collectively controlling the shortest wavelengths and only a single, or a few, longer wavelengths. In a specific example illustrated in Figure 3.19, four short wavelength pumps are grouped and collectively controlled and the longest wavelength pump is adjusted independently of the others.

Using this pumping method, three examples are given in which the gain is adjusted to three different values, 6.1 dB, 4.0 dB, and 2.0 dB, respectively, by adjusting the setting of the controller for Group #1 and the power of Group #2. The flatness achieved is ±0.46 dB, ±0.34 dB, and ±0.21 dB for the three settings.

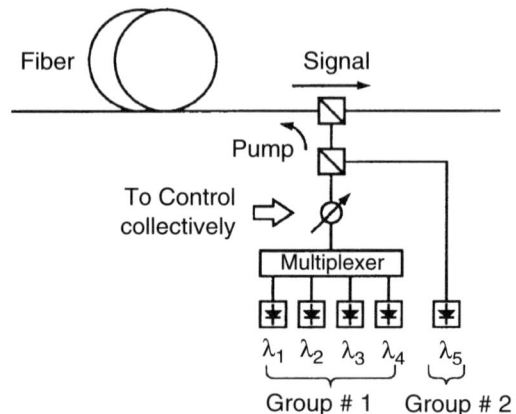

**Figure 3.19:** Schematic drawing of proposed method of simple gain control. Group #1 are shorter wavelength pumps while Group #2 are higher wavelength pumps. In Group #1, $\lambda_1 = 1420$ nm, $\lambda_2 = 1435$ nm, $\lambda_3 = 1450$ nm, and $\lambda_4 = 1465$ nm. In Group #2, $\lambda = 1495$ nm. (From [55].)

## 3.4.2  Broadened Pumps

The use of multiple pump wavelengths as described may be less attractive from a commercial point of view because of complexity and cost. To address this, a solution has been proposed [56, 57] in which the spectral width of the applied pump is broadened before it is coupled into the Raman amplifier.

In [56] a 50-nm broadband pump was constructed by using the spontaneously emitted light from an erbium-doped fiber amplifier (broadband ASE source) followed by a gain equalization filter and a booster amplifier. The emitted pump power was 23.5 dBm. By using this pump together with a fiber with a strong Raman efficiency, a 30-nm broadband amplifier was constructed with a 13-dB average net gain and peak-to-peak gain variations of less than 2 dB.

One of the constraints in the approach of using the above-mentioned broadband erbium-based ASE source as the pump is the wavelength of the pump [56]. To address this, an approach based on self-phase modulation of a pump laser [57] may be utilized. In this case spectral broadening is obtained by launching the pump laser first into a fiber with its zero dispersion wavelength matched to the wavelength of the pump laser, and then

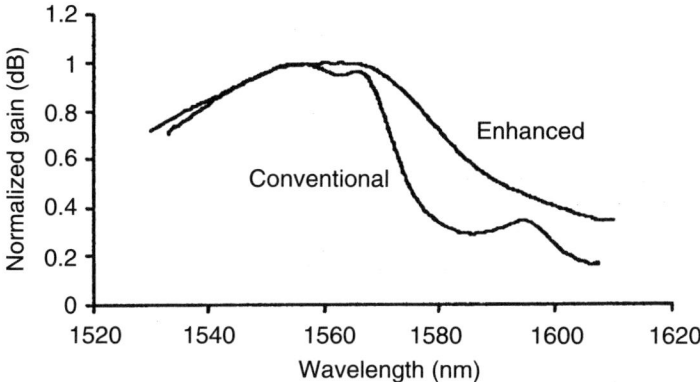

**Figure 3.20:** Measured Raman gain spectra of a 25-km-long fiber. The trace labeled Conventional refers to the use of a fiber Raman laser pump source, linewidth 0.6 nm. The trace labeled Enhanced is the gain spectrum obtained from a broadened pump laser, linewidth 14.23 nm. (From [57].)

into the amplifier fiber [57]. In this scheme the pump spectrum is defined by the fiber design of the pump fiber and the seed laser, as opposed to the wavelength band of the ASE source, typically an EDFA.

In [57] the pump laser had an initial 3-dB bandwidth of 0.63 nm at 1455 nm. After propagation through a 25-km zero-dispersion fiber, the 3-dB bandwidth of the pump was 14.23 nm when the pump power was 800 mW. Comparing the Raman gain spectra with and without the broadening, a significant change is found; see, Figure 3.20. From Figure 3.20 it is noted that the peak to peak variations in the gain spectrum has been significantly reduced, and in addition, the achieved gain spectrum is broadened as was expected.

The experimental results on both of the methods described above, using either an ASE source or a broadened pump laser, have illustrated that it is possible to strongly reduce the gain ripples and at the same time broaden the gain spectrum by optimizing the pump laser.

For completeness it is noted that conventional backward-pumped Raman amplifiers have been used in many other configurations and demonstrated great potential in hybrid amplifiers when combing the gain from EDFAs with Raman gain [58], to amplify wavelength outside the more conventional C and L band, for example, amplifiers around 1400 nm, and the so-called L and U band [59]. Finally, Raman gain has also found its application in dispersion compensation spans [60] to counteract their loss.

### 3.4.3   Time-Division-Multiplexed Pumps

As described previously, several pump lasers with various wavelengths are typically combined in a Raman amplifier either to flatten the spectral gain profile or to increase the available bandwidth for transmission. The composite pump beam is then injected into the optical fiber in the opposite direction of the signals to be amplified. In addition to this, high pump powers are required for the shorter wavelength pump to account for the Raman interaction among the pumps where pump power is transferred from the shorter pump wavelengths to the longer pump wavelengths.

As a result of the multiple pumps at relatively high power levels, hundreds of milliwatts, propagating in the same direction, four-wave mixing components may be generated. The strength of the four-wave mixing components depends particularly on the fiber dispersion and the actual wavelengths of the pumps. Figure 3.21 illustrates eight pump wavelengths that are combined to ensure a flat gain covering the C and L band [61]. As a result of four-wave mixing among the pumps, power is transferred to new frequency components, the four-wave mixing components. Unfortunately in the example displayed in Figure 3.21, these components end up in the signal band and result in degradation of the signal-to-noise ratio. In the example in Figure 3.21 the worst-case channel experiences an 8-dB degradation in the signal-to-noise ratio, from 27 dB to 19 dB. This degradation is measured within a 0.1-nm bandwidth. A 50-km-long dispersion shifted OFS TrueWave TW type fiber was used in the experiment.

To highlight the impact due to the four-wave mixing components, an experiment was performed in which the highest wavelength pumps and the L-band signals were removed; see Figure 3.22. The consequence of this is that the four-wave mixing wavelengths shift and do not interfere with the signals as strong as in Figure 3.21. The power spectrum of the C band only system is illustrated in Figure 3.22.

It is clear that the magnitude of the four-wave mixing components and hence the system performance is greatly impacted by the fiber type, in particular the fiber dispersion at the pump wavelengths, that is, the walk-off and the phase match between the different pump wavelengths. In Figures 3.21 and 3.22 no four-wave mixing was observed when using,

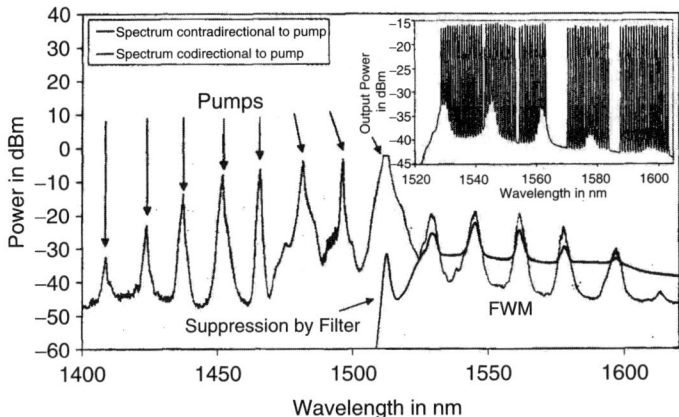

**Figure 3.21:** A measured power spectrum from a 50-km-long Raman amplifier in which 80 signal channels propagate in the C and L band. The fiber type was a TrueWave fiber and an on–off gain of 16 dB was achieved using a total pump power of 650 mW. The inset figure is a close-up of the signal channels counter propagating to the pumps. (From [61].)

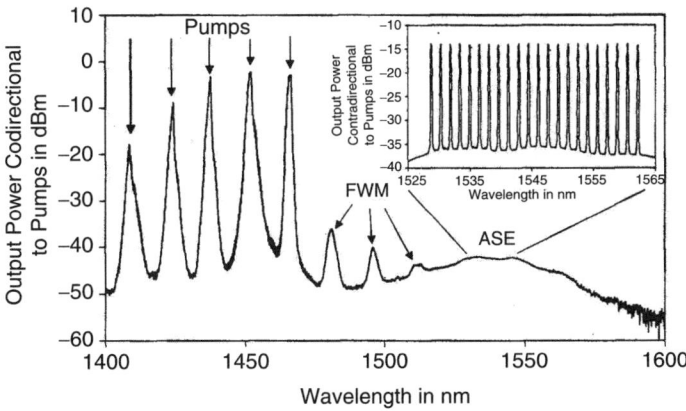

**Figure 3.22:** A measured power spectrum from a 50-km-long Raman amplifier in which 40 signal channels propagate in the C band. Similar to Figure 3.21, the fiber type was a TrueWave fiber but in this case an on–off gain of 20 dB was achieved using a total pump power of 650 mW. The inset figure is a close-up of the signal channels counter propagating to the pumps. (From [61].)

for example, a standard single mode-fiber [61] with a relatively large dispersion.

As opposed to changing the fiber dispersion at the pump wavelengths, four-wave mixing may be avoided and pump-to-pump interactions eliminated by sweeping the frequency of one pump laser [62–66] or by time division multiplexing (TDM) of a multiple of pump lasers each at different wavelengths.

Instead of using a single-frequency laser, a single but frequency-tunable laser is employed as a pump source. The tunable laser is periodically swept according to some wavelength pattern with the time spent at a particular wavelength determining the amount of Raman gain arising from that wavelength. If the pattern repetition rate is chosen high enough, the (counterpropagating) signal does not experience any significant temporal gain variations, but does experience the composite Raman gain of all pump wavelengths making up the sweep pattern [62–65], in analogy to a multiwavelength continuous Raman system. This method only works for backward pumping.

In [65] a time-division-multiplexed frequency swept pump is demonstrated as a means to dynamically flatten the spectral response of a Raman amplifier. An example is realized with a grating coupled sampled reflector semiconductor laser. This laser has a tuning range from 1515 nm to 1565 nm. The output power from this laser is then launched into a high-power EDFA after which a control signal is fed back to the laser. This setup ensures a pump laser with a controllable output frequency that may be swept at a preset sweeping pattern. As a result a gain of 22 dB $\pm$ 0.3 dB over 15 nm is achieved when the pump is launched into a 15-km-long fiber.

A second approach would be to TDM a large array of fixed-wavelength lasers [66]. The TDM would have to be performed with a high repetition rate, near 1 MHz, such that there is a sufficient averaging effect from the scan of pump wavelengths. This still suffers from the problem of efficiently wavelength multiplexing of the array of lasers and the cost associated with applying many lasers.

# 3.5  Advanced Pumping Configurations

The desire to increase the capacity in communication systems is one of the principal reasons for applying Raman gain. In an ideal configuration of a distributed amplifier, the intrinsic fiber attenuation is counterbalanced along every point of a transmission span resulting in no signal power variations along the span. A realistic approach to an ideal distributed Raman amplifier is obtained by using a bidirectionally Raman pumped amplifier, that is, the combined use of forward and backward pumping.

In Sections 3.2 and 3.3 it has been demonstrated that bidirectional pumping would result in a reduced accumulation of spontaneous emission, in addition to reduced penalties from Rayleigh scattering. On the other hand the path average power will increase when compared to a backward-pumped Raman amplifier and fixed-signal input power. In addition, but practically most challenging, an extra noise contribution arises from pump–signal cross talk and signal–pump–signal cross talk when comparing bidirectional pumping against backward pumping.

For completeness, an advantage of bidirectional pumping compared to backward or forward pumping is the ability to transmit signals in both the forward and the backward direction, that is, bidirectional pumping and bidirectional signal transmission. Finally, rather than launching very high power from one end, now the power requirement is split between two launch sites.

The above benefits and drawbacks have to be accounted for in a final evaluation of a pumping configuration.

One of the major obstacles in using forward pumping is the pump–signal cross talk as well as signal–pump–signal cross talk as discussed in Section 3.3. Both phenomena result in an additional noise contribution that may exceed the benefits otherwise obtained from the uniform gain distribution. However, two methods have been proposed to realize forward pumping. One is the use of so-called higher order pumping or cascaded pumping, which will be described in Section 3.5.1. The other method is the use of special pump modules that are designed with minimum noise, so-called quiet pumps. This will be described in Section 3.5.2.

## 3.5.1 Higher Order Pumping

In Section 3.2 and in Section 3.3 it has been shown that there are significant benefits in moving the gain from the end of a fiber span into the span, for example in improved signal-to-noise ratio and in reduced Rayleigh cross talk. One way to push the gain into the span is by amplifying the pump

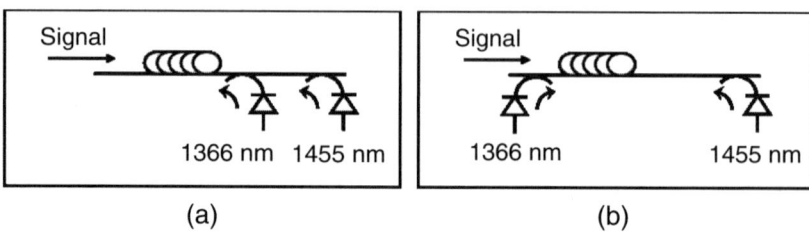

**Figure 3.23:** Configurations of second-order pumping. In (a) the first- and second-order pumps are copropagating but both are counterpropagating against the signal, whereas in (b) the first- and second-order pumps are counterpropagating against each other.

within the span using distributed amplification of the pump. This method thus relies on using another pump wavelength shifted another 13 THz from the first pump wavelength. This method is called second-order pump or cascaded Raman pumping.

The second-order pump may be launched from the same end as the first pump, displayed in Figure 3.23a. Alternatively it may be launched from the same end as the signal; see Figure 3.23b. In the latter case the remaining first pump is amplified close to the signal input end and as a result the first pump now provides gain for the signal both at the input and at the output end of the fiber span. In this case the signal power evolves, as in a bidirectionally pumped amplifier. However, when compared to a conventionally bidirectionally pumped amplifier, pump–signal and signal–pump–signal cross talk are avoided.

In the following we concentrate on the case in which the first and the second pump are counterpropagating. However, second-order pumping has been successfully demonstrated both in copropagating and counter-propagating pump configurations [67]. In addition efficient pump lasers have been demonstrated in which a single cavity provides both first- and second-order pump beams [68].

In a first demonstration, an 80-km-long transmission span was considered [69]. The noise figure for such a span is 19 dB if the intrinsic fiber loss is counterbalanced by an ideal EDFA. Replacing the EDFA with a distributed Raman amplifier, the noise figure is improved to about 14 dB. However, applying a second-order pump in the forward direction (Figure 3.23b), the noise figure is further improved. The improvement depends on the power of the applied second-order pump. For a pump power at 800 mW an improvement of 3 dB was measured.

The 3-dB improvement in noise figure does not translate directly into a 3-dB improvement in BER in a system experiment. Several factors influence the BER including the increased path average signal power and the direct pump signal coupling that occurs since the Raman spectrum is broad and in the above example extends from 1366 nm and much beyond 1550 nm. In the example discussed above, the 3-dB improvement in noise figure translates into a 1-dB improvement in receiver sensitivity.

More recently, second-order pumping has been improved further by applying yet another shorter pump wavelength [70]. In one configuration, a low-power 1455-nm pump is launched in the opposite direction as the signal. This weak pump is launched together with a second low-power pump at 1356 nm. A high-power third pump at 1257 nm amplifies the 1356-nm pump, which peaks in power about 15 km from the input end. The 1455-nm pump is amplified by the 1356-nm pump and assumes a maximum value approximately 25 km from the input end. As compared to a conventional and a second-order pumping configuration, the Raman gain is pushed further into the span, leading to a further improvement of the noise performance. A system demonstration using a 100-km span shows an improvement of 2.5 dB compared to conventional Raman pumping configuration when using third-order pumping [70]. This improvement is achieved on the cost of applying a 3-W pump laser at 1276 nm.

## 3.5.2  Quiet Pumps

For the purpose of forward pumping it is essential to have both a high degree of wavelength stabilization and a stable and low noise output intensity, the latter characterized by RIN. These requirements need to be fulfilled while retaining a high-output power level and multimode operation. The latter is desired to avoid stimulated Brillouin scattering. Finally the output of the pump should preferably be unpolarized.

As mentioned earlier, there are several options when choosing a pump, including cascaded fiber Raman lasers and Fabry Perot semiconductor lasers with an external fiber Bragg grating to lock the wavelength. However, both the cascaded Raman laser and the external Bragg grating locked laser

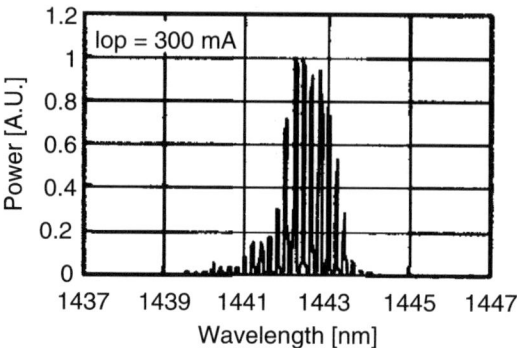

**Figure 3.24:** Output spectrum of internal-grating stabilized multimode pump laser. (From [55].)

are characterized by a high RIN, typically $-110$ dB/Hz for the Raman fiber laser and $-135$ dB/Hz for the fiber grating laser. Thus none of these is practical for forward-pumping Raman amplifier configurations.

As opposed to the external grating, a multimode pump laser with an internal grating incorporated into the laser cavity has been demonstrated [55, 71]. With this method only a few longitudinal modes are contained within the output beam, and the mode spacing in this laser is relatively large: tens of GHz as opposed to hundreds of MHz for an external grating about 1 m from the cavity. Figure 3.24 illustrates the power spectrum of a laser with an internal grating, and the relative large mode spacing is clear.

The RIN of the laser with the output power spectrum in Figure 3.24 is displayed in Figure 3.25. From the figure, the RIN is as low as $-160$ dB/Hz, which is about 30 dB lower than the RIN of a Fabry Perot laser with an external grating [55, 72].

The benefit of forward pumping is that gain may be obtained close to the input end of the fiber, which as a consequence prevents the signal-to-noise ratio from dropping as much as it otherwise does in a counterpumped Raman amplifier. However, the path average signal power is higher in the forward-pumped amplifier compared to the counterpumped amplifier. Taking both the spontaneous emission and the nonlinear effects evaluated through the path average power effects into account leads to the conclusion that bidirectional pumping provides the optimum conditions; see Sections 3.2 and 3.3.1 for further details.

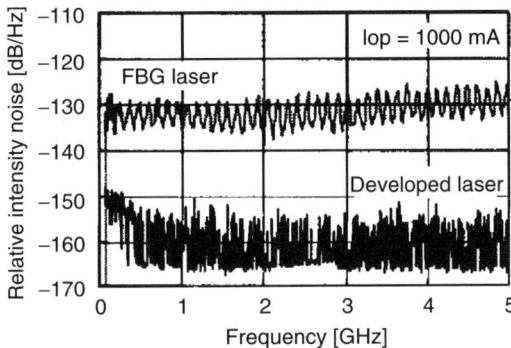

**Figure 3.25:** Comparison of relative intensity noise between a usual external fiber Bragg grating stabilized laser and the internal-grating stabilized laser; cf. Figure 3.24. (From [55].)

The highlighted benefits of bidirectional pumping were used in an experimental demonstration of a 1.28 Tbit/s ($32 \times 40$ Gbit/s) unrepeatered transmission through a 240-km-long fiber span consisting of conventional nondispersion shifted fiber [73]. In the transmission, the losses were compensated solely by bidirectionally pumped distributed Raman amplification; that is, the pump power in the transmission span was chosen exactly so that the Raman gain was sufficient to counterbalance the 24-dB intrinsic fiber losses. The forward pump provided approximately 8 dB of gain.

The Raman pumping modules were multiplexed depolarized semiconductor lasers operating at wavelengths of 1427 and 1455 nm. The RIN values of the pump modules were around $-140$ dB/Hz. The bidirectional pumping configuration resulted in a 6-dB improvement in the optical signal-to-noise ratio when compared against either co- or counterpumping. In the system there was no significant transfer of RIN from the pump to the signal.

Forward pumping may also be used to flatten the spectral dependence of the noise figure. In conventional counterpumped Raman amplifiers, pump power is transferred from the short pump wavelengths to the longer wavelengths during propagation. Thus, the gain for the longer signal wavelengths is pushed further into the transmission span. As a result the longer signal wavelengths perform the best. To eliminate this difference in performance between the longest and shortest wavelength channels, bidirectional pumping will, in addition to improved noise performance discussed in Section 3.2, also lead to the possibility of flattening the spectral dependence of the noise figure by applying forward pumping for the poorest performing channels.

This has been demonstrated on a 76-km-long Raman amplifier. The fiber in the experiment was a standard single-mode fiber. By using five pump wavelengths in the backward direction, totaling 580 mW, and three wavelengths in the forward direction, totaling 90 mW, a noise figure of 15.2 dB $\pm$ 0.4 dB was achieved over an 80-nm bandwidth. For the same configuration a gain of 9.5 dB $\pm$ 0.3 dB was achieved [72].

## 3.6  Summary

It is without question that one of the most promising applications of Raman scattering is as a means to achieve distributed amplification. The use of distributed Raman amplification has already been demonstrated in ultra-high-capacity optical communication systems as the enabling method to transmit 40 Gbit/s per channel in a wavelength-division-multiplexed transmission system.

Among the benefits of Raman amplification as opposed to, for example, discrete erbium-doped fiber amplification is that the gain mechanism is intrinsic to the silica fiber; that is, no specialty fiber is needed; that gain at any wavelength may be achieved by choosing an appropriate pump wavelength; and that the useable bandwidth may be extended simply by combining multiple pump wavelengths. These benefits are all of great advantage also in distributed Raman amplifiers; however, the fact that the gain is distributed throughout the entire fiber span is specific and unique to Raman amplification, and as a result of this the signal-to-noise ratio at the output of the fiber span is significantly improved.

To evaluate the performance of the distributed amplifier, the improved noise performance has to be balanced against the challenges in using distributed Raman amplification. The most severe challenges are caused by Rayleigh reflections that are critical since the gain is accumulated over tens of kilometers, nonlinear effects that may be enhanced in distributed Raman amplifiers since the path average signal power is higher along the transmission fiber when compared to an unpumped fiber. In addition to these challenges, the Raman process also has a very fast response time on the order of tens of femtoseconds. These represent a major challenge, especially if the pump beam is propagating in the same direction as the signal.

In this chapter the application of Raman scattering as a means to achieve distributed amplification was described. Focus was directed toward the specific advantages and challenges in using distributed Raman amplification. In the last part of the chapter conventional and more advanced methods used to pump distributed Raman amplifiers were described.

## Acknowledgements

Much of the work reported in this chapter was done at Bell Laboratories, Lucent Technologies. Colleagues at the Fiber Research department are greatly acknowledged and especially Andrew Stentz, Jake Bromage and Mei Du are thanked, without them I would not have been able to write this chapter. I would also like to thank former colleagues at Tyco Submarine Systems, especially Howard Kidorf, with whom I did much initial work on Raman amplification and in addition Henrik Smith from the Copenhagen University and Jørn Hedegaard Povlsen, Research Center COM, Technical University of Denmark for numerous fruitful technical discussions.

## References

[1] L. F. Mollenauer, R. H. Stolen, and M. N. Islam, *Opt. Lett.*, **10**, 229 (1985).

[2] P. B. Hansen, L. Eskildsen, S. G. Grubb, S. K. Korotky, T. A. Strasser, J. E. J. Alphonsus, J. J. Veselka, D. J. DiGiovanni, D. W. Peckham, and D. Truxal, *Electron. Lett.* **32**, 1018 (1996).

[3] T. N. Nielsen, A. J. Stentz, K. Rottwitt, D. S. Vengsarkar, Z. J. Chen, P. B. Hansen, J. H. Park, K. S. Feder, S. Cabot, S. Stulz, D. W. Peckham, L. Hsu, C. K. Kan, A. F. Judy, S. Y. Park, L. E. Nelson, and L. Gruner-Nielsen, *IEEE Photon. Technol. Lett.* **12**, 1079 (2000).

[4] B. Chu, L. Leng, L. E. Nelson, L. Gruner-Nielsen, Y. Qian, J. Bromage, S. Stultz, S. Kado, Y. Emori, S. Namiki, P. Gaarde, A. Judy, B. Palsdottir, and R. L. Lingle, Jr, 3.2 Tb/s (80 × 42.7 Gb/s) transmission over 20 × 100 km of non-zero dispersion fiber with simultaneous C + L band dispersion compensation (Optical Fiber Communication Conferences, Anaheim, CA, March 2002), postdeadline paper FC8.

[5] M. Nissov, C. R. Davidson, K. Rottwitt, R. Menges, P. C. Corbett, D. Innis, and N. S. Bergano, 100 Gb/s (10 × 10 Gb/s) WDM transmission over 7200 km using distributed Raman amplification, European Conference on Optical Communication (ECOC, Edinburgh, Scotland, September 1997), postdeadline paper.

[6] B. Mikkelsen, G. Raybon, R. J. Essiambre, A. J. Stentz, T. N. Nielsen, D. W. Peckham, L. Hsu, L. Gruner-Nielsen, K. Dreyer, and J. E. Johnson, *IEEE Photon. Technol. Lett.* **12**, 1400 (2000).

[7] See, for example: C. Xu, X. Liu, L. Mollenauer, and X. Wei, *IEEE Photon. Technol. Lett.* **15**, 617 (2003) S. Wabnitz, and G. L. Meur, *Opt. Lett.* **26**, 777 (2001).

[8] E. Desurvire, *Erbium Doped Fiber Amplifiers, Principles and Applications* (Academic Press, San Diego, 1995).

[9] C. Lester, K. Rottwitt, J. H. Povlsen, P. Varming, M. Newhouse, and A. J. Antos, *Opt. Lett.* **20**, 1250 (1995).

[10] R. Stolen, J. P. Gordon, W. J. Tomlinson, and H. A. Haus, *J. Opt. Soc. Am. B* **6**, 1159 (1989).

[11] S. G. Grubb, T. A. Strasser, W. Y. Cheung, W. A. Reed, V. Mizrahi, T. Erdogan, P. J. Lemaire, A. M. Vengsarkar, and D. J. DiGiovanni, High power 1.48 mm cascaded Raman laser in germanosilicate fibers (Optical Amplifiers and Their Applications, Davos, Switzerland, 1995), p. 197.

[12] P. B. Hansen, V. L. da Silva, G. Nykolak, J. R. Simpson, D. L. Wilson, J. E. J. Alphonsus, and D. J. DiGiovanni, *IEEE Photon. Technol. Lett*, **7**, 588 (1995).

[13] P. B. Hansen, L. Eskildsen, S. G. Grubb, A. J. Stentz, T. A. Strasser, J. Judkins, J. J. DeMarco, R. Pedrazzini, and D. J. DiGiovanni, *IEEE Photon. Technol. Lett.* **9**, 262 (1997).

[14] P. B. Hansen, L. Eskildsen, A. J. Stentz, T. A. Strasser, J. Judkins, J. J. DeMarco, R. Pedrazzani, and D. J. DiGiovanni, *IEEE Photon. Technol. Lett.* **10**, 159 (1998).

[15] M. X. Ma, H. D. Kidorf, K. Rottwitt, F. Kerfoot, III, and Davidson, C. *IEEE Photon. Technol. Lett.* **10**, 893 (1998).

[15a] T. N. Nielson, A. J. Stentz, P. B. Hansen, C. J. Chen, D. S. Vengsarkar, T. A. Strasser, K. Rottwitt, O. H. Park, S. Stulz, S. Cabot, K. S. Feder, P. S. Westbrook, K. G. Kosinski, "1.6 Tb/s (40 × 40 Gb/s) transmission over 4–100 km nonzero-dispersion fiber using hybrid Raman/Er doped inline amplifiers," ECOC, Nice, France, 1999, Paper pd2-2.

[16] K. Rottwitt and H. D. Kidorf, A 92-nm bandwidth Raman amplifier (Optical Fiber Communication Conference, San Jose, CA, 1998), postdeadline paper PD6.

[17] Y. Emori and S. Namiki, 100 nm bandwidth flat gain Raman amplifiers pumped and gain equalized by 12 wavelength channel WDM high power laser diodes (Optical Fiber Communication Conference, San Diego, CA, February 1999), paper PD19.

[18] A. K. Srivastava, D. L. Tzeng, A. J. Stentz, J. E. Johnson, M. L. Pearsall, O. Mizuhara, T. A. Strasser, K. F. Dreyer, J. W. Sulhoff, L. Zhang, P. D. Yeates, J. R. Pedrazzani, A. M. Sergent, R. E. Tench, J. M. Freund, T. V. Nguyen, H. Manar, Y. Sun, C. Wolf, M. M. Choy, R. B. Kummer, D. Kalish, and A. R. Chraplyvy, High speed WDM transmission in AllWave fiber in both the 1.4 mm and 1.55 mm bands (Optical Amplifiers and Their Applications, Vail, CO, July 1998), PD2.

[19] J. Bromage, K. Rottwitt, and M. E. Lines, *IEEE Photon. Technol. Lett.* **14**, 24 (2002).

[20] H. Kidorf, K. Rottwitt, M. Nissov, M. Ma, and E. Rabarijaona, *IEEE Photon. Technol. Lett.* **11**, 530 (1999).

[21] E. Desurvire, M. J. F. Digonnet, and H. J. Shaw, *J. Lightwave Technol.* **4**, 426 (1986).

[22] K. Rottwitt, J. Bromage, A. J. Stentz, L. Leng, M. E. Lines, and H. Smith, *J. Lightwave Technol.* **21**, 1452 (2003).

[23] K. Rottwitt, J. H. Povlsen, A. Bjarklev, O. Lumholt, B. Pedersen, and T. Rasmussen, *IEEE Photon. Technol. Lett.* **5**, 218 (1993).

[24] K. Rottwitt and A. J. Stentz, in *Optical Fiber Telecommunications,* Vol. IVA (Academic Press, San Diego, CA, 2002), chap. 5.

[25] R. W. Hellwarth, *Prog. Quantum Electron.* **5**, 1 (1977).

[26] M. J. Adams, *An Introduction to Optical Waveguides* (Wiley, New York, 1981).

[27] G. P. Agrawal, *Nonlinear Fiber Optics* (Academic Press, San Diego, CA, 1995).

[28] R. J. Essiambre, B. Mikkelsen, and G. Raybon, *Electron. Lett.* **35**, 1576 (1999).

[29] S. N. Knudsen, B. Zhu, L. E. Nelson, M. Ø. Pedersen, D. W. Peckham, and S. Stulz, *Electron. Lett.* **37**, 965 (2001).

[30] See A. R. Chraplyvy, *J. Lightwave Technol.* **8**, 1548 (1990). R. J. Essiambre, P. Winzer, J. Bromage, and C. H. Kim, *Photon. Technol. Lett.* **14**, 914 (2002).

[31] See J. Stark, P. Mitra, and A. Sengupta, *Opt. Fiber Technol.* **7**, 275 (2001). P. Mitra and J. Stark, *Nature*, **411**, 1027 (2001).

[32] G. P. Agrawal, *Fiber Optic Communication Systems* (Wiley, New York, NY, 2002).

[33] D. Marcuse, *J. Lightwave Technol.* **8**, 1816 (1990).

[34] A. F. Evans, J. Grochocinski, A. Rahman, C. Raynolds, and M. Vasilyev, Distributed amplification: How Raman gain impacts other fiber nonlinearities (Optical Fiber Communication Conference, Anaheim, CA, 2001), paper MA7.

[35] F. Forghieri, R. W. Tkach, and A. R. Chraplyvy. *IEEE Photon. Technol. Lett.* **7**, 101 (1995).

[36] J. L. Gimlett, M. Z. Iqbal, L. Curtis, N. K. Cheung, A. Righetti, F. Fontana, and G. Grasso, *Electron. Lett.* **25**, 1393 (1989).

[37] J. L. Gimlett and N. K. Cheung, *J. Lightwave Technol.* **7**, 888 (1989).

[38] H. Takahashi, K. Oda, and H. Toba, *J. Lightwave Technol.* **14**, 1097 (1996).

[39] P. Wan and J. Conradi, *J. Lightwave Technol.* **14**, 288 (1996).

[40] P. Wan and J. Conradi, *Electron. Lett.* **31**, 383 (1995).

[41] M. Nissov, K. Rottwitt, H. D. Kidorf, and M. X. Ma, *Electron. Lett.* **35**, 997 (1999).

[42] M. O. van Deventer, *J. Lightwave Technol.* **11**, 1895 (1993).

[43] J. Bromage, C. H. Kim, R. M. Jopson, K. Rottwitt, and A. J. Stentz, Dependence of double Rayleigh back-scatter noise in Raman amplifiers on gain and pump depletion (Optical Amplifiers and Their Applications, Stresa, Italy, July 2001).

[44] C. R. S. Fludger, Y. Zhu, V. Handerek, and R. J. Mears, *Electron. Lett.* **37**, 970 (2001).

[45] S. Wang and C. Fan, *Opt. Comm.* **210**, 355 (2002).

[46] J. Bromage and L. E. Nelson, Relative impact of multiple-path interference and amplified spontaneous emission noise on optical receiver performance (Optical Fiber Communication Conference, Anaheim, CA, March 2002), paper TuR3.

[47] D. Hollenbeck and C. D. Cantrell, *J. Opt. Soc. Am. B* **19**, 2886 (2002).

[48] A. J. Stentz, S. G. Grubb, C. E. Headley, III, J. R. Simpson, T. Strasser, and N. Park, Raman amplifier with improved system performance (Optical Fiber Communication, San Jose, CA, 1996), paper TuD3.

[49] See H. F. Mahlein, *Opt. Quantum Electron.* **16**, 409 (1984). M. D. Mermelstein, C. Headley, and J.-C. Bouteiller, *Electron. Lett.* **38**, 403 (2002).

[50] C. R. S. Fludger, V. Handerek, and R. J. Mears, *J. Lightwave Technol.* **19**, 1140 (2001).

[51] W. Jiang and P. Ye, *J. Lightwave Technol.* **7**, 1407 (1989).

[52] F. Forghieri, R. W. Tkach, and A. R. Chraplyvy, Bandwidth of cross talk in Raman amplifiers (Optical Fiber Communication Conference, San Jose, CA, Feb. 20–25, 1994), paper FC6.

[53]  S. V. Chernikov, Y. Zhu, R. Kashyap, and J. R. Taylor, *Electron. Lett.* **31**, 472 (1995).

[53a] H. Wandel, P. Krishnan, T. Veng, Y. Qian, L. Quay, L, Grüner-Nielsen, "Dispersion compensating fibers for non-zero dispersion fibres," (Optical Fiber Communication Conference, Anaheim, California, March 2002), paper WU1.

[54]  See, for example, J. D. Ania-Castanon, S. M. Kobtsev, A. A. Pustovskikh, and S. K. Turitsyn, Simple design method for gain-flattened three-pump Raman amplifiers (Lasers and Electro-Optics Society, 2002), p. 500. V. E. Perlin and H.G. Winful, *J. Lightwave Technol.* **20**, 409 (2002). M. Yan, J. Chen, W. Jiang, J. Li, J. Chen, and X. Li, *IEEE Photon. Technol. Lett.* **13** (2001).

[55]  Y. Emori, S. Kado, and S. Namiki, *Opt. Fiber Technol.* **8**, 107 (2002).

[56]  T. Tsuzaki, M. Kakui, M. Hirano, M. Onishi, Y. Nakai, and M. Nishimura, Broadband discrete fiber Raman amplifier with high differential gain operating over 1.65 μm-band (Optical Fiber Communication Conference, Anaheim, CA, March 2001), paper MA3.

[57]  T. J. Ellingham, L. M. Gleeson, and N. J. Doran, Enhanced Raman amplifier performance using non-linear pump broadening (ECOC'02, Copenhagen), paper 4.1.3.

[58]  H. Musada, S. Kawai, and K. Aida, *IEEE Photon. Technol. Lett.* **10**, 516 (1998).

[59]  P. C. Reeves-Hall, D. A. Chestnut, C. J. S. De Matos, and J. R. Taylor, *Electron. Lett.* **37**, 883 (2001).

[60]  P. B. Hansen, G. Jacobovitz-Veselka, L. Gruner-Nielsen, and A. J. Stentz, *Electron. Lett.* **34**, 1136 (1998).

[61]  R. E. Neuhauser, P. M. Krummrich, H. Bock, and C. Glingener, Impact of nonlinear pump interactions on broadband distributed Raman amplification (Optical Fiber Communication Conference, Anaheim, CA, Mar. 17–22, 2001), paper MA4-1.

[62]  L. F. Mollenauer, A. R. Grant, and P. V. Mamyshev, *Opt. Lett.* **27**, 592 (2002).

[63]  A. R. Grant, *IEEE J. Quantum Electron.* **38**, 1503 (2002).

[64]  P. J. Winzer, J. Bromage, R. T. Kane, P. A. Sammer, and C. Headley, Temporal gain variations in time-division multiplexed Raman pumping schemes (CLEO, Long Beach, CA, May 2002), paper CW01.

[65]  P. J. Winzer, K. Sherman, and M. Zirngibl, *IEEE Photon. Technol. Lett.* **14**, 789 (2002).

[66]  C. R. S. Fludger, V. Handerek, N. Jolley, and R. J. Mears, Novel ultra-broadband high performance distributed Raman amplifier employing pump

modulation (Optical Fiber Communication Conference, Anaheim, CA, March 2002), paper WB4.

[67] D. A. Chestnut, C. J. S. de Matos, P. C. Reeves-Hall, and J. R. Taylor, *Opt. Lett.* **27**, 1708 (2002).

[68] J.-C. Bouteiller, K. Bar, S. Radic, J. Bromage, Z. Wang, and C. Headley, Dual-order Raman pump providing improved noise figure and large gain bandwidth (Optical Fiber Communication Conference, Anaheim, CA, March 2002), postdeadline paper FB3.

[69] K. Rottwitt, A. Stentz, T. Nielson, P. Hansen, K. Feder, and K. Walker, Transparent 80 km bi-directionally pumped distributed Raman amplifier with second-order pumping (ECOC, Nice, France, 1999), paper II-144.

[70] S. B. Papernyi, V. J. Karpov, and W. R. L. Clements, Third-order cascaded Raman amplification (Optical Fiber Communication Conference, Anaheim, CA, March 2002), postdeadline paper FB4.

[71] R. P. Espindola, K. L. Bacher, K. Kojima, N. Chand, S. Srinivasan, G. C. Cho, F. Jin, C. Fuchs, V. Milner, and W. C. Dautremont-Smith, High power, low RIN, spectrally-broadened 14xx DFB pump for application in co-pumped Raman amplification (ECOC, Amsterdam, Netherlands, Sep. 30–Oct. 4, 2001), paper PD.F.1.7.

[72] S. Kado, Y. Emori, S. Namiki, N. Tsukiji, J. Yoshida, and T. Kimura, Broadband flat-noise Raman amplifier using low-noise bi-directionally pumping sources (ECOC, Amsterdam, Netherlands, Sep. 30–Oct. 4, 2001), paper PD.F.1.8.

[73] Y. Zhu, W. S. Lee, C. R. S. Fludger, and A. Hadjifotiou, All-Raman unrepeatered transmission of 1.28 Tbit/s (32×40 Gbit/s) over 240 km conventional 80 mm$^2$ NDSF employing bi-directional pumping (Optical Fiber Communication Conference, Anaheim, CA, March 2002), paper ThFF2.

# Chapter 4

# Discrete Raman Amplifiers

**Shu Namiki, Yoshihiro Emori,** and **Atsushi Oguri**

This chapter focuses on the issues surrounding Raman amplifiers in a discrete module, or discrete Raman amplifiers. Because of the small scattering cross-section, Raman amplification may better fit in a distributed amplifier rather than a discrete one. Therefore, in designing discrete Raman amplifiers, several challenges such as increasing efficiency and solving fundamental trade-offs are required. However, discrete Raman amplifiers have many attractive aspects over rare-earth-doped fiber amplifiers such as an erbium-doped fiber amplifier (EDFA) including arbitrary gain band, better adjustability of gain shape, and better linearity. It is thus important to understand the design issues of discrete Raman amplifiers, which are dependent on each other, so that one can optimize the design for a particular application. This chapter unveils the trade-offs and identifies the design issues through understanding the physics of Raman amplification in discrete configurations. It also discusses dispersion characteristics as well as wide-band operation of discrete Raman amplifiers.

## 4.1  Basic Configuration and Its Model

Figure 4.1 shows the basic configurations of discrete Raman amplifiers. It generally comprises a gain fiber, a wavelength-division-multiplexed

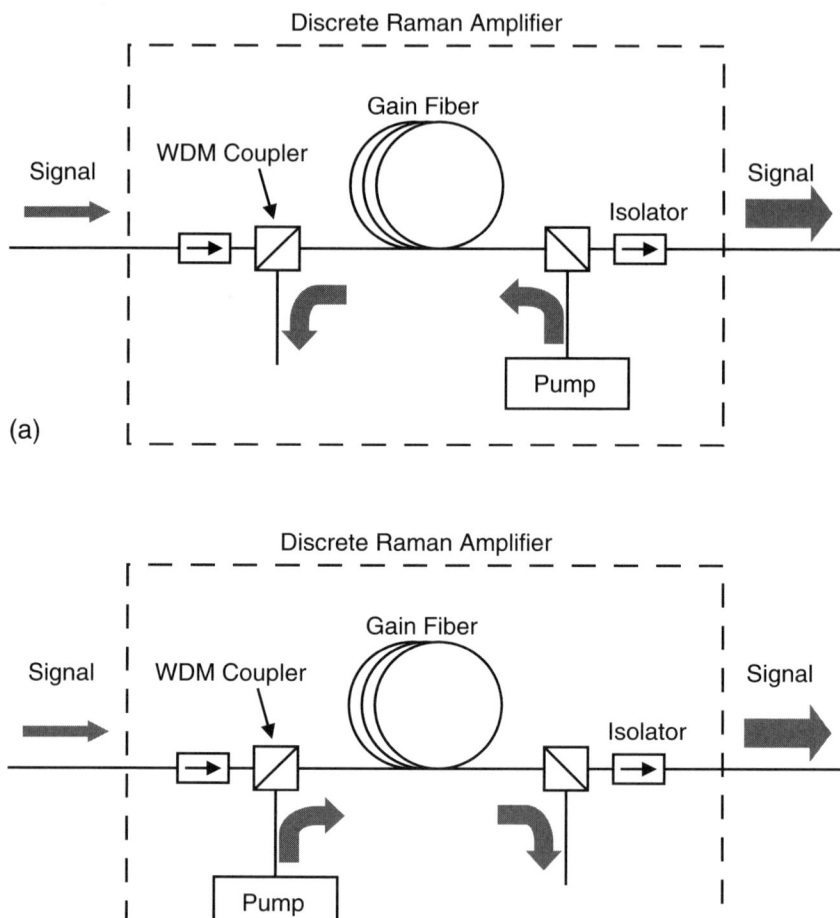

**Figure 4.1:** Schematic configurations of (a) single-stage counterpumped, (b) single-stage copumped, (c) single-stage bidirectional, and (d) dual-stage discrete Raman amplifiers.

(WDM) coupler for combining the pump and the signal, and isolators at the input and output ends. The orientation of the pump can be either forward or backward with respect to the signal propagation. Whereas the counterpropagating one is called *counterpumping*, the copropagating pumping scheme is called *copumping*. There is also an option of *bidirectional pumping*, in

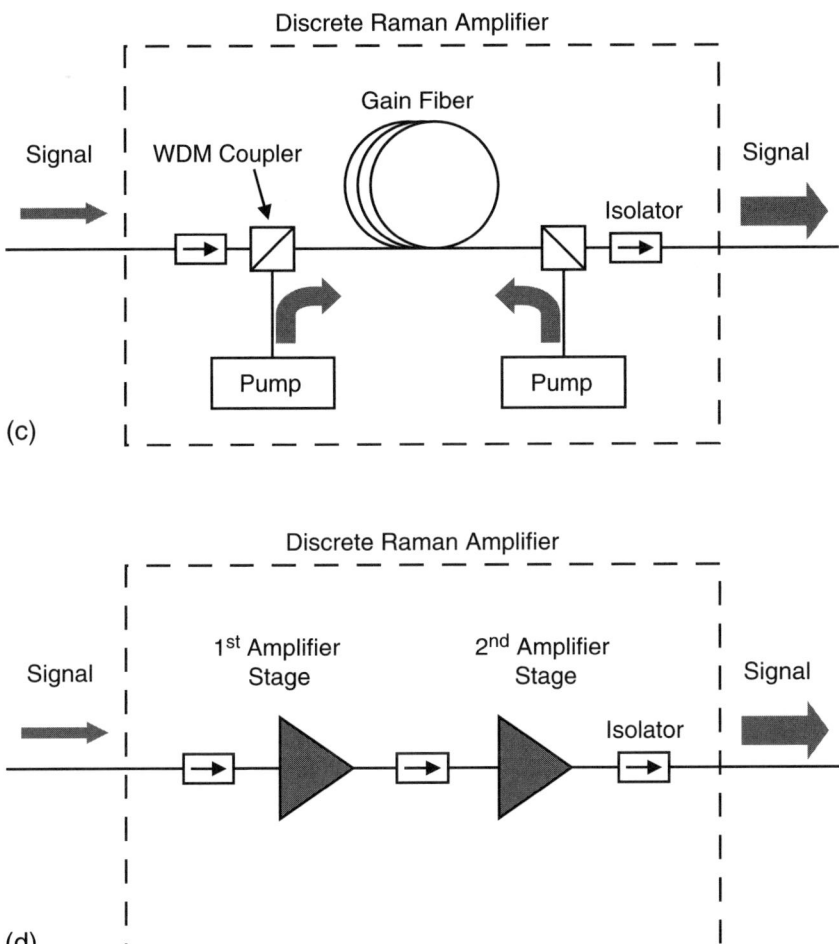

**Figure 4.1:** Continued.

which the gain fiber is pumped in both directions. In general, there are several derivatives or combinations of these simple configurations for various applications. In a dual-stage Raman amplifier, two amplifier stages are concatenated.

The important parameters representing discrete Raman amplifiers are (i) the wavelength and input power level of the signal, (ii) the wavelength and input power level of the pump, and (iii) the type and length of the gain fiber. As for the gain fiber, the following properties in signal and pump wavelength bands are required to design the amplifier in detail: (i) the attenuation coefficient, (ii) the Raman gain coefficient for the given pump wavelengths, (iii) the Rayleigh backscattering coefficient, and (iv) the nonlinear coefficient. The wavelength dependence of each property should be given as precisely as possible for the accurate design over the entire signal and pump band.

An optimal discrete Raman amplifier is designed and found by changing these parameters for various configurations. The targeted optical characteristics of a discrete Raman amplifier are usually gain, noise figure, output signal power level, optical signal-to-noise ratio (OSNR), double-Rayleigh backscattering (DRBS) noise power, nonlinear phase shift, and pump-to-signal power conversion efficiency. For WDM signals, the dependence of the above characteristics on wavelength or signal channel should also be optimized.

### 4.1.1   Single-Pump Amplification

The simplest case of single pump amplification is reviewed here, though it has already been described in Chapter 2. The power evolutions of signal and pump waves along an optical fiber are described by the following coupled equations [1] (refer to Eqs. (2.1.7) and (2.1.8)),

$$\frac{dP_s}{dz} = g_R P_p P_s - \alpha_s P_s \qquad (4.1)$$

$$\eta \frac{dP_p}{dz} = -\frac{\omega_p}{\omega_s} g_R P_p P_s - \alpha_p P_p, \qquad (4.2)$$

where $P_s$ and $P_p$ are the signal and pump powers, respectively, $g_R$ is the Raman gain efficiency for the wavelengths of the signal and pump, $\omega_s$ and $\omega_p$ are the frequencies of signal and pump waves, respectively, and $\alpha_s$ and $\alpha_p$ are the attenuation coefficients of the optical fiber at the signal and pump wavelengths, respectively. The parameter $\eta$ takes the plus sign for the copumping and the minus sign for the counterpumping.

Equations (4.1) and (4.2) show that the signal receives gain proportional to the pump power with a proportionality constant given by the Raman gain efficiency and loss due to the attenuation of optical fiber, while the pump power receives loss due to the energy transfer to the signal and the attenuation of optical fiber. The first term on the right-hand side of Eq. (4.2) describes the effect of pump depletion from which the gain saturation results. The frequency ratio in Eq. (4.2) appears to conserve the photon number in the amplification process.

In the case that the signal power is low enough that the amount of pump depletion is negligible as compared with the high pump power, one can drop the first term on the right-hand side of Eq. (4.2) and then solve Eqs. (4.1) and (4.2) to obtain the power of signal at distance $z$ as (refer to Eq. (2.1.11))

$$P_s(z) = P_s(0) \exp(g_R P_0 L_{\text{eff}} - \alpha_s z), \qquad (4.3)$$

where $P_0$ is the input pump power at $z = 0$ and $L_{\text{eff}}$ is called the effective interaction length and defined as (refer to Eq. (2.1.12))

$$L_{\text{eff}} = \int_0^z \frac{P_p(z')}{P_p(0)} dz' = \frac{1 - \exp(-\alpha_p z)}{\alpha_p}. \qquad (4.4)$$

The relation between the on–off Raman gain and the Raman gain efficiency is given as (refer to Eq. (2.1.17))

$$G_A \equiv \frac{P_s(L) \text{ with pump on}}{P_s(L) \text{ with pump off}} = \exp(g_R P_0 L_{\text{eff}}), \qquad (4.5)$$

where $L$ is the physical length of the optical fiber and $P_s(L)$ with pump on is assumed to be the amplified signal power without the amplified spontaneous emission (ASE) and thermal noise. By comparing the signal output between the pump source turned on and off, the Raman gain efficiency is easily measured. However, note that the ASE and thermal noise are neglected in Eq. (4.5) while they are contained in the actually measured power of the amplified signal. In the small signal regime, the ASE and thermal noise can be directly measured by turning off the input signal.

## 4.1.2   Multiple-Pump Amplification

A more general equation for stimulated Raman scattering in the presence of multiple wavelengths and noise is written as [2–5] (refer to Eq. (2.1.23))

$$
\pm\frac{dP_v^{\pm}}{dz} = -\alpha_v P_v^{\pm} + \varepsilon_v P_v^{\mp}
$$

$$
+ \sum_{\mu > v} g_R\,(\mu, v)\left[P_{\mu}^{+} + P_{\mu}^{-}\right]P_v^{\pm}
$$

$$
+ \sum_{\mu > v} g_R\,(\mu, v)\left[P_{\mu}^{+} + P_{\mu}^{-}\right]2h v \Delta v\,[1 + \Theta\,(\mu - v, T)]
$$

$$
- \sum_{\mu > v} \frac{v}{\mu} g_R\,(v, \mu)P_v^{\pm}\left[P_{\mu}^{+} + P_{\mu}^{-}\right]
$$

$$
- \sum_{\mu > v} \frac{v}{\mu} g_R\,(v, \mu)P_v^{\pm} 4h\mu \Delta\mu\,[1 + \Theta\,(v - \mu, T)], \qquad (4.6)
$$

where subscripts $\mu$ and $v$ denote optical frequencies, superscripts $+$ and $-$ denote forward- and backward-propagating waves, respectively, $P_v$ is the optical power within infinitesimal bandwidth $\Delta v$ around frequency $v$, $\alpha_v$ is the attenuation coefficient, $\varepsilon_v$ is the Rayleigh backscattering coefficient that is the product of the Rayleigh scattering and the Rayleigh capture coefficients, $g_R(\mu, v)$ is the Raman gain efficiency at frequency $v$ due to pump at frequency $\mu$, $h$ is Planck's constant, $T$ is temperature, and $\Delta\mu$ is the bandwidth of each frequency component around frequency $\mu$. The state of polarization (SOP) of all lights is assumed to be uncorrelated and randomly changing with respect to each other. The function $\Theta(\mu - v, T)$ is the mean number of phonons in thermal equilibrium at temperature $T$, or the Bose-Einstein factor, where $h$ and $k$ are Planck and Boltzmann constants, respectively [6, 7]:

$$
\Theta(\mu - v, T) = \frac{1}{\exp\left[\dfrac{h\,(\mu - v)}{kT}\right] - 1} \qquad (4.7)
$$

**Table 4.1 List of parameters used in the numerical simulations conducted in this chapter**

| | |
|---|---|
| Pump wavelength [nm] | 1451.779 |
| Signal wavelength [nm] | 1550.116 |
| Pump power [mW] | 400 |
| Input signal power [mW] | 0.01 |
| Attenuation coefficient at 1452 nm [dB/km] | 0.64 |
| Attenuation coefficient at 1550 nm [dB/km] | 0.45 |
| Peak of Raman gain efficiency pumped at 1452 nm [1/W/km] | 3.3 |
| Fiber length [km] | 5 |

In this model, pump-to-pump, pump-to-signal, and signal-to-signal Raman interactions, Rayleigh backscattering, fiber loss, spontaneous Raman emission noise, and noise due to thermal phonons are included. The first term of Eq. (4.6) is fiber loss, whereas the next one represents Rayleigh backscattering from the light propagating in the opposite direction. The third and fourth terms represent the gain and the spontaneous scattering caused by the light at frequency $\mu$, respectively. The fifth and sixth terms correspond to the depletion of the light at frequency $\nu$ due to the amplification and the spontaneous scattering to the light at frequency $\mu$. The factor of two in the fourth term is because the spontaneous scattering is independent of the polarization of the signal light and hence has twice the efficiency of the stimulated emission under our assumption. In addition to this, another factor of 2 enters in the sixth term because of the spontaneous emission in both directions.

In the following sections, various numerical simulations based on the above-mentioned models will be conducted. The default set of parameters used for the simulations in this chapter is listed in Table 4.1, unless otherwise mentioned.

## 4.1.3  Nonlinear Phase Shift

Discrete Raman amplifiers usually have a kilometers-long amplifier fiber, which is roughly from 10 to 100 times longer than that of EDFA. Furthermore, the gain fibers of Raman amplifiers tend to have larger nonlinearity than common transmission fibers, because a smaller mode field diameter is desirable for a higher Raman gain efficiency. These two aspects lead us to consider the nonlinear effect in a discrete Raman amplifier. Nonlinear phase shift is a useful quantity to evaluate how much the nonlinear effect is. After solving the above numerical model (Eq. 2.6), the nonlinear phase shift $\Phi_{NL}$ of the signal is calculated as the path-integrated power times the nonlinear coefficient (refer to Eq. (2.1.15))

$$\Phi_{NL} = \gamma \int_0^L P_s(z)\,dz, \tag{4.8}$$

where $\gamma$ is the nonlinear coefficient describing the self-phase modulation term in a nonlinear Schrödinger equation [1] and is defined as

$$\gamma \equiv \frac{2\pi n_2}{\lambda_s A_{\text{eff}}}, \tag{4.9}$$

where $n_2$ is the nonlinear refractive index. Effective area $A_{\text{eff}}$ in the nonlinear coefficient is defined as in Eq. (4.10) in contrast to Eq. (2.1.5) that is used for Raman gain efficiency:

$$A_{\text{eff}} = \frac{\left[\int \int_{-\infty}^{\infty} I_s(x, y, z)\,dxdy\right]^2}{\int \int_{-\infty}^{\infty} [I_s(x, y, z)]^2\,dxdy}, \tag{4.10}$$

where $I_s$ is the intensity of the signal wave in the optical fiber and $\int \int_{-\infty}^{\infty} dxdy$ denotes the integral over the entire transverse plane. The nonlinear coefficient scales inversely with the wavelength and effective area. This scaling is the same as the Raman gain efficiency because the Raman process is closely related to the self-phase modulation (a real part of the third-order nonlinear susceptibility) [8]. (Refer to Chapter 2.)

## 4.2 Gain Fibers and Material

Before designing discrete Raman amplifiers, it is necessary to know the list of available fibers as the gain medium. Although there are extensive activities in developing a novel fiber specifically for discrete Raman amplifiers [9, 10], it is important to recognize that there is a limited number of fiber types that are commercially available. Almost all of them are based on germano-silicate glass. The designs and fabrications of optical fibers are limited by many fundamental trade-offs of the design issues and the basic material properties. The Raman properties of other glass materials have also been investigated.

## 4.2.1   Raman Properties of Germano-Silicate Fibers

For thorough design, the properties of gain fibers associated with Raman amplification need to be known accurately. Figure 4.2 plots the measured Raman gain efficiency spectra of standard single-mode fiber (SMF), nonzero dispersion shifted fiber (NZDSF), and dispersion-compensating fiber (DCF). In this plot, a 1511-nm pump laser is used for the measurement.

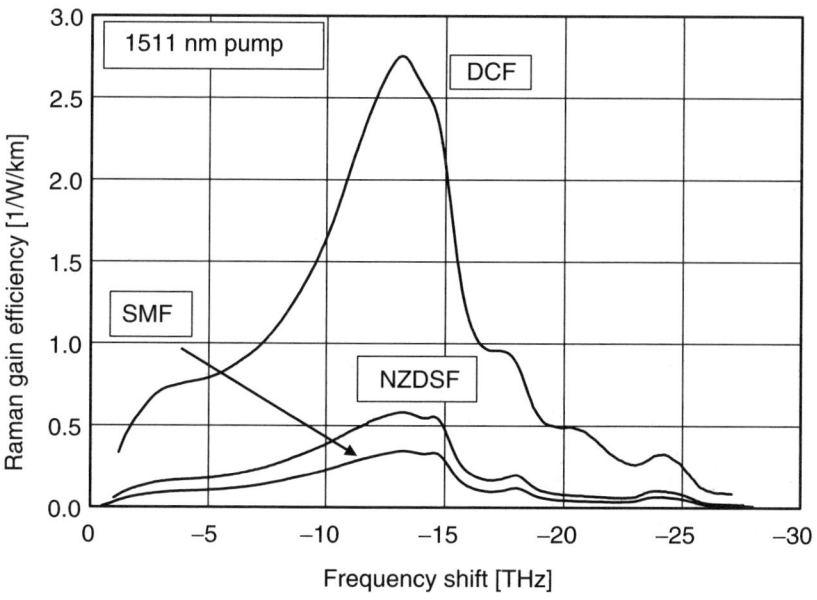

**Figure 4.2:** The measured Raman gain efficiency spectra of single-mode fiber (SMF), nonzero dispersion shifted fiber (NZDSF), and dispersion compensating fiber (DCF).

Vibrational modes such as bending, stretching, and rocking motions of vitreous $SiO_2$ and $GeO_2$ and so forth were extensively studied in order to explain the spectral structures of Raman gain efficiency [11–13]. The spectral profile depends on the concentration of $GeO_2$ [14, 15]. The peak value is proportional to the inverse effective area according to Eq. (2.1.4) and the concentration of $GeO_2$ in the host glass of silica [15–17]. Detailed studies on Raman gain spectra for transmission fibers can be found, for example, in [18].

To perform the full numerical computation using Eq. (4.6), one also needs to know the spectra of the attenuation coefficient and the Rayleigh backscattering coefficient, as well as the gain spectrum [19]. Figure 4.3 shows an example of the spectra of attenuation and Rayleigh backscattering coefficients for different fibers. In general, DCF has a higher concentration of germanium and a smaller core area than SMF and NZDSF. This leads to higher attenuation and Rayleigh backscattering coefficients. In Figure 4.3a, the attenuation of DCF increases rapidly toward 1400 nm. This comes from the OH absorption peak around 1385 nm. One also needs to know the value of the nonlinear coefficient for the fiber of interest in order to calculate the nonlinear phase shifts. In most cases, the nonlinear coefficient can be treated as a constant over the entire low loss wavelength range of the fiber.

## 4.2.2    Raman Properties of Other Fiber Materials

In order to reduce the pump power requirement and also expand the gain bandwidth, more efficient and broadbandwidth gain medium than conventional germano-silicate fibers have been desired. Several research studies on Raman properties of fiber material other than germano-silicate, such as phosphate fiber [20, 21] and tellurite fiber [22–25], have been reported for these objectives. Because of the different vibrational modes of different materials, these fibers exhibit different Raman gain spectra from that of germano-silicate fibers. Silica-based phosphate-doped fiber has an additional Raman shift frequency at 40 THz and the efficiency is about two times as large as that at 13.2 THz of silica (see Fig. 7.8). Tellurite fiber is completely different material from silica-based fiber. Its frequency peaks of Raman shift are located at 21 and 13 THz and the efficiency is about 160 times as large as that of SMF. Compared with germano-silicate fiber, it is more difficult to obtain flat gain due to their plural peaks, though these

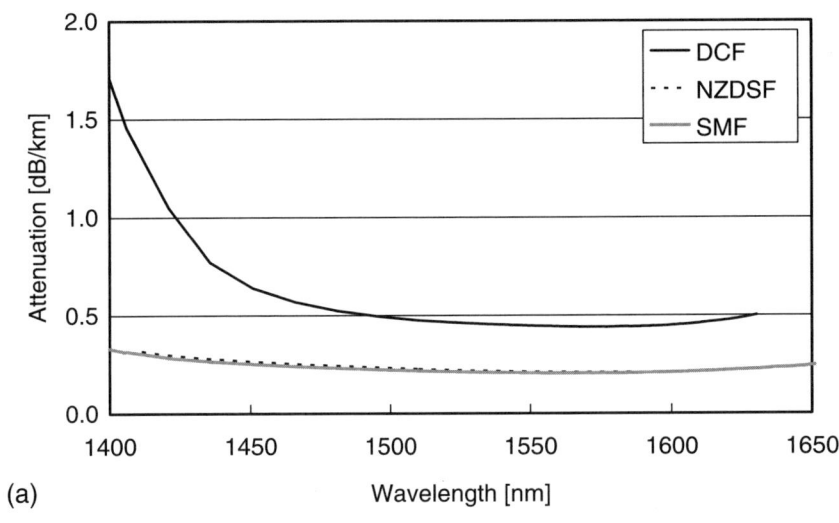

(a)

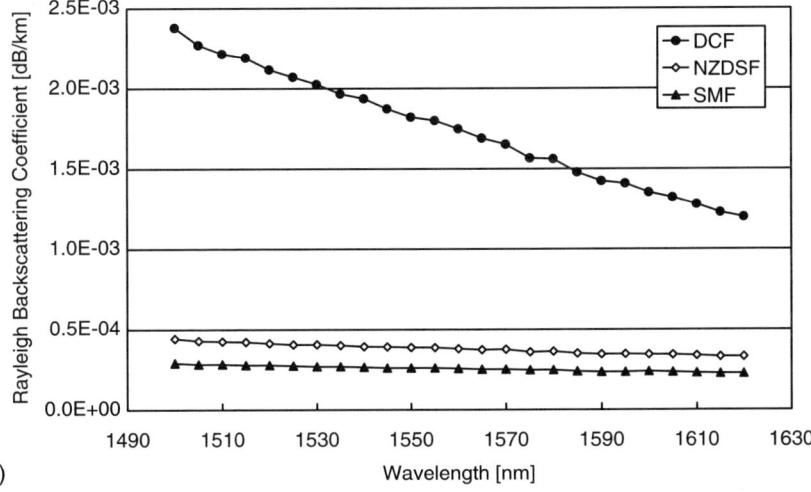

(b)

**Figure 4.3:** An example of the (a) attenuation spectra and (b) Rayleigh backscattering coefficients for different fibers. (Courtesy of authors of Ref. [19].)

fibers have the advantage of large Raman gain efficiency and broadband-width gain. Raman gain spectra for fluorine-doped silica fiber have been also studied [26], where it was reported that the Raman gain efficiency at 13.2 THz decreases with increasing fluorine concentration. Table 4.2 shows the comparison of the gain efficiency and loss coefficients among several types of fibers mentioned so far.

## 4.3   Design Issues of Discrete Raman Amplifier

### 4.3.1   Maximum Raman Gain as a Function of Fiber Length

Once the gain fiber and pump power are given, the net gain, $G$, can be written explicitly from Eq. (4.3) as a function of the fiber length $z$ (refer to Eq. (2.1.11))

$$G(z) = \exp\left(g_R P_0 \frac{1 - \exp(-\alpha_p z)}{\alpha_p} - \alpha_s z\right) \qquad (4.11)$$

though it is valid only under the small signal approximation. In the net gain, there are two contributions. One is the on–off Raman gain and the other is the fiber attenuation, which correspond to the first and second terms on the right-hand side of Eq. (4.11), respectively. Figure 4.4 plots the net gain versus the fiber length for a fixed pump power, based on Eq. (4.6) and the parameters in Table 4.1. As can be seen from the figure, the Raman gain is dominant for shorter fiber length and the net gain increases with increasing fiber length, while the fiber attenuation plays a more important role for longer fiber length and the net gain decreases. In between, the net gain has a maximum.

The maximum of the net gain in terms of the fiber length $z$ can be easily found by taking the derivative of Eq. (4.11) with respect to $z$ and seeking the condition for $z$ in which the derivative becomes zero. The net gain becomes maximal when

$$z = -\frac{1}{\alpha_p} \ln\left(\frac{\alpha_s}{g_R P_0}\right), \qquad (4.12)$$

where the small signal approximation is used as well as Eq. (4.3).

**Table 4.2 Comparison between the fiber parameters of several types of fibers**

| Fiber | Main Peak | | Second Main Peak | | $\lambda_{pump}$ [nm] | $\alpha$ [dB/km] | References |
| | Frequency Shift [THz] | $g_R$ [1/W/km] | Frequency Shift [THz] | $g_R$ [1/W/km] | | | |
|---|---|---|---|---|---|---|---|
| SMF | 13.2 | 0.34 | | | 1511 | 0.208 (1550 nm) | Figure 4. 2, 4.3a |
| NZDSF | 13.2 | 0.58 | | | 1511 | 0.214 (1550 nm) | |
| DCF | 13.1 | 3.04 | | | 1511 | 0.447 (1550 nm) | |
| HNLF | (13) | 6.5 | | | | 0.49 (1550 nm) | [10] |
| Phosphate-doped silica fiber | 40 | 1 | 13 | 0.5 | | | [20] |
| Tellurite-based fiber | 21 | 55 | 13 | 38 | 1460 | 20.4 (1560 nm) | [22] |
| Fluorine-doped silica fiber[a] | 13 | <SMF | | | 1452 | 0.19 (1550 nm) | [26] |

[a]The Raman gain coefficient around 13.2 THz frequency shift decreases with increasing fluorine concentration.

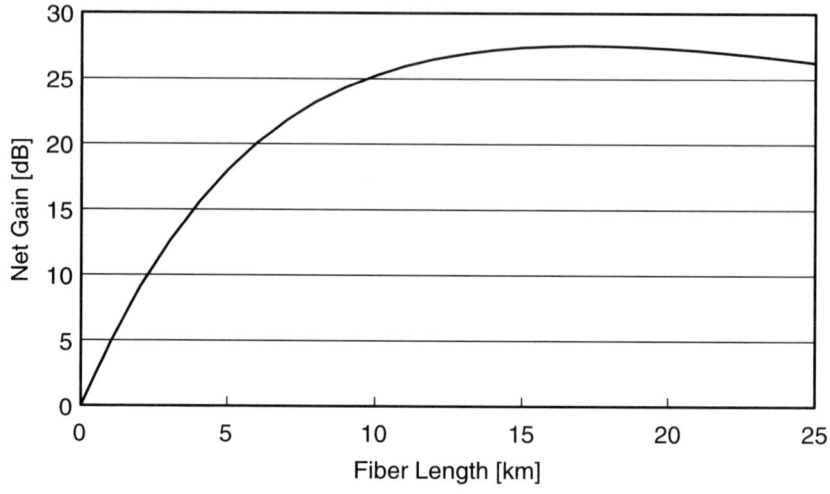

**Figure 4.4:** Net gain versus fiber length.

## 4.3.2   Figure of Merit of Gain Fiber

The effective length (Eq. (4.4)) asymptotically becomes for long enough fiber

$$L_{\text{eff}}(z \to \infty) = \frac{1}{\alpha_p}. \tag{4.13}$$

In this limit, the on–off gain will be

$$G_A(z \to \infty) = \exp\left(\frac{g_R P_0}{\alpha_p}\right) \equiv \exp\left(\text{FOM} \cdot P_0\right), \tag{4.14}$$

where the figure of merit (FOM) for a fiber with regard to the applicability to a discrete Raman amplifier is defined as

$$\text{FOM} \equiv \frac{g_R}{\alpha_p}. \tag{4.15}$$

By knowing FOM, the efficiency of the fiber for discrete Raman amplifiers can be estimated and compared. As example values of FOM, those of SMF,

NZDSF (TrueWave RS), and DCF are shown in [27] as 1.498, 2.423, and 3.667 [1/dB/W], respectively. Note that the unit of FOM used in [27] is 1/dB/W whereas it is 1/W in Eq. (4.15).

### 4.3.3 Efficiency and Linearity

Let us turn to the relation between the pump-to-signal power conversion efficiency (PCE) and the linearity of amplification. PCE is defined as follows.

$$\text{PCE} = \frac{P_s(L) - P_s(0)}{P_0} \times 100 = \frac{(G(L) - 1)\, P_s(0)}{P_0} \times 100\, [\%], \quad (4.16)$$

where $P_s(L)$, $P_s(0)$, and $P_0$ are the output and input signal power and the launched pump power, respectively. If the net gain is sufficiently large, $P_s(0)$ is negligible. Figure 4.5 plots the relation between PCE and Raman gain efficiency. The dashed line in Figure 4.5 indicates the PCE obtained by using Eq. (4.3). Here the counterpumped configuration (a) in Figure 4.1 is used, and it is assumed that an arbitrary Raman gain efficiency can be

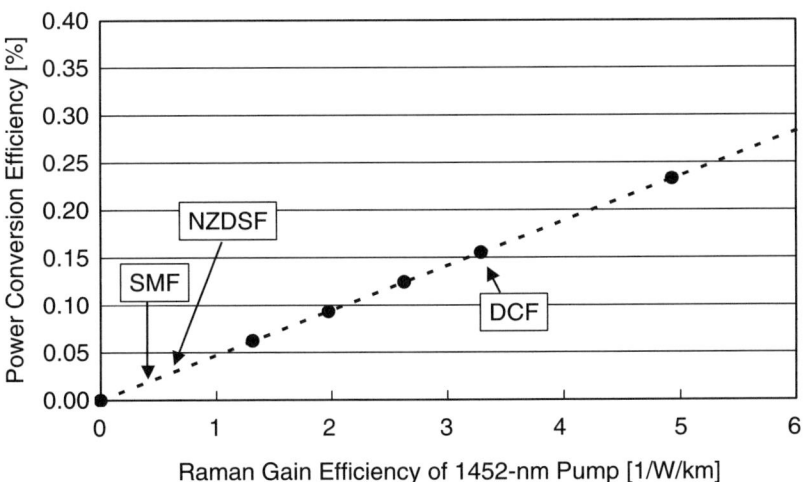

**Figure 4.5:** Power conversion efficiency versus Raman gain efficiency.

used. For a given Raman gain efficiency, the pump power is adjusted so that the net gain of the amplifier is 18 dB. The simulation parameters other than the efficiency and the pump power are fixed at the values in Table 4.1. As can be seen from the figure, the PCE increases with increasing Raman gain efficiency.

The PCE of Raman amplifiers appears very low as compared with EDFAs, which can easily achieve several tens percent of PCE [28]. However, the PCE of a Raman amplifier can be increased by using higher input power, because the gain of the Raman amplifier has a larger saturation power than EDFA. Therefore, for example, PCE increases with the number of WDM signal channels. Figure 4.6 shows the PCE as a function of the total input signal power while the net gain is kept at 18 dB by changing the pump power. The dashed line indicates the PCE obtained by using Eq. (4.3). The other values of the simulation parameters are as shown in Table 4.1. Incidentally, it is noted that the small signal approximation should not be used to investigate how to increase PCE beyond 0-dBm input signal power, as the pump depletion term is not negligible under the large PCE. The discrepancy between the results using Eqs. (4.3) and (4.6) is also shown in Figure 4.6. It is interesting to see the PCE increases monotonically with

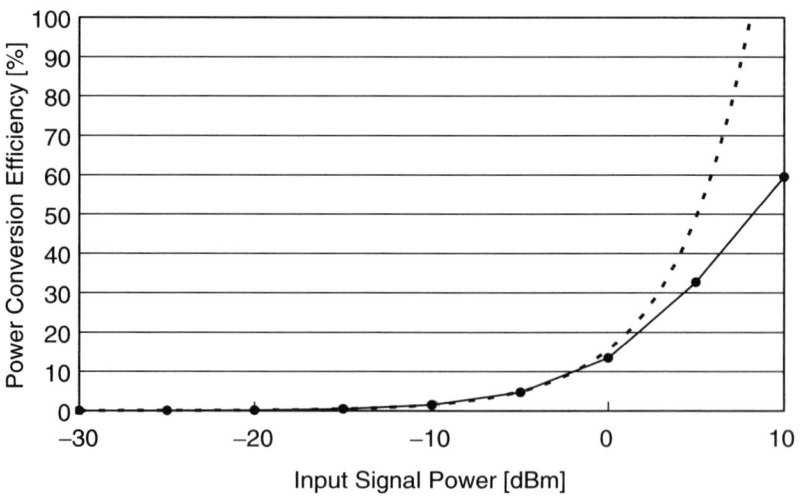

**Figure 4.6:** The power conversion efficiency as a function of the total input signal power while the net gain is kept constant.

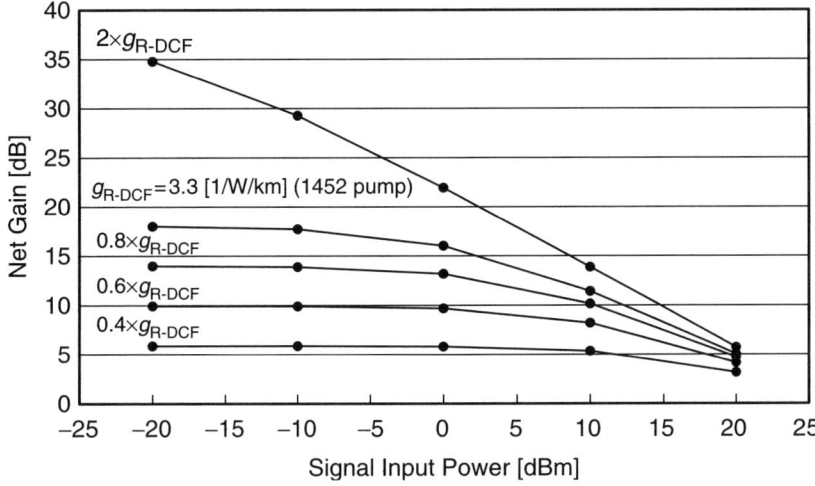

**Figure 4.7:** Net gain versus signal input power for various Raman gain efficiencies.

increasing input signal power. Due to gain saturation, larger input signal means larger pump power. Indeed, PCE could increase up to 13% if the input signal power were 0 dBm, or 32 channels of −15 dBm. Experimental results of high PCE have been actually reported: One exceeded 20% for the output power of 213 mW [29] and more than 70% of internal PCE for output powers more than 1 W was demonstrated [30, 31].

Figure 4.7 plots gain versus input signal power for various Raman gain efficiencies. In this plot, the simulation parameters other than input signal power and Raman gain efficiency are fixed at the values in Table 4.1. From the figure, it is found that the larger Raman gain efficiency results in the more gain saturation under a certain constant pump power. This is because larger efficiency makes the output signal power closer to the pump power.

The gain saturates differently for different pumping configurations. Figure 4.8 compares the difference for the fiber lengths of 5 km and 10 km. In this plot, the other values of the simulation parameters are as shown in Table 4.1, while 200 mW of pump power is used for each direction in a bidirectionally pumped configuration. At the higher input power, the co-pumping exhibits the largest gain saturation in case of 10 km, while the difference is not noticeable for 5 km. This is because the path-average

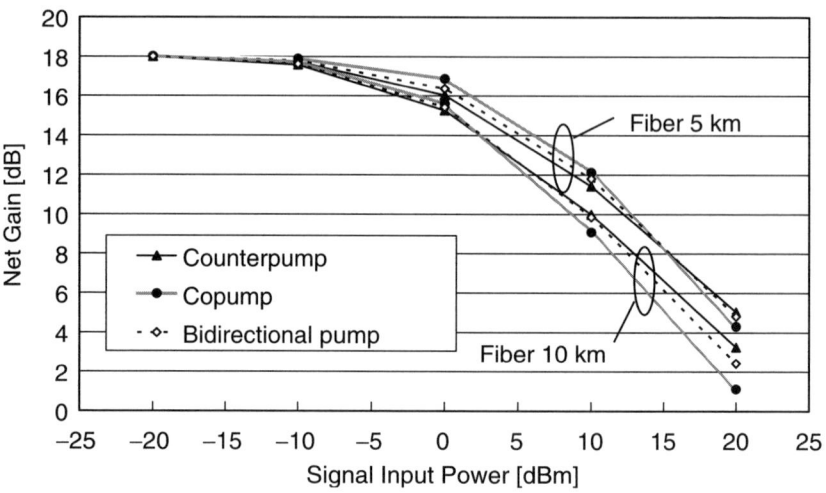

**Figure 4.8:** Net gain versus signal input power for different pumping orientations and fiber lengths.

power of the signal is significantly higher for the copumping configuration than for the other pumping configurations in the case of 10 km, whereas it is not significantly higher in the case of 5 km. This is seen in Figure 4.9, which shows the signal power evolution for the different pumping orientations for the 5- and 10-km lengths. If the path-average power increases, the coupling between the pump and the signal becomes larger, and therefore, more pump energy is exhausted in the amplifier fiber, resulting in a lower gain. Referring back to the amount of gain, saturation increases also as the fiber length increases because the more interaction or coupling due to the longer length, the more energy is cumulatively transferred from the pump to the signal.

## 4.3.4   Pump-Mediated Noise

Because the fluorescence lifetime of the stimulated Raman scattering is on the order of 10 fs, the temporal fluctuations of the pump laser will reflect on the temporal shape of the gain, as discussed in Chapter 2. As a result,

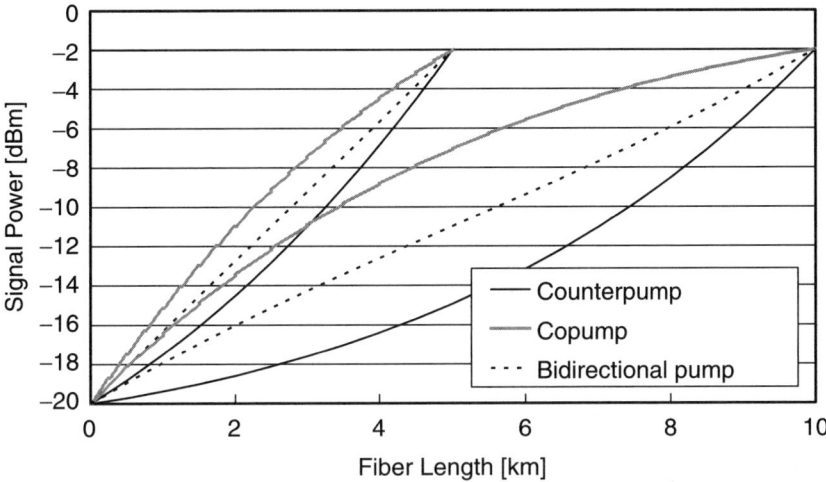

**Figure 4.9:** The signal power evolutions for different pumping orientations.

the noise of the pump laser will increase the noise of the signal through the Raman gain. Therefore, care should be taken particularly for the copumping scheme so as not to deteriorate the signal quality. On the other hand, in the counterpumping scheme, the pump-mediated noise will not be as serious an issue as copumping. Then, it is desirable to avoid using the copumping as much as possible. Or the pump lasers with sufficiently low noise should be used for copumping configurations [32–34].

The relative intensity noise (RIN) transfer function consists of several parameters such as on–off gain, loss and dispersion coefficients of fiber, and the wavelength difference between pump and signal [35]. Although the amount of Raman gain determines the maximum transfer, the other parameters affect the shape of the function. Figure 4.10 shows examples of RIN transfer functions in which the characteristics of the gain fibers listed in Table 4.3 are used. In general, the shorter gain fiber leads to a significant chirped oscillation as shown in the case of DCF in Figure 4.10. To put it more precisely, the amplitude of the oscillation is determined by the total attenuation of the fiber at the pump wavelength. Thus, the frequency at which the RIN transfer is 3 dB lower than the maximum is raised when the gain fiber is shortened.

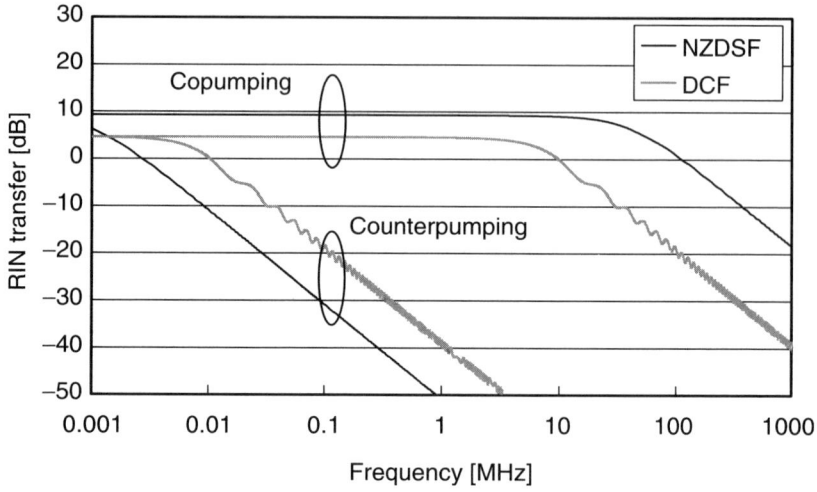

**Figure 4.10:** RIN transfer characteristics for NZDSF and DCF. (Courtesy of authors of Ref. [35].)

**Table 4.3 Parameters used for the calculation of RIN transfer characteristics in Figure 4.10**

| Fiber Type | NZDSF (TrueWave RS) | DCF |
|---|---|---|
| Length [km] | 81.1 | 6.8 |
| Added gain [dB] | 12.6 | 7.4 |
| Dispersion @ 1500 nm [ps/km/nm] | 2.3 | −97.6 |
| Attenuation @ 1455 nm [dB/km] | 0.25 | 1.52 |

## 4.3.5  ASE Noise Figure

Let us discuss ASE and the thermal noise of discrete Raman amplifiers. In general, in order to calculate such effects, Eq. (4.6) needs to be numerically solved. For simplicity, only one pump and one signal frequency are assumed, neglecting gain saturation effects and Rayleigh backscattering. Equation (4.6) can then be reduced to the following equation as modified from Eq. (4.1) with the pertinent noise term.

$$\frac{dP_s}{dz} = g_R P_p P_s - \alpha_s P_s + 2h\nu \Delta\nu g_R P_p, \tag{4.17}$$

where the thermal term is also neglected for simplicity. In the small signal regime, the pump power $P_p$ has a simple exponential form in the copumping scheme as

$$P_p(z) = P_0 \exp(-\alpha_p z), \tag{4.18}$$

while in the counterpumping scheme,

$$P_p(z) = P_0 \exp\{-\alpha_p (L - z)\}, \tag{4.19}$$

where the signal inputs at $z = 0$ and outputs at $z = L$.

The noise figure can be calculated based on Eqs. (4.17), (4.18), and (4.19) through the following definition:

$$\mathrm{NF\,(dB)} = 10\log\left(\frac{S_{\mathrm{in}}/N_{\mathrm{in}}}{S_{\mathrm{out}}/N_{\mathrm{out}}}\right), \tag{4.20}$$

where $S$ and $N$ denote the signal and noise parts in optical power at the given frequency, respectively, and $P_s(0) = S_{\mathrm{in}} + N_{\mathrm{in}}$ and $P_s(L) = S_{\mathrm{out}} + N_{\mathrm{out}}$.

Figure 4.11 plots the ASE noise figure as a function of the fiber length, for (a) counterpumping and (b) copumping, respectively. These figures also plot the net gain as a function of the fiber length. Although the full set of Eq. (4.6) in these calculations was used, Eqs. (4.17)–(4.19) should also provide a reasonable approximation [35]. The values of the simulation parameters are as shown in Table 4.1, except for the fiber length. From the figures, it is found that the difference between the two schemes is not significant for relatively short fiber length. On the other hand, that for longer fiber length becomes remarkable. This is because the accumulation of noise along the fiber is different in the two cases. By comparing the plots in Figure 4.9, the signal power in the counterpumping scheme experiences a lower level than that in the copumping scheme. This results in a worse noise figure in the counterpumping scheme for longer fiber lengths.

Discrete Raman amplifiers with noise figures as low as 4.2 dB have been reported [37]. In this paper, such a low NF was achieved by means

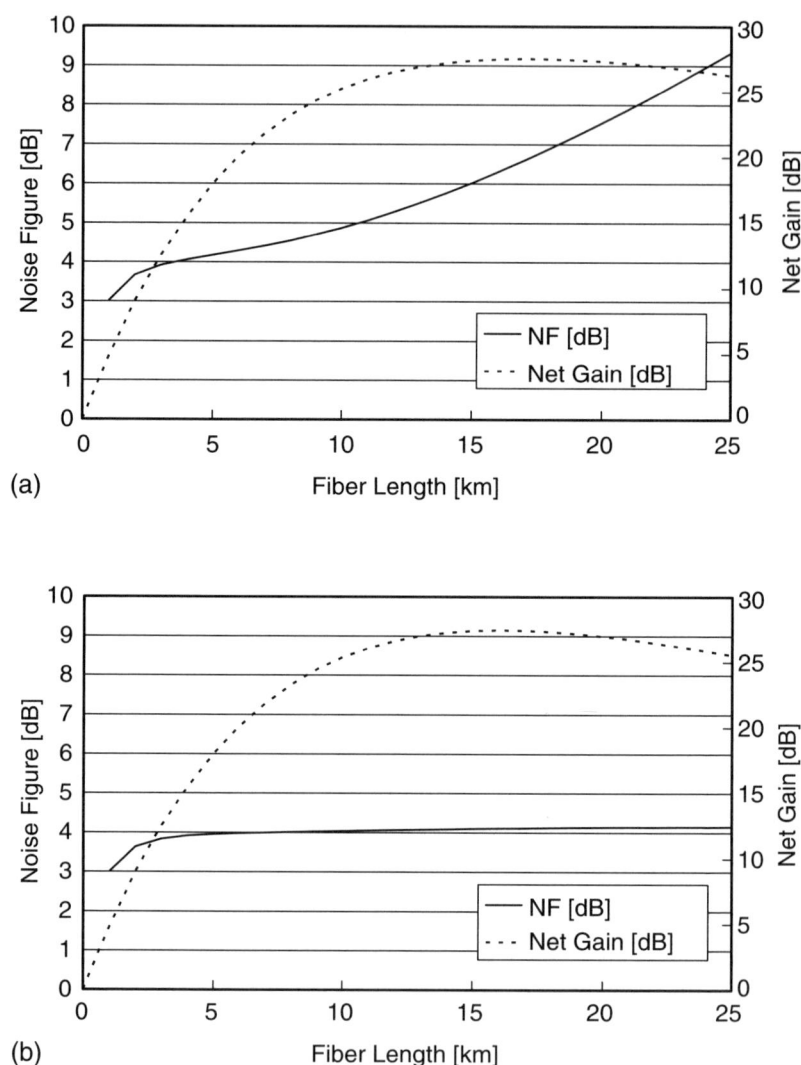

**Figure 4.11:** NF and net gain versus fiber length for (a) a counterpumped configuration and (b) a copumped configuration.

of the following techniques: (i) a short length and high efficiency of the gain fiber, which is 2.1 km of depressed cladding fiber, (ii) a two-stage amplifier configuration with an intermediate isolator to block the Rayleigh backscattered signal, and (iii) the high power intracavity pump in which a part of a ring cavity for the pump light is used to amplify the signal light. In addition, studies of noise figures and other properties of discrete Raman amplifiers are reported for various gain fibers [38]. In this paper, it is indicated that the desirable characteristics of a fiber are high Raman gain coefficient, small effective area, low attenuation coefficient, and short fiber length to obtain high net gain and low noise figure.

## 4.3.6    Nonlinear Effects and Double Rayleigh Backscattering Noise

Because a long fiber length is required for a Raman amplifier compared with EDFAs, it is necessary to account for the nonlinear effects and the DRBS effects that occur inside the Raman amplifier. These effects are particularly important for Raman amplifiers, as they tend to persist because the gain is distributed over a long length of the fiber.

The nonlinear effects are complicated and difficult to thoroughly examine, as they depend on many system parameters such as the fiber dispersion, the signal pulse parameters, and the modulation formats of the signal. However, at least if the total nonlinear phase shift of the signal defined as Eq. (4.8) inside the Raman amplifier could be sufficiently reduced, it would be desirable in general. The total nonlinear phase shift can be reduced by decreasing at either the path average power of the signal, the length of fiber, or the nonlinear coefficient.

The DRBS effects will be discussed in detail in Chapter 5. It mainly causes multipath interference (MPI) of the signal, which is interference between two divided lights going through different lengths of light path, and results in an intensity noise via conversion of phase noise (see Chapter 3). The DRBS of the signal is the phenomenon that a small portion of the signal reflects back due to Rayleigh backscattering, is amplified through the Raman gain, and reflects back again due to Rayleigh backscattering, acting as noise after traveling over indefinite length. This is shown schematically in Figure 1.11. The points at which the signal is backscattered are equally distributed over the fiber, and the effect is cumulative and not decaying out due to the gain distributed over the fiber length. The DRBS signal is amplified twice. Therefore, the MPI noise due to DRBS scales as the length of the fiber and the Raman gain [39].

Using Eq. (4.6), the signal power evolution can be calculated from which the integral of Eq. (4.8) can be performed to obtain the total non-linear phase shift. Also, the power ratio of the output signal to the DRBS noise can be calculated if their terms in Eq. (4.6) are separated in the simulation. Figure 4.12 plots the nonlinear phase shift and the signal-to-DRBS noise ratio versus fiber length for (a) a counterpumping scheme and (b) a copumping scheme, while the net gain was kept constant at 18 dB by adjusting the pump power. The other conditions are as shown in Table 4.1. The nonlinear phase shift is larger and incremental with the length for the copumping case. The effect of MPI is almost the same for both cases. Both the nonlinear and the MPI effects become larger for longer fiber length. Therefore, shorter fiber length would be desirable in order to avoid these effects.

The relationship between MPI cross talk and the Raman gain for co-, counter-, and bidirectional pumping schemes is shown in Figure 4.13. The fiber lengths used here were 5 km and 10 km, and the Raman gain was changed by changing the pump power. The MPI cross talk grows as the Raman gain increases. The bidirectional pumping shows slightly smaller MPI cross talk [40]. This difference becomes larger when the net loss of the gain fiber is larger, for example, when the fiber is longer.

### 4.3.7   Optimum Fiber Length and Number of Stages

From the above discussions, a shorter fiber length would be desirable for various reasons. The nonlinear effects, the MPI cross talk, the noise figure, and the linearity of amplification are better for shorter fiber length. One serious drawback, however, is efficiency. In any case, it is important to determine the length of the given fiber for a certain purpose by compromising the efficiency for the limits of the other adverse effects.

One way to circumvent this fundamental dilemma is to adopt a multistage scheme. By inserting an optical isolator, the MPI effects will be mitigated, as the DRBS portions will be isolated between the stages and not cumulative over the entire length [41]. This scheme therefore can realize a larger Raman gain than the single stage scheme within the same limit of the MPI cross talk.

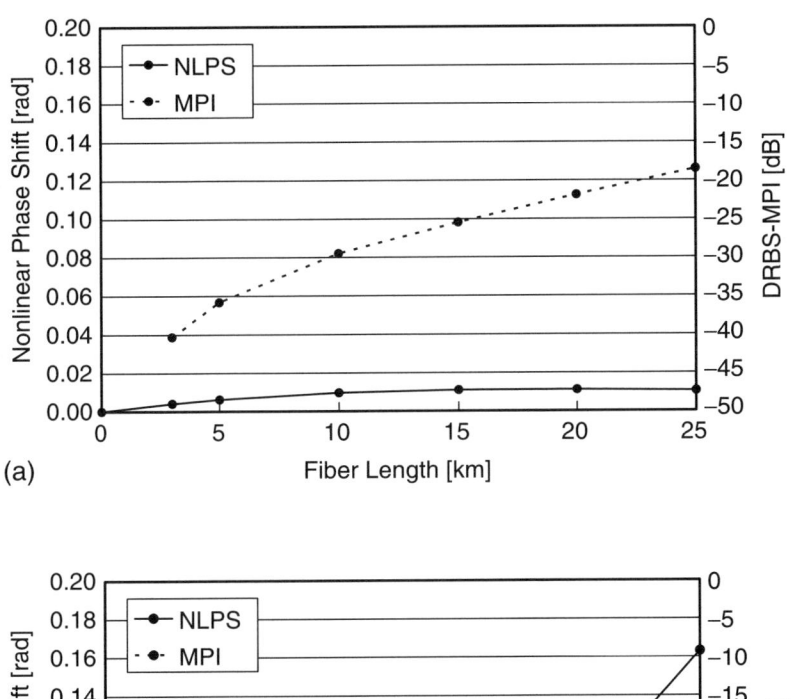

**Figure 4.12:** Nonlinear phase shift and MPI cross talk versus fiber length for (a) counterpumping and (b) copumping.

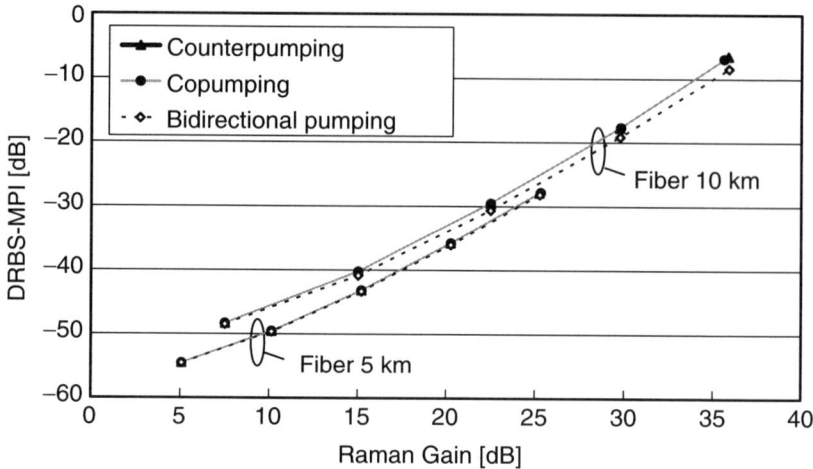

**Figure 4.13:** MPI cross talk versus Raman gain for co-, counter-, and bidirectional pumping.

It is also desirable to have a multistage scheme in which the first stage consists of a relatively short length of fiber in the sense that the noise figure becomes better if only counterpumping is allowed [42, 43]. As discussed previously, the short length of fiber tends to provide a linear gain at the expense of lower efficiency.

A drawback of the multistage configurations is the risk of increasing nonlinear effects [44]. Because of the additional losses due to isolators and couplers, the multistage amplifiers tend to require larger on–off Raman gain than the single stage one for the same net gain. This results in the larger path average power and hence the larger nonlinear phase shift as shown in Figures 4.14 and 4.15. It is therefore important to carefully choose and optimize the configuration of the amplifier by understanding the priority and trade-offs in the performance.

## 4.3.8   Transient Effects

Suppose that the total input power would suddenly decrease by a large fraction of power, say 3 dB, for example. In such a situation, if the amplifier

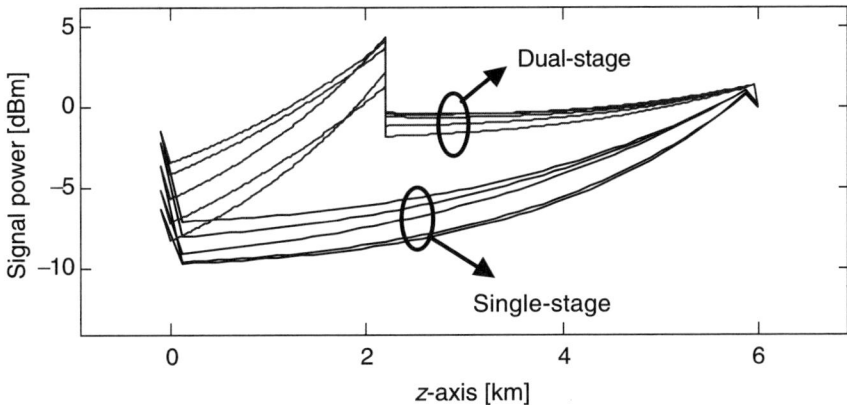

**Figure 4.14:** Power evolution in both DCRA configurations calculated for different channels, assuming perfect gain flattening of the GFF [44].

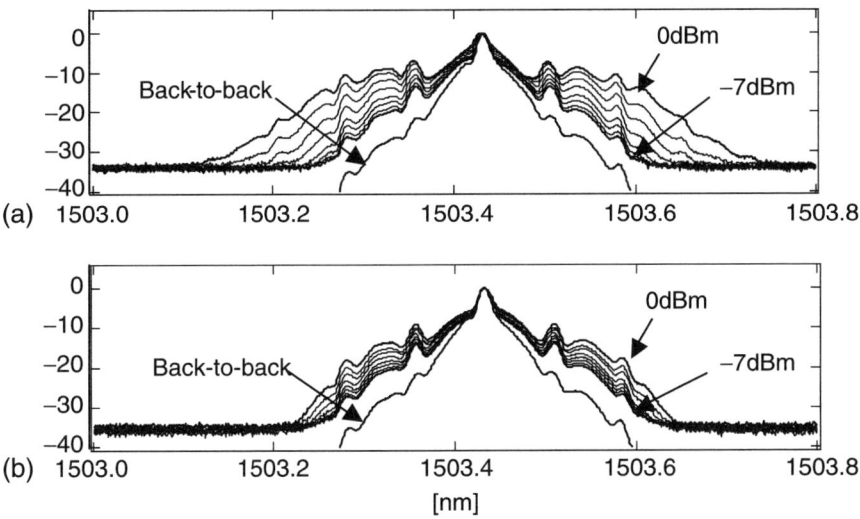

**Figure 4.15:** Spectral broadening due to SPM measured in (a) double- and (b) single-stage DCRA for input powers ranging from −7 to 0 dBm per channel in 1-dB steps [44].

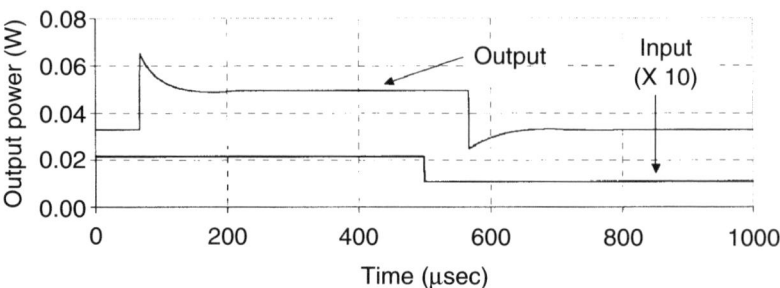

**Figure 4.16:** Simulated transient effect with a 50% amplitude modulation on the input signal. In addition to the overshoot at the leading edge, an undershoot at the trailing edge of the signal pulse appears [45].

operates at a saturated regime, the gain has to change accordingly before and after the sudden decrease in the input power. A finite response time of the amplifier may then result in a transient response in the gain. Although the stimulated Raman scattering is nearly instantaneous, the time delay due to the propagation of the signal and pump through the kilometers-long gain fiber causes a relatively slow response as a whole system of amplification. Reference [44] measured the transient response of a counterpumped discrete Raman amplifier comprising 13.9 km of DCF as the gain medium. They showed that the transient response to 50% modulation of input signal power lasted for approximately 50 $\mu$s as shown in Figure 4.16.

The magnitude of the gain change in the transient response depends on the degree of gain saturation. In other words, if the amplifier operates in a linear gain regime, the transient change in the gain will not be as large. Therefore, in order to avoid the transient effect, it is again desirable to use a short length of fiber to make the gain as linear as possible. There were also proposed some countermeasures to suppress the transients. One employs a fast electronic feedback to control the level of pump power as quickly as it can suppress the change in the gain [46], and another employs a co-pump to suppress the transients [47]. As the copumping light propagates with the signal, the gain change can be offset by changing the power of copumping without a significant time delay. It should be noted that the transients in EDFAs are reported to be much larger than those in Raman amplifiers [48].

# 4.4 Dispersion-Compensating Raman Amplifiers

## 4.4.1 Dispersion-Compensating Fiber

In high bitrates and long-haul optical transmissions, the effect of the group velocity dispersion (GVD) becomes nonnegligible. The signal pulse having a finite bandwidth tends to spread because each frequency component has a different group velocity due to GVD. In practice, it is necessary to compensate for the GVD of most commercially available fibers in the transmissions at the bit rates of 10 Gb/s or higher. As this is a linear process, the pulse spreading due to GVD can be offset by applying the same magnitude but opposite sign of the GVD. One of the most practical and effective ways to compensate for the GVD is to use DCFs [49]. DCFs are usually spooled and put in a module box so that they can be mounted on the circuit board together with other optical regenerating components such as EDFAs. DCFs in WDM applications usually have an opposite sign of both dispersion and dispersion slope to compensate for the GVD over a wide band [50, 51]. For such a purpose, the ratio of the dispersion coefficient to the dispersion slope coefficient has to be almost the same as that of the transmission fibers. Table 4.4 shows typical characteristics of DCF [49].

One critical drawback of the use of DCF is the loss added to the system due to the attenuation of the DCF. In optically amplified transmission links, the loss always turns out to be the noise. In order to mitigate the noise due to the loss of the DCF module, the in-line EDFAs are usually of two-stage configuration and the DCF is placed between the EDFA stages as depicted in Figure 4.17. In spite of such a configuration, the noise figure increases with the larger loss of the DCF.

**Table 4.4 Typical characteristics of DCF**

| | |
|---|---|
| RDS (relative dispersion slope) [nm$^{-1}$] | 0.0035 |
| FOM (figure of merit) [ps/(nm dB)] | 190 |
| Attenuation at 1550 nm [dB/km] | 0.50 |
| Dispersion at 1550 nm [ps/(nm km)] | −95 |
| Mode field diameter [μm] | 5.1 |
| PMD [ps km$^{-0.5}$] | 0.08 |

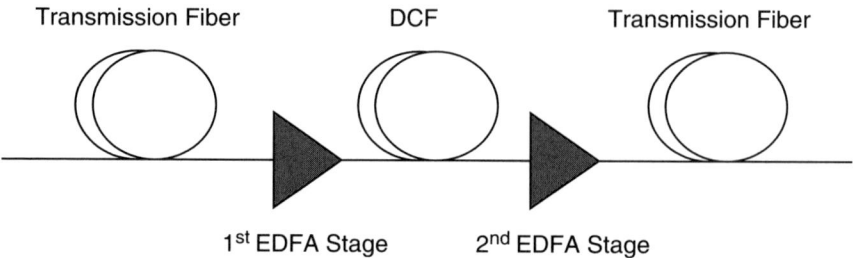

**Figure 4.17:** Two-stage EDFA with DCF module as an interstage device.

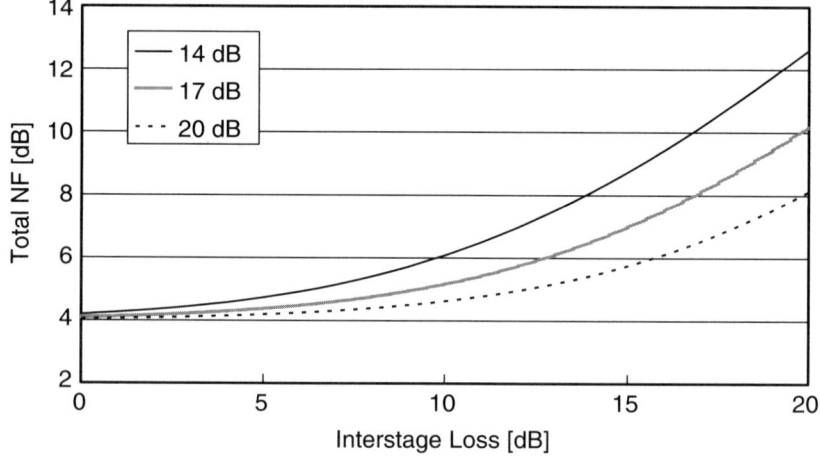

**Figure 4.18:** Noise figure versus interstage loss of the two-stage EDFA.

Figure 4.18 shows the noise figure of a two-stage EDFA versus the interstage loss for several gain levels of the EDFA's first stage. The noise figures of the EDFA's first and second stages used in this calculation are 4 dB and 6 dB, respectively. For a loss larger than 10 dB, the noise figure increases significantly. If the dispersion and loss of a DCF are assumed $-150$ ps/nm/km dispersion and 0.5 dB/km, respectively, about 10 dB of DCF loss is required to compensate 170 km of standard single-mode fiber. In EDFAs for WDM applications, the DCF is not the only interstage device in general, but a gain flattening filter, a variable optical attenuator, add–drop components, and so on are also used, adding their losses as interstage devices. Since only the DCF among these interstage devices can be gain medium, it could be inevitable to reduce DCF loss by using Raman amplification.

## 4.4.2  DCF as a Raman Gain Fiber

As shown in Figure 4.2, the peak value of the Raman gain efficiency of DCF is approximately 10 times as large as that of SMF. This means that the DCF is a very efficient Raman gain medium. Therefore, it would be a good idea to pump the interstage DCF to compensate for the loss of the DCF [52]. As discussed in the previous sections, a modest amount of Raman gain makes design much easier, suppressing the adverse effects such as nonlinearity, DRBS-MPI, and transient effects. In this case, only the noise figure increase in the in-line EDFA needs to be avoided. For this purpose, the Raman gain in the DCF need not be as large.

Figure 4.19 compares the noise figure spectra with and without the Raman gain in the DCF. In this experiment, a counterpumped DCF Raman

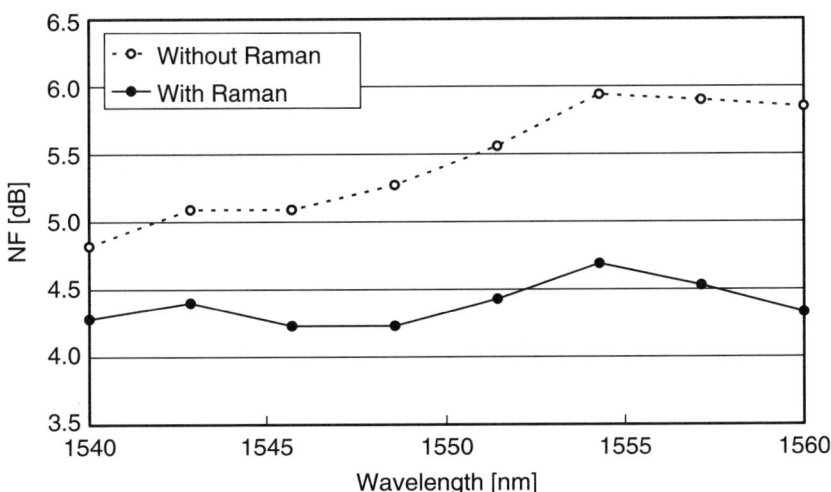

**Figure 4.19:** Noise figure spectra with and without lossless Raman gain in the DCF.

amplifier is placed between two EDFAs. Two pump wavelengths, 1435 nm and 1465 nm, are used in the Raman pump source, so that less than 0.5 dB of gain flatness is achieved. The noise figure of the first stage of the EDFA is less than 4 dB at any signal wavelength, whereas the gain is varied from 14 dB to 20 dB depending on the wavelength. The Raman gain is set to cancel out the excess loss due to the insertion of the DCF Raman amplifier, which is about 10 dB. In the case of the presence of Raman gain in the figure, the noise figure is improved by at most 1.5 dB [53].

The DCF is also attractive as a Raman gain medium for more versatile discrete Raman amplifiers with larger net gains, because in addition to high Raman gain efficiency, it has a large absolute magnitude of dispersion. As discussed before, nonlinear effects in discrete Raman amplifiers need to be sufficiently suppressed. Nonlinear effects appear less detrimental in the presence of large dispersion. For example, the four-wave mixing and cross-phase modulation hardly occur due to the dephasing and walk-off effects of dispersion [1], and the effect of self-phase modulation can be turned into an advantage in dispersion-managed solitons [54].

The design issues of a dispersion-compensating Raman amplifier (DCRA) are not just the same as those discussed in the previous sections. Because the DCF is usually used to compensate for the dispersion of the transmission fiber, the length of the DCF is determined by the amount of the cumulated dispersion in the transmission line. On the other hand, a discrete Raman amplifier has to be carefully designed to optimize the overall performances with respect to efficiency, ASE noise, nonlinearity, MPI, and so on. And the length of fiber is a critical parameter. Therefore, the DCF used in a DCRA might be slightly different from the passive DCF. Ideally speaking, the length of the DCF should be simultaneously optimized as a discrete Raman amplifier and a dispersion-compensating element. In general, the length of the DCF should be as short as possible. However, it is noted that the design of fiber characteristics is not straightforward; for example, the dispersion characteristics and effective area are not independent, and the attenuation coefficient is not independent either. The issues of optimization are different for different system configurations. One can find some references referring to C+L-band DCRAs for SMF transmission links [55, 56], S-band DCRAs for SMF [57], and 100-nm-bandwidth DCRAs for SMF [58]. Because the DCF is an efficient Raman gain medium, only 200 mW of pump power can be enough for the lossless operation of the DCF [52].

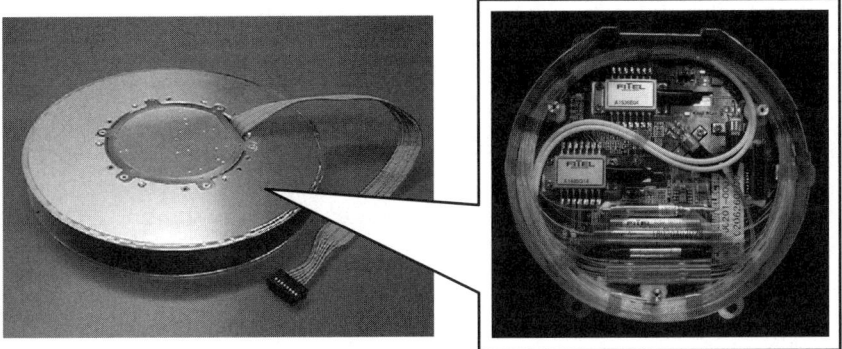

**Figure 4.20:** Photograph of a DCRA. (Courtesy of Fitel Products Division, Furukawa Electric Co., Ltd.)

Figure 4.20 shows a photograph of a DCRA. This module is designed to be mechanically compatible with the conventional passive DCF modules. All necessary optics and electronics for pumping DCF are housed in the center part, which is usually empty in the DCF modules.

## 4.5 Wideband Operation by WDM Pumping

The previous sections have mainly been focused on using one pump wavelength. This section attempts to extend the previous discussions to the case of multiwavelength pumping, or WDM pumping, for wide and flat gain operation of discrete Raman amplifiers. The design issues are basically the same, but each of the issues has now to be understood in conjunction with the homogeneity over the entire wavelength range of interest. In other words, the wavelength dependence of each property has to be optimally designed as well.

WDM pumping is a technique in which the multiwavelength pumps are combined to simultaneously launch into a single optical fiber, creating a composite Raman gain [2]. The composite Raman gains realized by WDM pumping can be designed so as to be suitable for the applications to WDM signal transmissions. Let us investigate the behavior of discrete Raman amplifiers using the WDM pumping technique and discuss the optimum designs of such amplifiers.

## 4.5.1  Wide Flat Composite Gain

First, let us define a WDM-pumped discrete Raman amplifier in order to facilitate our following discussions. Figure 4.21 shows the schematic diagram of the discrete Raman amplifier comprising five-wavelength WDM pumping and DCF as the gain fiber. The spectral data of the Raman gain efficiency, attenuation coefficient, and Rayleigh backscattering coefficient of the DCF are those shown in Figures 4.2 and 4.3, respectively. The number of WDM signals is assumed to be 92 channels in the range from 1530 nm to 1605 nm with 100-GHz spacing.

Figure 4.22 plots the on–off Raman gain spectra for the total input signal powers of −30, −10, 0, +10 dBm, respectively. The launched

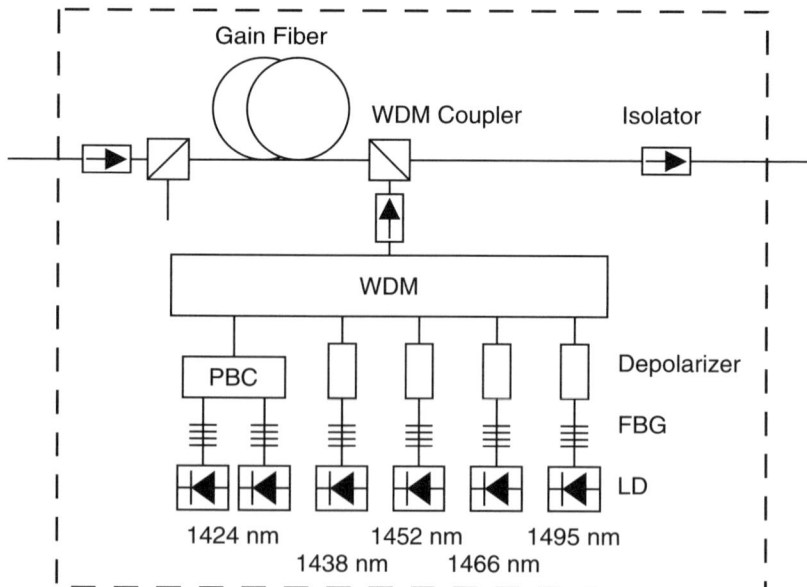

**Figure 4.21:** Schematic diagram of the discrete Raman amplifier comprising five-wavelength WDM pumping and DCF as the gain fiber.

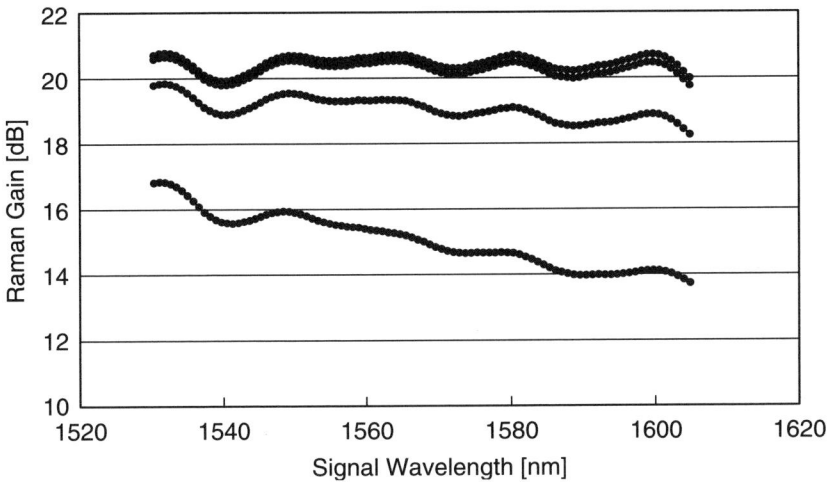

**Figure 4.22:** The on–off Raman gain spectra for the total input signal powers of −30, −10, 0, and +10 dBm, respectively.

pump powers of the pumping wavelengths are fixed for all plots, which are 375 mW, 167 mW, 93 mW, 99 mW, and 71 mW for 1424-nm, 1438-nm, 1452-nm, 1466-nm, and 1495-nm pump wavelengths, respectively. The change in the gain spectra derives mainly from the gain saturation and the stimulated Raman scattering among signals [59–61]. The average net gain is 18 dB and the flatness is better than 1 dB over the C and L bands for total input signal power of −30 dBm.

Figure 4.23 plots the Raman gain and output signal spectra under the same conditions as Figure 4.22. Comparing the two curves shown in Figure 4.23, it is found that the signal output spectra are not of the same profile as that of the on–off Raman gain, where the difference around 1600 nm for +10 dBm input looks larger than −30-dBm input. This is due to not only the spectral profile of fiber attenuation coefficients but also Raman interactions between signals. As described in Eq. (4.6), the interactions between the WDM signals and WDM pumps are complicated. Indeed, the on–off Raman gain spectra are inseparable from the signal power level spectra through saturation and they in turn influence the degree of Raman interactions between the signals. In this sense, when one designs wideband

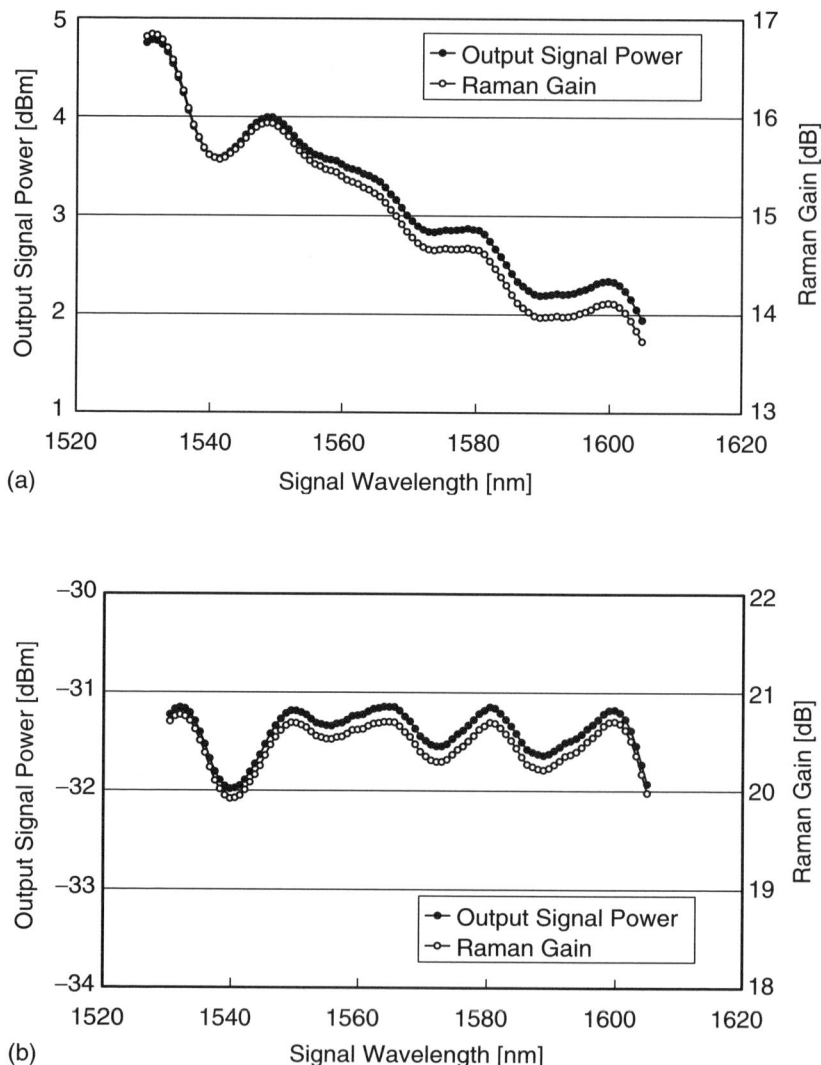

**Figure 4.23:** The Raman gain and output signal spectra for the total input signal power of (a) +10 dBm and (b) −30 dBm under the same conditions as Figure 4.22.

**Table 4.5 Comparison of path-average pump powers**

| Pump wavelength [nm] | 1424 | 1438 | 1452 | 1466 | 1495 |
|---|---|---|---|---|---|
| Input pump power [mW] | 375 | 167 | 93 | 99 | 71 |
| Path average power [mW] (Total signal power = −30 dBm) | 86 | 65 | 59 | 96 | 248 |
| Path average power [mW] (Total signal power = 10 dBm) | 89 | 61 | 48 | 71 | 150 |

Raman amplifiers for WDM signals, it is important to optimize the relevant parameters with regard to the signal power level spectra rather than the on–off Raman gain spectra.

## 4.5.2  Pump SRS Tilt—Effect of Saturation

As shown in Figure 4.22, the on–off gain tends to have a tilt in the spectrum when the composite gain is saturated. This is because each of the WDM pumps is differently depleted through the amplification process, resulting in different contributions from each pump wavelength. Table 4.5 lists the path-average powers of each pump wavelength when the total input powers are −30 and +10 dBm. The path-average pump power is in proportion to the amount of the contribution in the composite gain [2]. Looking at these values closely, the balance among the contributions from the pump wavelengths changes depending on the different total input signal powers. This results in the tilt of the composite on–off Raman gain when saturated.

## 4.5.3  Signal SRS Tilt—How to Define Gain

The system would become even more complicated when the so-called signal stimulated Raman scattering (SRS) tilt is taken into account. The signal SRS tilt is the phenomenon in which the WDM signals interact with each other through the stimulated Raman scattering process and as a result the energy in the shorter wavelength signals is transferred to the longer wavelength signals. The energy transfer causes a tilt in the optical spectra of the WDM signals.

Because this process is via the stimulated Raman scattering, the amount of the tilt, or in other words the on–off Raman gain, depends on the power and the number of the signals. This fact also suggests that the signal output power spectrum is as important as the on–off Raman gain. Furthermore, in the presence of signal SRS tilt, care should be taken to clearly define the Raman gain with appropriate parameters given when discussing such Raman amplifiers.

### 4.5.4   Control of Gain

In the WDM pumping, the pump power allocation is an important parameter to determine the composite Raman gain spectra. In practice, each of the pump powers has to be controlled possibly through some feedback. In order to control the gain, it is necessary to precisely know how much pump power is required to achieve the desirable gain. One proposed method is to measure the spontaneous Raman scattering by each pump laser beforehand and prepare a table of the pump-power allocations for the reference to realize desirable gain shapes [62]. Another approach is to predict the necessary pump powers based on a simple linearized model with numerical iterations [63].

### 4.5.5   Flattening Other Parameters

For wideband operations, not only the gain and output signal power but also the other characteristics such as noise figure, OSNR, DRBS-MPI, and nonlinear phase shift should be flattened. Figure 4.24 plots examples of the spectra of the ASE noise figure, OSNR, DRBS-MPI, and nonlinear phase shift, where the input signal power is −20 dBm/ch. In this example, the following characteristics are obtained in terms of wavelength dependence: (i) Net gain is flattened within 18 ± 0.5 dB by offsetting the wavelength dependence of fiber loss and signal SRS tilt with that of pump-to-signal Raman gain. (ii) ASE noise figure at the shorter wavelength is worse than at the longer, primarily because spontaneous Raman scattering becomes large toward the pump wavelength according to Eq. (4.7). (iii) OSNR at the shorter wavelength is smaller than at the longer due to the wavelength dependence of ASE noise figure and the signal SRS, which is equivalent

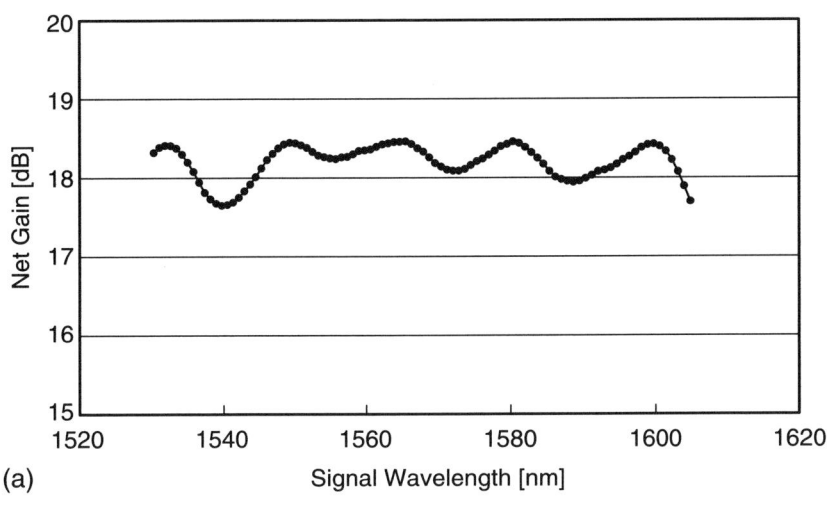

(a)

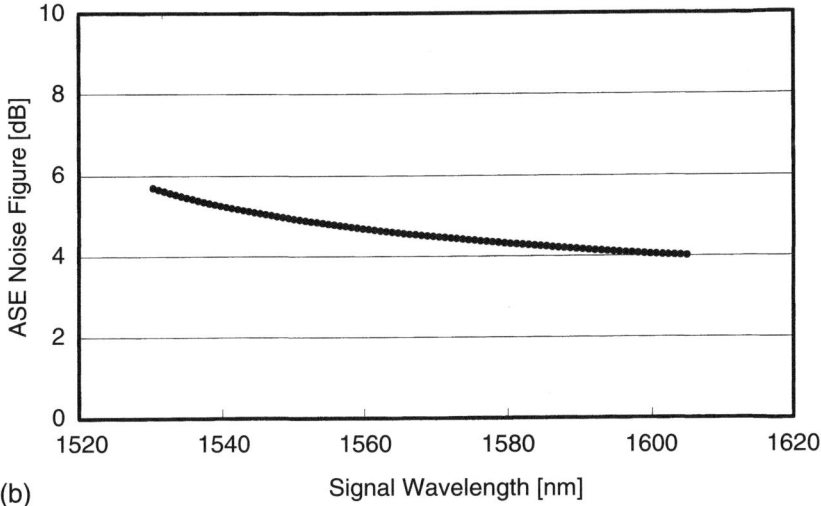

(b)

**Figure 4.24:** The spectrum of (a) net gain, (b) ASE noise figure, (c) OSNR, (d) DRBS-MPI, and (e) nonlinear phase shift.

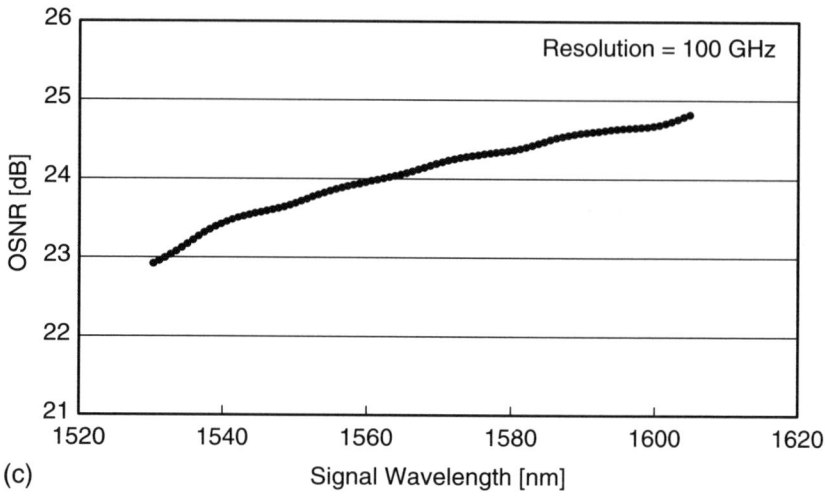

(c)

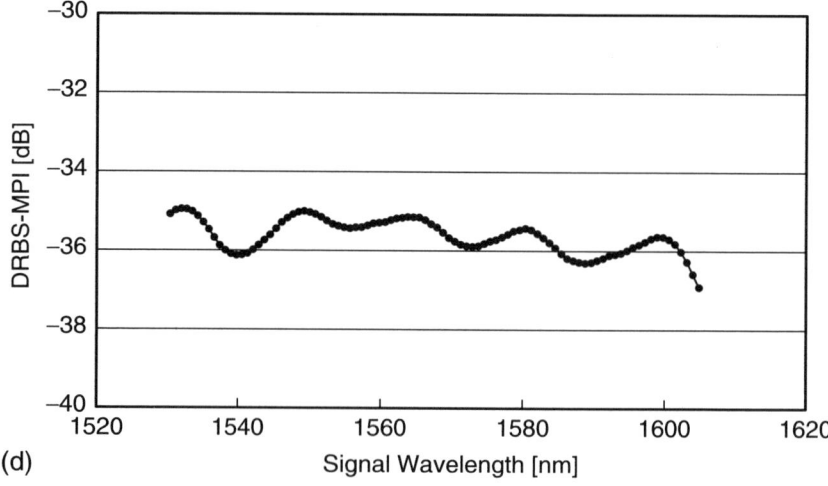

(d)

**Figure 4.24:** Continued.

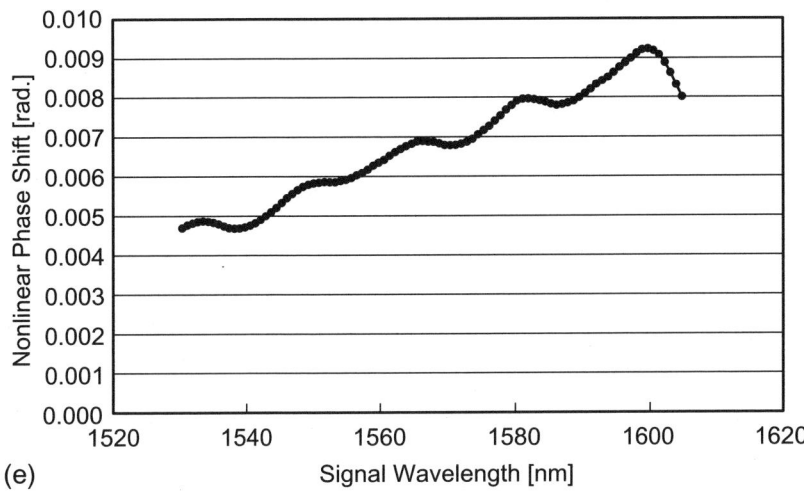

**Figure 4.24:** Continued.

to the fiber loss increase at the shorter wavelength. (iv) DRBS-MPI at the shorter wavelength is larger than at the longer because differential Raman gain at the shorter wavelength is larger. (v) Nonlinear phase shift at the longer wavelength is larger than at the shorter because the signal power is kept higher through the gain fiber. In designing the wideband discrete Raman amplifiers, one has to optimize all the parameters over the entire signal range while watching the trade-off among them.

# References

[1] G. P. Agrawal, *Nonlinear Fiber Optics*, 3rd ed. (Academic Press, San Diego, CA, 2001).

[2] S. Namiki and Y. Emori, *IEEE J. Selected Top. Quantum Electron.* **7**, 3 (2001).

[3] H. Kidorf, K. Rottwitt, M. Nissov, M. Ma, and E. Rabarijaona, *IEEE Photon. Technol. Lett.* **11**, 530 (1999).

[4] B. Min, W. J. Lee, and N. Park, *IEEE Photon. Technol. Lett.* **12**, 1486 (2000).

[5]   M. Achtenhagen, T. G. Chang, B. Nyman, and A. Hardy, *Appl. Phys. Lett.*
      **78**, 1322 (2001).

[6]   R. H. Stolen, S. E. Miller, and A. G. Chynoweth, eds., *Nonlinear Properties
      of Optical Fibers: Optical Fiber Telecommunications* (Academic Press,
      New York, 1979).

[7]   R. Eisberg and R. Resnick, *Quantum Physics of Atoms, Molecules, Solids,
      Nuclei, and Particles*, 2nd ed. (Wiley, New York, 1985).

[8]   R. H. Stolen, J. P. Gordon, W. J. Tomlinson, and H. A. Haus, *J. Opt. Soc.
      Am. B* **6**, 1159 (1989).

[9]   Y. Qian, J. H. Povlsen, S. N. Knudsen, and L. Grüner-Nielsen, in *Optical
      Amplifiers and Their Applications, OSA Technical Digest Series* (Optical
      Society of America, Washington, DC, 2000), paper OMB6.

[10]  T. Tsuzaki, M. Kakui, M. Hirano, M. Onishi, Y. Nakai, and M. Nishimura,
      in *Optical Fiber Communication Conference, OSA Technical Digest Series*
      (Optical Society of America, Washington, DC, 2001), paper MA3.

[11]  F. L. Galeener and G. Lucovsky, *Phys. Rev. Lett.* **37**, 1474 (1976).

[12]  F. L. Galeener, *Phys. Rev. B* **19**, 4292 (1979).

[13]  N. Shibata, M. Horiguchi, and T. Edahiro, *J. Non-Crystalline Solids*, **45**,
      115 (1981).

[14]  F. L. Galeener, J. C. Mikkelsen, Jr., R. H. Geils, and W. J. Mosby, *Appl.
      Phys. Lett.* **32**, 34 (1978).

[15]  S. T. Davey, D. L. Williams, B. J. Ainslie, W. J. M. Rothwell, and B.
      Wakefield, *IEEE Proc.* **136**, 301 (1989).

[16]  V. L. da Silva and J. R. Simpson, in *Optical Fiber Communication
      Conference, OSA Technical Digest Series* (Optical Society of America,
      Washington, DC, 1994), paper WK13.

[17]  J. Bromage, K. Rottwitt, and M. E. Lines, *IEEE Photon. Technol. Lett.* **14**,
      24 (2002).

[18]  C. Fludger, A. Maroney, N. Jolly, and R. Mears, in *Optical Fiber Com-
      munication Conference, OSA Technical Digest Series* (Optical Society of
      America, Washington, DC, 2000), paper FF2.

[19]  P. B. Gaarde, Y. Qian, S. N. Knudsen, and B. Pálsdóttir, in *Symposium on
      Optical Fiber Measurements*, 143–146 (2002).

[20]  Y. Akasaka, S. Kado, K. Aiso, T. Yagi, T. Suzuki, H. Koaizawa, and
      T. Kamiya, in *27th European Conference on Optical Communications,
      Proceedings* (Amsterdam, The Netherlands, 2001), paper Mo.B.3.2.

[21]  E. M. Dianov, *J. Lightwave Technol.* **20**, 1457 (2002).

[22]  A. Mori, H. Masuda, and M. Shimizu, in *28th European Conference on
      Optical Communications, Proceedings* (Copenhagen, Denmark, 2002),
      paper Symposium 3.1.

[23] A. Mori, H. Masuda, K. Shikano, K. Oikawa, K. Kato, and M. Shimizu, *Electron. Lett.* **37**, 1442 (2001).

[24] H. Masuda, A. Mori, K. Shikano, K. Oikawa, K. Kato, and M. Shimizu, *Electron. Lett.* **38**, 867 (2002).

[25] A. Mori, H. Masuda, K. Shikano, and M. Shimizu, *J. Lightwave Technol.* **21**, 1300 (2003).

[26] C. Fukai, K. Nakajima, and M. Ohashi, in *7th OptoElectronics and Communications Conference, Proceedings* (Kanagawa, Japan, 2002), paper 10D2-6.

[27] Y. Qian, J. H. Povlsen, S. N. Knudsen, and L. Grüner-Nielsen, in *Optical Amplifiers and Their Applications, OSA Technical Digest Series* (Optical Society of America, Washington, DC, 2000), paper OMD18.

[28] Y. Tashiro, S. Koyanagi, K. Aiso, and S. Namiki, in *Optical Amplifiers and Their Applications, OSA Technical Digest Series* (Optical Society of America, Washington, DC, 1998), paper WC2.

[29] D. Hamoir, D. Bayart, J.-Y. Boniort, and C. Le Sergent, in *Optical Amplifiers and Their Applications, OSA Technical Digest Series* (Optical Society of America, Washington, DC, 1999), paper ThD23.

[30] S. A. E. Lewis, S. V. Chernikov, and J. R. Taylor, *Electron. Lett.* **35**, 923 (1999).

[31] A. Oguri, Y. Taniguchi, R. Sugizaki, T. Yagi, and S. Namiki, *Optical Amplifiers and Their Applications*, OSA Technical Digest Series (Optical Society of America, Washington, DC, 2004), paper OTuA6.

[32] R. P. Espindola, K. L. Bacher, K. Kojima, N. Chand, S. Srinivasan, G. C. Cho, F. Jin, C. Fuchs, V. Milner, and W. C. Dautremont-Smith, in *27th European Conference on Optical Communication, Proceedings* (Amsterdam, The Netherlands, 2001), paper PD.F.1.7.

[33] S. Kado, Y. Emori, S. Namiki, N. Tsukiji, J. Yoshida, and T. Kimura, in *27th European Conference on Optical Communications, Proceedings* (Amsterdam, The Netherlands, 2001), paper PD.F.1.8.

[34] N. Tsukiji, N. Hayamizu, H. Shimizu, Y. Ohki, T. Kimura, S. Irino, J. Yoshida, T. Fukushima, and S. Namiki, in *Optical Amplifiers and Their Applications, OSA Technical Digest Series* (Optical Society of America, Washington, DC, 2002), paper OMB4.

[35] C. R. S. Fludger, V. Handerek, and R. J. Mears, *J. Lightwave Technol.* **19**, 1140 (2001).

[36] K. Mochizuki, N. Edagawa, and Y. Iwamoto, *J. Lightwave Technol.* **4**, 1328 (1986).

[37] A. J. Stentz, T. Nielsen, S. G. Grubb, T. A. Strasser, and J. R. Pedrazzani, in *Optical Fiber Communication Conference, OSA Technical*

*Digest Series* (Optical Society of America, Washington, DC, 1996), paper PD16.

[38] Y. Akasaka, I. Morita, M.-C. Ho, M. E. Marhic, and L. G. Kazovsky, in *25th European Conference on Optical Communications, Proceedings*, Vol. 1, p. 288 (Munich, Germany, 1999).

[39] C. R. S. Fludger and R. J. Mears, *J. Lightwave Technol.* **19**, 536 (2001).

[40] M. Nissov, K. Rottwitt, H. D. Kidorf, and M. X. Ma, *Electron. Lett.* **35**, 997 (1999).

[41] A. Bononi and M. Papararo, in *Optical Amplifiers and Their Applications, OSA Technical Digest Series* (Optical Society of America, Washington, DC, 2001), paper OTuA2.

[42] D. Hamoir, J.-Y. Boniort, L. Gasca, and D. Bayart, in *Optical Amplifiers and Their Applications, OSA Technical Digest Series* (Optical Society of America, Washington, DC, 2000), paper OMD8.

[43] J. Bromage, J.-C. Bouteiller, J. J. Thiele, K. Brar, J. H. Park, C. Headley, L. E. Nelson, Y. Qian, J. DeMarco, S. Stulz, L. Leng, B. Zhu, and B. J. Eggleton, in *Optical Fiber Communication Conference, OSA Technical Digest Series* (Optical Society of America, Washington, DC, 2001), paper PD4.

[44] H. J. Thiele, J. Bormage, and L. Nelson, in *28th European Conference on Optical Communications, Proceedings* (Copenhagen, Denmark, 2002), paper Symposium 3.9.

[45] C.-J. Chen and W. S. Wong, in *Optical Amplifiers and Their Applications, OSA Technical Digest Series* (Optical Society of America, Washington, DC, 2001), paper OMC2.

[46] C.-J. Chen, J. Ye, W. S. Wong, Y.-W. Lu, M.-C. Ho, Y. Cao, M. J. Gassner, J. S. Pease, H.-S. Tsai, H. K. Lee, S. Cabot, and Y. Sun, *Electron. Lett.* **37**, 1304 (2001).

[47] Y. Sugaya, S. Muro, Y. Sato, and E. Ishikawa, in *28th European Conference on Optical Communications, Proceedings* (Copenhagen, Denmark, 2002), paper 5.2.3.

[48] E. Schulze, M. Malach, and F. Raub, in *28th European Conference on Optical Communications, Proceedings* (Copenhagen, Denmark, 2002), paper Symposium 3.8.

[49] L. Gruner-Nielsen, S. Nissen Knudsen, B. Edvold, T. Veng, D. Magnussen, C. Christian Larsen, and H. Damsgaard, *Opt. Fiber Technol.* **6**, 164 (2000).

[50] Y. Akasaka, R. Sugizaki, and T. Kamiya, in *European Conference on Optical Communication, Proceedings* (Brussels, Belgium, 1995), paper We.B.2.4.

[51] K. Mukasa, Y. Akasaka, Y. Suzuki, and T. Kamiya, *Proc. 23rd European Conf. Opt. Comm.* **1**, 127 (Edinburgh, Scotland, 1997).

[52] P. B. Hansen, G. Jacobovitz-Veselka, L. Gruner-Nielsen, and A. J. Stentz, *Electron. Lett.* **34**, 1136 (1998).

[53] Y. Emori, Y. Akasaka, and S. Namiki, in *Optical Amplifiers and Their Applications, OSA Technical Digest Series* (Optical Society of America, Washington, DC, 1998), paper PD3.

[54] N. J. Smith, N. J. Doran, F. M. Knox, and W. Forysiak, *Opt. Lett.* **21**, 1981 (1996).

[55] L. Grüner-Nielsen, Y. Qian, B. Pálsdóttir, P. B. Gaarde, S. Dyrbøl, and T. Veng, in *Optical Fiber Communication Conference, OSA Technical Digest Series* (Optical Society of America, Washington, DC, 2002), paper TuJ6.

[56] Y. Qian, S. Dyrbøl, J. S. Andersen, P. B. Gaarde, C. G. Jørgensen, B. Pálsdóttir, and L. Grüner-Nielsen, in *28th European Conference on Optical Communications, Proceedings* (Copenhagen, Denmark, 2002), paper 4.1.6.

[57] R. Sugizaki, Y. Emori, S. Namiki, and T. Yagi, in *Optical Amplifiers and Their Applications, OSA Technical Digest Series* (Optical Society of America, Washington, DC, 2001), paper OTuB6.

[58] T. Miyamoto, T. Tsuzaki, T. Okuno, M. Kakui, M. Hirano, M. Onishi, and M. Shigematsu, in *Optical Fiber Communication Conference, OSA Technical Digest Series* (Optical Society of America, Washington, DC, 2002), paper TuJ7.

[59] C. Fludger, in *Optical Amplifiers and Their Applications, OSA Technical Digest Series* (Optical Society of America, Washington, DC, 2000), paper OMD7.

[60] A. Berntson, S. Popov, E. Vanin, G. Jacobsen, and J. Karlsson, in *Optical Fiber Communication Conference, OSA Technical Digest Series* (Optical Society of America, Washington, DC, 2001), paper MI2.

[61] A. Oguri and S. Namiki, in *29th European Conference on Optical Communications, Proceedings* (Rimini, Italy, 2003), paper Tu3.2.2.

[62] M. Sobe and Y. Yano, in *Optical Fiber Communication Conference, OSA Technical Digest Series* (Optical Society of America, Washington, DC, 2002), paper TuJ5.

[63] K. Fujimura, M. Sakano, T. Nakajima, and S. Namiki, in *Optical Amplifiers and Their Applications, OSA Technical Digest Series* (Optical Society of America, Washington, DC, 2003), paper MC3.

# Chapter 5

# System Impairments

**Morten Nissov** and **Howard Kidorf**

## 5.1 Introduction

Creating a Raman amplifier in an optical fiber has been shown to be a relatively simple feat. Managing the impairments that arise due to the nature of the optical fiber and the Raman pumps has been a far more daunting task. In this chapter the two main impairment mechanisms strongly emphasized or specific to Raman-based systems, transfer of relative intensity noise (RIN) from pumps to signals and double Rayleigh backscattering, are analyzed.

An analytical model for RIN transfer is derived for the case in which pump depletion can be ignored and it is shown that this model agrees qualitatively very well with a direct numerical solution of the coupled equations taking pump depletion into account. The usage of the analytical model only leads to slightly conservative specifications on tolerable pump laser RIN. Experimental verification of the RIN transfer models is presented. Also discussed is a type of laser that achieves both low RIN and a sufficiently broad linewidth suitable for use as forward Raman pump. With these lasers, bidirectional or forward pumping can be fully exploited by the Raman amplifier designer.

Double Rayleigh scattering (DRS) is studied, and simple analytical models are presented. Double Rayleigh scattering is the mechanism responsible for multipath interference (MPI) noise, which can severely affect transmission performance for a long Raman-based system. With enough Raman gain in the span and little enough amplified spontaneous emission (ASE) noise from the amplifier, MPI presents the next important limit to long distance transmission using Raman amplification. The noise is studied theoretically and numerical results are presented. Techniques for measuring this noise and for suppressing it are also presented.

## 5.2  Pump Noise Transfer

Initially, only backward Raman pumping was a viable option, which was therefore used in all Raman experiments [1, 2]. Being able to use bidirectional pumping has many benefits especially for long span lengths. However, due to the fast gain dynamics of the Raman amplification process (on the order of fs) any RIN on the pump lasers couples quite efficiently to the signals when forward pumping is employed. The typically used cladding pumped fiber lasers and fiber Bragg grating stabilized Fabry-Perot (FBG-FP) lasers are generally too noisy to be useful for forward Raman pumping (see Section 7.2.4). Distributed feedback (DFB) lasers exhibit low enough RIN, but due to the narrow linewidth, the stimulated Brillouin threshold is too low to be useful as Raman pumps [3]. To enable the use of forward pumping, low noise pump lasers with linewidths appropriate for high power forward Raman pumping were developed. Two solutions are spectrally broadened DFBs (SB-DFBs) [4, 5] and inner-grating-stabilized multimode (IGM) laser [6]. Such lasers have been used in system experiments demonstrating good performance from long bidirectionally pumped Raman amplifiers [7]. However, even though such devices exist, it is still important to be able to adequately specify RIN requirements for the pump lasers to avoid penalties from RIN transfer.

First the definitions of RIN will be described. Then models for RIN transfer for undepleted amplifiers are derived both excluding and including pump signal walk-off due to dispersion. The impact of pump depletion is then analyzed and measurements of RIN transfer are described. Finally pump lasers suitable for forward pumping are described.

### 5.2.1  Relative Intensity Noise

The output power of a laser $P(t)$ can be described as

$$P(t) = P_0 + \Delta P(t), \tag{5.1.1}$$

where the power, $P_0$, is constant in time and $\Delta P(t)$ is a random fluctuation with a time averaged value of zero,

$$\langle \Delta P(t) \rangle = 0, \tag{5.1.2}$$

where $\langle \ldots \rangle$ denotes time averaging as defined by

$$\langle X \rangle = \lim_{T \to \infty} \frac{1}{T} \int_T X \, dt. \tag{5.1.3}$$

The time averaging is performed by integrating a time-dependent function $X$ over a period of time $T$. For a periodic function, integrating over a single period is sufficient. The time-averaged power of the laser is therefore equal to $P_0$:

$$\langle P(t) \rangle = P_0 \tag{5.1.4}$$

Another useful term is the relative fluctuation, $\delta P(t)$, defined as

$$\delta P(t) = \frac{\Delta P(t)}{P_0}, \tag{5.1.5}$$

which allows the power of the laser to be expressed as

$$P(t) = P_0 \left( 1 + \delta P(t) \right). \tag{5.1.6}$$

This expression will allow the following equations to be written more conveniently.

### 5.2.1.1 Fundamental RIN Definition

The RIN of a laser can now be defined (Eq. (11.9-5) in Ref. [8]) as the fluctuation power in 1 Hz of bandwidth relative to the average power squared

$$\mathrm{RIN}(f) = \frac{S_{\Delta P}(f) \cdot \Delta f_0 (\equiv 1\mathrm{Hz})}{P_0^2}, \tag{5.1.7}$$

where $S_{\Delta P}$ is the power spectral density of the fluctuation [W$^2$/Hz], $f$ is the frequency, and $\Delta f_0 \equiv 1\text{Hz}$ is the bandwidth. Since the power spectral density function and the autocorrelation function $C_{\Delta P}$ form a Fourier transform pair, the RIN can also be expressed as

$$\text{RIN}(f) = \frac{\Im\{f; C_{\Delta P}(\tau)\}}{P_0^2} \Delta f_0, \tag{5.1.8}$$

where $\Im\{f; \ldots\}$ denotes the Fourier transform for a frequency $f$. Definitions for both the Fourier transform and the autocorrelation are described in the Appendix.

Using properties of autocorrelations and Fourier transforms (see Appendix) the RIN can also be expressed as

$$\text{RIN}(f) = \Im\{f; C_{\delta P}(\tau)\} \Delta f_0, \tag{5.1.9}$$

where $C_{\delta P}$ is the autocorrelation of the relative fluctuation $\delta P$.

### 5.2.1.2 Simplified RIN Definition

The fundamental RIN equations (5.1.7) and (5.1.9) are not always convenient for studying RIN transfer. Often the correlation functions cannot be easily derived in closed form, and a simplified approach is warranted. In the following sections, the RIN transfer in an amplifier without walk-off between the signal and the pump is studied using the fundamental definition. However, when pump–signal walk-off is introduced, the equations are not easily solved. The following equations allow much more complex situations to be easily analyzed.

The power in the fluctuation, which can be expressed as the time-averaged mean square fluctuation, is equal to the integral of the spectral density of the fluctuation for all frequencies (Eq. (11.9-2) in Ref. [8]):

$$\left\langle \Delta P^2(t) \right\rangle = \int_0^\infty S_{\Delta P}(f)df. \tag{5.1.10}$$

If the fluctuation is bandwidth limited to a small bandwidth, $\Delta f$, over which the spectral density is constant the integral can be simplified:

$$\left\langle \Delta P^2(t) \right\rangle = S_{\Delta P} \cdot \Delta f. \tag{5.1.11}$$

Using the RIN definition in Eq. (5.1.7) the RIN for such a bandwidth limited signal can now be expressed as

$$\text{RIN} = \frac{\left\langle \Delta P^2(t) \right\rangle}{P_0^2} \cdot \frac{\Delta f_0}{\Delta f}. \tag{5.1.12}$$

Further, using Eq. (5.1.5) for a fluctuation with a bandwidth of $f = \Delta f_0$, the RIN expression can be further simplified as

$$\text{RIN} = \left\langle \delta P^2(t) \right\rangle. \tag{5.1.13}$$

Using this definition, RIN can be represented as a sinusoidal fluctuation. In this way the integrations can be performed and also more complex situations can be analyzed.

As an example, a pump signal with a modulation index, $m$, at frequency, $f$, and random phase is considered:

$$P_p(t) = P_{p,0} \left( 1 + m \sin \left( 2\pi f t + \phi \right) \right), \tag{5.1.14}$$

where $P_{p,0}$ is the time-averaged pump power.

The relative pump fluctuation is therefore expressed according to (5.1.6) as

$$\delta P_p(t) = m \sin \left( 2\pi f t + \phi \right). \tag{5.1.15}$$

The pump laser RIN can be found by performing the integration in (5.1.13):

$$\text{RIN}_p = \frac{2}{\pi} \int_{\pi/2} m^2 \sin^2 t \cdot dt = \frac{1}{2} m^2. \tag{5.1.16}$$

This result will be used later when pump–signal walk-off is treated. In the following sections the fundamental RIN definition, Eq. (5.1.9), as well as the simplified RIN definition, Eq. (5.1.13), are used to discuss the theory of RIN transfer.

## 5.2.2 Undepleted Model

First the theory is presented for undepleted Raman amplifiers. The advantage of first studying the undepleted case is that analytical equations can be derived, whereas the depleted case can only be solved numerically. In Chapter 2, the gain of an undepleted Raman amplifier is described by Eq. (2.1.11):

$$G = \exp\left(g_R P_p L_{\text{eff}} - \alpha_s L\right), \tag{5.1.17}$$

where $g_R$ is the Raman gain efficiency (see Section 2.1.1), $P_p$ is the injected pump power, $\alpha_s$ is the fiber loss at the signal wavelength, $L$ is the amplifier length, and the effective length, $L_{\text{eff}}$, is defined as in Eq. (2.1.12):

$$L_{\text{eff}} = \frac{1}{\alpha_p}\left(1 - e^{-\alpha_p L}\right), \tag{5.1.18}$$

where $\alpha_p$ is the fiber loss at the pump wavelength.

### 5.2.2.1  Forward Pumping with No Walk-off

To further simplify the theory, equations are first presented for the case in which a signal and a pump copropagate with the same speed of propagation (i.e., forward pumping with no dispersion). This is also the worst case for RIN transfer, since any walk-off between pump and signal reduces the efficiency of the RIN transfer [9].

Similar to Eq. (5.1.6), the time-varying pump signal $P_p(t)$ is described by

$$P_p(t) = P_{p,0}\left(1 + \delta P_p(t)\right), \tag{5.1.19}$$

where the subscript $p$ is used to denote the pump signal. According to Eq. (5.1.9) the pump RIN can now be expressed as

$$\text{RIN}_p(f) = \Im\left\{f; C_{\delta P_p}(\tau)\right\}\Delta f_0. \tag{5.1.20}$$

The gain, $G(t)$, of the Raman amplifier with a time-varying pump signal can be expressed by combining the equations for the pump (5.1.19) and the Raman gain (5.1.17):

$$G(t) = \exp\left(g_R P_{p,0} L_{\text{eff}} \left(1 + \delta P_p(t)\right) - \alpha_s L\right) = G_0 \exp\left(g_R P_{p,0} L_{\text{eff}} \delta P_p(t)\right),$$

$$(5.1.21)$$

where $G_0$ is the gain in the absence of fluctuations, defined by

$$G_0 = \exp\left(g_R P_{p,0} L_{\text{eff}} - \alpha_s L\right). \tag{5.1.22}$$

For most pump lasers the RIN is very small ($< -100$ dB/Hz) and the exponential function in Eq. (5.1.21) can therefore be approximated by a first-order Taylor expansion (Eq. (20.15) in Ref. [10]) as shown below:

$$\exp(x) \cong 1 + x. \tag{5.1.23}$$

Applying Eqs. (5.1.21) and (5.1.23), an approximate equation for the gain can now be written:

$$G(t) \cong G_0 \left(1 + g_R P_{p,0} L_{\text{eff}} \delta P_p(t)\right). \tag{5.1.24}$$

From this equation it can also be seen that for small fluctuations (i.e., where the Taylor expansion is valid) the gain is equal to the average gain in the absence of fluctuations:

$$\langle G \rangle = G_0. \tag{5.1.25}$$

In Chapter 2 the Raman on–off gain was defined by Eq. (2.1.17) as

$$G_A = \exp\left(g_R P_{p,0} L_{\text{eff}}\right). \tag{5.1.26}$$

The time-varying gain can therefore be expressed by combining Eqs. (5.1.24), (5.1.25), and (5.1.26):

$$G(t) = \langle G \rangle \left(1 + \ln G_A \cdot \delta P_p(t)\right). \tag{5.1.27}$$

By multiplying the input signal (denoted by subscript $s$) by the Raman gain and expressing the input signal in a form similar to the pump signal equation (5.1.19), the output signal $P_s^{\text{out}}(t)$ can now be written as

$$P_s^{\text{out}}(t) = G \cdot P_{s,0}^{\text{in}}\left(1 + \delta P_s^{\text{in}}(t)\right). \tag{5.1.28}$$

Inserting the equation for the gain (5.1.27) results in the following:

$$P_s^{\text{out}}(t) = \langle G \rangle P_{s,0}^{\text{in}} \cdot \left(1 + \ln G_A \cdot \delta P_p(t)\right)\left(1 + \delta P_s^{\text{in}}(t)\right). \tag{5.1.29}$$

The time invariant output power $P_{s,0}^{\text{out}}$ is the product of the average gain and the time invariant input power:

$$P_{s,0}^{\text{out}} = \langle G \rangle P_{s,0}^{\text{in}}. \tag{5.1.30}$$

Inserting this into Eq. (5.1.29) and performing the multiplication of the terms within the parentheses (ignoring second-order fluctuations, which are much smaller than the first order), results in

$$P_s^{\text{out}}(t) \cong P_{s,0}^{\text{out}} \cdot \left(1 + \ln G_A \cdot \delta P_p(t) + \delta P_s^{\text{in}}(t)\right). \tag{5.1.31}$$

Using Eqs. (5.1.1) and (5.1.5) the relative signal fluctuation at the output after Raman amplification can therefore be expressed as follows:

$$\delta P_s^{\text{out}}(t) = \ln G_A \cdot \delta P_p(t) + \delta P_s^{\text{in}}(t). \tag{5.1.32}$$

The autocorrelation of the relative fluctuation of the output signal $C_{\delta P_s^{\text{out}}}$ can now be described using Eqs. (5.A.3) and (5.A.5) (listed in the Appendix) as

$$C_{\delta P_s^{\text{out}}} = \ln^2 G_A \cdot C_{\delta P_p} + C_{\delta P_s^{\text{in}}}, \tag{5.1.33}$$

where $C_{\delta P_p}$ and $C_{\delta P_s^{\text{in}}}$ are the autocorrelations of the relative pump fluctuation and input signal fluctuation, respectively. The RIN of the signal after Raman amplification [9] can therefore be expressed by using the RIN definition of Eq. (5.1.9):

$$\text{RIN}_s^{\text{out}} = \ln^2 G_A \cdot \text{RIN}_p + \text{RIN}_s^{\text{in}}. \tag{5.1.34}$$

This equation shows that in the worst case (no walk-off between signal and copropagating pump) the RIN coupled to the signal can significantly exceed the pump RIN. As an example, typical terrestrial span lengths are $\sim$100 km. If all amplification is provided by Raman amplification with an on–off gain of $\sim$20 dB the pump RIN would be enhanced by 13 dB. This illustrates the importance of using low RIN pump lasers for forward pumping.

### 5.2.2.2 Pump–Signal Walk-off

The case in which the pump and the signal do not propagate with the same speed (walk-off) will now be treated. This case is more complex than the case with no walk-off, since the temporal properties of the pump and signal must be taken into account. Before the special cases of either forward or backward pumping are described in detail, general equations will be described and solutions will be derived that will make the subsequent treatments clearer.

A set of partial differential equations can be used to describe the propagation of pump and signal in time and space. The signal will be considered to be forward propagating (i.e., from $z = 0$ to $z = L$). The propagation of the signal can be described as

$$\left(\frac{\partial}{\partial z} + \frac{1}{V_s} \cdot \frac{\partial}{\partial t}\right) P_s(z, t) = -\alpha_s P_s(z, t) + g_R P_p(z, t) P_s(z, t), \quad (5.1.35)$$

where $V_s$ is the group velocity of the signal.

Similarly, the propagation equations for a forward (upper signs) or backward (lower sign) propagating pump (ignoring pump depletion) can be written as

$$\left(\frac{\partial}{\partial z} \pm \frac{1}{V_p} \cdot \frac{\partial}{\partial t}\right) P_p(z, t) = \mp \alpha_p P_p(z, t), \quad (5.1.36)$$

where $V_p$ is the group velocity of the pump.

Since pump depletion is ignored, the propagation equation for the pump does not depend on the signal, and the equation can therefore be solved analytically for each propagation direction. First the solution for a forward-propagating pump is described. The solution is of the form of an attenuated traveling wave:

$$P_p^+(z, t) = F\left(t - \frac{z}{V_p}\right) e^{-\alpha_p z}, \qquad (5.1.37)$$

where $F$ is an arbitrary function that describes the pump signal launched into the amplifier.

By substituting $z$ with $L - z$, the solution for the backward-propagating pump can also be written:

$$P_p^-(z, t) = F\left(t - \frac{L - z}{V_p}\right) e^{-\alpha_p (L-z)}. \qquad (5.1.38)$$

These equations can be further simplified by changing to a frame of reference moving with the signal ($z = V_s t$). The time-dependent solutions for the forward- and backward-propagating pumps can then be simplified as shown below:

$$P_p^+(t) = F\left(t - \frac{V_s}{V_p} t\right) e^{-\alpha_p V_s t} \qquad (5.1.39)$$

$$P_p^-(t) = F\left(t - \frac{L - V_s t}{V_p}\right) e^{-\alpha_p (L - V_s t)} \qquad (5.1.40)$$

The partial differential equation describing the signal propagation can also be simplified by choosing a frame of reference moving with the signal and by using the chain rule for partial derivatives [11, p. 198] as

$$\frac{d P_s}{dt} = \frac{\partial P_s}{\partial z} \cdot \frac{dz}{dt} + \frac{\partial P_s}{\partial t} \cdot \frac{dt}{dt} = V_s \cdot \frac{\partial P_s}{\partial z} + \frac{\partial P_s}{\partial t}. \qquad (5.1.41)$$

A time dependent differential equation can therefore be written for the signal:

$$\frac{1}{V_s} \cdot \frac{d P_s(t)}{dt} = -\alpha_s P_s(t) + g_R P_p(t) P_s(t). \qquad (5.1.42)$$

Using the simplified RIN treatment as described above (5.1.14) and introducing a sinusoidal modulation on the pump, the pump power launched into the amplifier can now be described as

$$F(x) = P_{p,0} \left[ 1 + m \sin(2\pi f x + \phi) \right], \qquad (5.1.43)$$

where $m$ is the modulation index and $x$ is the argument given either by (5.1.39) or by (5.1.40).

The variation of the pump as seen by the signal can be calculated by combining Eq. (5.1.43) with either Eq. (5.1.39) for forward pumping or Eq. (5.1.40) for backward pumping. This is illustrated in Figure 5.1 for forward pumping and Figure 5.2 for backward pumping, which demonstrate the effect of averaging over the transit time of the amplifier due to the dispersion-induced pump–signal walk-off. The calculations were performed for a 100-km-long Raman amplifier with a loss coefficient at the pump wavelength of 0.22 dB/km, a pump modulation index of 1, and wavelength separation of pump and signal of 95 nm. The transit time through this 100-km amplifier was $\sim$0.5 ms, and because a frame of reference moving with the signal was chosen, the horizontal axis therefore also indicates position. Thus the exponential decay of the pump envelope with time for the forward-pumped case is caused by fiber loss. Similarly, for the backward-pumped case the pump envelope increases with time as the signal gets closer to the output of the amplifier where the pump was injected.

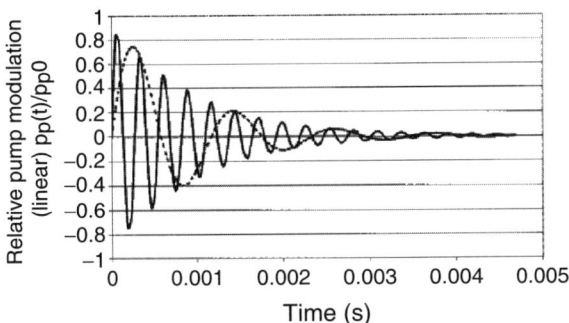

**Figure 5.1:** Variation of the pump as seen by the signal for forward pumping [9, Fig.1]. Pump modulation frequency of 100 MHz, dispersion of 17 ps/nm/km (solid line) and 4 ps/nm/km (dashed line). © 2004 IEEE.

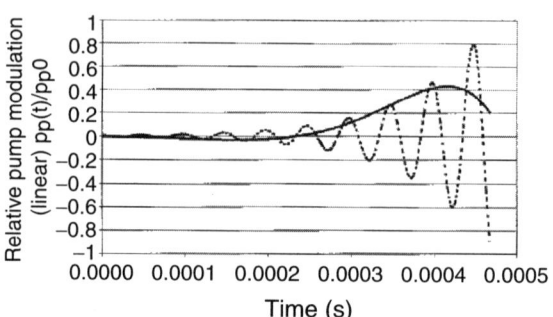

**Figure 5.2:** Variation of the pump as seen by the signal for backward pumping [9, Fig. 2]. Pump modulation frequency of 1 kHz (solid line) and 10 kHz (dashed line). © 2004 IEEE.

For the forward-pumping case, it is seen that the rate of oscillations is increased with dispersion. Thus the signal will experience more oscillations in gain, which will tend to average out and therefore reduce the RIN transfer. For the backward-pumping case the counterpropagation dominates over the dispersion walk-off. It is worth noting that similar pump variations are seen by the signal in these two examples, even though the modulation frequencies differ by four orders of magnitude. This already illustrates the large difference in sensitivity to pump RIN in the forward- and backward-pumping cases.

The formalism is now ready to treat the two distinct cases of RIN transfer from forward- and backward-propagating pumps to the signal.

*Forward pumping with walk-off:* The fundamental equation for analyzing the RIN transfer from the forward-propagating pump with noise to the signal is formed by combining (5.1.39), (5.1.42), and (5.1.43). The resultant first-order time-dependent differential equation for the signal power is

$$\frac{1}{V_s} \cdot \frac{dP_s(t)}{dt} = -\alpha_s P_s(t) + g_R P_s(t) P_{p,0} e^{-\alpha_p V_s t}$$

$$\times \left[ 1 + m \sin \left( 2\pi f t \left( 1 - \frac{V_s}{V_p} \right) + \phi \right) \right]. \quad (5.1.44)$$

This equation can be simplified [9] by substituting $b = 2\pi \left(1 - V_s/V_p\right)$. Introducing this substitution and separating the variables results in

$$\frac{dP_s(t)}{P_s(t)} = V_s \left(-\alpha_s + g_R P_{p,0} e^{-\alpha_p V_s t} \left[1 + m \sin\left(ftb + \phi\right)\right]\right) dt$$

(5.1.45)

This equation can now be integrated over the propagation time:

$$\int_{P_s(0)}^{P_s(T)} \frac{dP}{P} = V_s \int_0^T \left(-\alpha_s + g_R P_{p,0} e^{-\alpha_p V_s t} \left[1 + m \sin\left(ftb + \phi\right)\right]\right) dt,$$

(5.1.46)

where $T$ is the transit time through the fiber. The solution can be expressed as

$$\ln\left(\frac{P_s(T)}{P_s(0)}\right) = -\alpha_s L + g_R P_{p,0} \left(\frac{1 - e^{-\alpha_p L}}{\alpha_p} + V_s m \int_0^T e^{-\alpha_p V_s t} \sin(ftb + \phi) dt\right),$$

(5.1.47)

where the amplifier length, $L$, was substituted for $V_s T$ outside the integral.

Because the gain is the ratio of the output to the input signal power, an equation for the signal gain can now be written

$$G = \langle G \rangle \exp\left(g_R V_s P_{p,0} m \int_0^T e^{-\alpha_p V_s t} \sin\left(ftb + \phi\right) dt\right), \quad (5.1.48)$$

where the average gain $\langle G \rangle$ as defined in Eq. (5.1.25) is introduced to further simplify the expression.

The fluctuation is only a small perturbation on the average gain and the exponential function can therefore be Taylor expanded according to Eq. (5.1.23):

$$G = \langle G \rangle \left(1 + g_R V_s P_{p,0} m \int_0^T e^{-\alpha_p V_s t} \sin\left(ftb + \phi\right) dt\right). \quad (5.1.49)$$

Integrals of this form are described in standard integral tables (combine Eq. (14.518) and (14.519) in Ref. [10]):

$$\int e^{ax} \sin(bx + c) \cdot dx = e^{ax} \frac{a \sin(bx + c) - b \cos(bx + c)}{a^2 + b^2}. \quad (5.1.50)$$

Based on the previous two equations, the relative gain fluctuation can therefore be expressed:

$$\delta G = \frac{\Delta G}{\langle G \rangle} = g_R P_{p,0} m \frac{V_s}{(\alpha_p V_s)^2 + (fb)^2}$$

$$\times \left[ \begin{array}{c} \alpha_p V_s \sin\phi + fb \cos\phi \\ -e^{-\alpha_p L} (\alpha_p V_s \sin(fTb + \phi) + fb \cos(fTb + \phi)) \end{array} \right].$$

$$(5.1.51)$$

The signal fluctuation and thus RIN can be expressed in terms of the gain fluctuation:

$$P_s^{out} = G P_{s,0}^{in} \left(1 + \delta P_s^{in}\right) = \langle G \rangle P_{s,0}^{in} \left(1 + \delta P_s^{in}\right)(1 + \delta G). \quad (5.1.52)$$

Neglecting second-order products of fluctuations this equation can be simplified to

$$P_s^{out} \cong \langle G \rangle P_{s,0}^{in} \left(1 + \delta P_s^{in} + \delta G\right) = \langle P_{s,0}^{out}\rangle \left(1 + \delta P_s^{in} + \delta G\right). \quad (5.1.53)$$

It follows that the relative signal fluctuation at the output of the Raman amplifier is

$$\delta P_s^{out} = \delta P_s^{in} + \delta G. \quad (5.1.54)$$

Using Eq. (5.1.13) the signal RIN after Raman amplification can be calculated (noting that the fluctuations on the input signal and gain are independent):

$$RIN_s^{out} = \left\langle \left(\delta P_s^{out}\right)^2 \right\rangle = \left\langle \left(\delta P_s^{in} + \delta G\right)^2 \right\rangle = \left\langle \left(\delta P_s^{in}\right)^2 \right\rangle + \left\langle (\delta G)^2 \right\rangle$$

$$= RIN_s^{in} + \left\langle (\delta G)^2 \right\rangle. \quad (5.1.55)$$

To calculate the signal RIN, the time-averaged gain fluctuation must be derived. Using Eq. (5.1.51) this equation can be expressed as

$$\left\langle (\delta G)^2 \right\rangle = 2 \ln^2 G_A \mathrm{RIN}_p \left( \frac{V_s/L_{\mathrm{eff}}}{\left(\alpha_p V_s\right)^2 + (fb)^2} \right)^2$$

$$\times \left\langle \left( \begin{array}{c} \alpha_p V_s \sin \phi + fb \cos \phi \\ -e^{-\alpha_p L} \left(\alpha_p V_s \sin \left( fTb + \phi \right) + fb \cos \left( fTb + \phi \right)\right) \end{array} \right)^2 \right\rangle .$$

$$\text{(5.1.56)}$$

Before calculating the time averages, some basic properties of time averaging of trigonometric functions will be listed. These equations will aid in the calculations of the time averages:

$$\left\langle \sin^2 x \right\rangle = \left\langle \cos^2 x \right\rangle = 1/2 \qquad \text{(5.1.57)}$$

$$\left\langle \sin x \cdot \cos x \right\rangle = 0. \qquad \text{(5.1.58)}$$

From these two equations and the basic trigonometric rules (see Eqs. (5.34) and (5.35) in Ref. [10]) it follows:

$$\left\langle \sin x \cdot \sin (x + a) \right\rangle = \left\langle \cos x \cdot \cos (x + a) \right\rangle = \frac{1}{2} \cos a \qquad \text{(5.1.59)}$$

$$\left\langle \cos x \cdot \sin (x + a) \right\rangle = - \left\langle \sin x \cdot \cos (x + a) \right\rangle = \frac{1}{2} \sin a. \qquad \text{(5.1.60)}$$

Using the above four equations as well as Eq. (5.1.55), the time-averaged gain fluctuation and thus the RIN of the signal can now be expressed [12]:

$$\mathrm{RIN}_s^{\mathrm{out}} = \mathrm{RIN}_s^{\mathrm{in}} + \mathrm{RIN}_p \ln^2 G_A \frac{(V_s/L_{\mathrm{eff}})^2}{\left(\alpha_p V_s\right)^2 + (fb)^2}$$

$$\times \left( 1 - 2e^{-\alpha_p L} \cos \left( fTb \right) + e^{-2\alpha_p L} \right). \qquad \text{(5.1.61)}$$

Substituting $2\pi D \Delta \lambda V_s$ for $b$ [9] and recognizing that $e^{-\alpha_p L} \ll 1$ for long lengths of fiber, typical of practical Raman amplifiers, this simplifies to

$$\text{RIN}_s^{\text{out}} = \text{RIN}_s^{\text{in}} + \text{RIN}_p \ln^2 G_A \frac{1/L_{\text{eff}}^2}{\alpha_p^2 + (2\pi f D \Delta \lambda)^2}, \tag{5.1.62}$$

where $D$ is the dispersion and $\Delta \lambda$ is the wavelength separation between the signal and the pump. It can be seen from this equation that the RIN transfer due to walk-off demonstrates a low-pass filtering effect [13, 14]. As the frequency of the noise increases the RIN transfer decreases. This also applies to the fiber dispersion, effective length, and pump–signal wavelength separation. The rate of pump oscillations increases with fiber dispersion (as discussed in the previous section) as well as with pump–signal wavelength separation. Thus the signal will experience more oscillations in gain, which will tend to reduce the RIN transfer, as the signal gain is averaged over an increasing number of these oscillations. The corner frequency (where the transferred RIN is reduced by a factor of two) of this low-pass characteristic can be expressed for a long Raman amplifier as follows [9]:

$$f_c = \frac{\alpha_p}{2\pi D \Delta \lambda} = \frac{\alpha_p}{2\pi S \left| \dfrac{\lambda_s - \lambda_p}{2} - \lambda_0 \right| (\lambda_s - \lambda_p)}, \tag{5.1.63}$$

where $S$ is the dispersion slope and $\lambda_0$ is the zero-dispersion wavelength of the fiber.

This corner frequency equation is illustrated versus fiber attenuation (Fig. 5.3) and fiber zero-dispersion wavelength (Fig. 5.4). Due to the large variations of corner frequency, logarithmic scales are used on both figures. Figure 5.3 shows that the corner frequency decreases with increased dispersion due to the increased walk-off and decreased fiber attenuation. A decrease in loss at the pump wavelength distributes the gain more evenly and thus allows the gain oscillations to better average out. Figure 5.4 shows that for fibers with a zero-dispersion wavelength in between the wavelengths of the signal and the pump, the RIN corner frequency (and thus the RIN transfer) can be dramatically increased, because the walk-off is reduced. The worst case happens when the fiber zero-dispersion wavelength is exactly in between the pump and the signal wavelength in

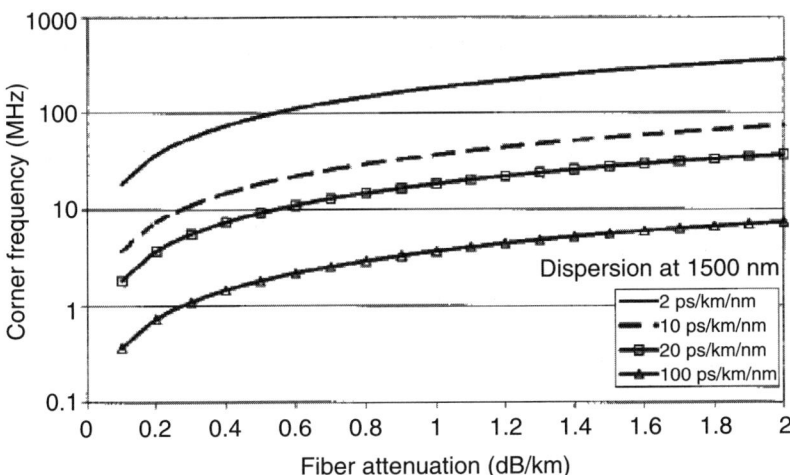

**Figure 5.3:** The corner frequency of a forward-pumped 100-km Raman amplifier as a function of fiber attenuation calculated for a pump–signal wavelength separation of 100 nm [9, Fig. 4]. © 2004 IEEE.

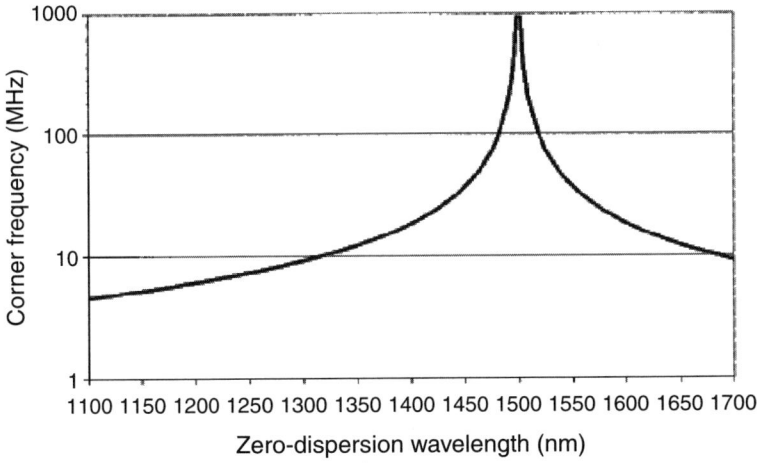

**Figure 5.4:** Corner frequency for a forward-pumped Raman amplifier as a function of zero-dispersion wavelength. Pump attenuation is 0.25 dB/km, dispersion slope 0.05 ps/nm/km, and the pump and signal wavelengths are 1450 nm and 1550 nm, respectively [9, Fig. 5]. © 2004 IEEE.

which case there is no walk-off. From the point of view of RIN transfer, terrestrial nonzero dispersion shifted fibers (NZDSFs), which often have zero-dispersion wavelength around 1500 nm, would be the worst choice for forward-pumped Raman amplifiers.

*Backward pumping with walk-off:* Backward pumping can now be treated using the same formalism as for forward pumping. The fundamental equation for analyzing a backward-propagating pump with noise is formed by combining Eqs. (5.1.40), (5.1.42), and (5.1.43). The same group velocity, $V_g$, is assumed for both signal and pump since walk-off due to counterpropagation dominates over dispersion [9]. This first-order, time dependent, differential equation is shown below:

$$\frac{1}{V_g} \cdot \frac{dP_s(t)}{dt} = -\alpha_s P_s(t) + g_R P_s(t) P_{p,0} e^{-\alpha_p(L-V_g t)}$$

$$\times \left[ 1 + m \sin\left( 2\pi f \left( 2t - \frac{L}{V_g} \right) + \phi \right) \right] \qquad (5.1.64)$$

After separating the variables and applying the relationship $V_g T = L$, this equation can be integrated over the propagation time.

$$\int_{P_s(0)}^{P_s(T)} \frac{dP(t)}{P} = V_g \int_0^T \left( -\alpha_s + g_R P_{p,0} e^{-\alpha_p(L-V_g t)} \right.$$

$$\left. \times [1 + m \sin(2\pi f(2t - T) + \phi)] \right) dt, \qquad (5.1.65)$$

where $T$ is the transit time through the fiber. The solution can be expressed as

$$\ln\left( \frac{P_s(T)}{P_s(0)} \right) = -\alpha_s L + g_R P_{p,0} \left( \frac{1 - e^{-\alpha_p L}}{\alpha_p} + V_g e^{-\alpha_p L} m \right.$$

$$\left. \times \int_0^T e^{\alpha_p V_g t} \sin(2\pi f(2t - T) + \phi) \, dt \right). \qquad (5.1.66)$$

Since the gain is the ratio of the output to the input signal power, an equation for the signal gain can now be written after Taylor expansion of the exponential function:

$$G = \langle G \rangle \left( 1 + g_R V_g P_{p,0} e^{-\alpha_p L} m \int_0^T e^{\alpha_p V_g t} \sin \left( 2\pi f \left( 2t - T \right) + \phi \right) dt \right),$$

(5.1.67)

where the average gain $\langle G \rangle$ was introduced to further simplify the expression. After integration using Eq. (5.1.50), the gain fluctuation can now be expressed:

$$\delta G = g_R P_{p,0} m \frac{V_g}{\left( \alpha_p V_g \right)^2 + (4\pi f)^2}$$

$$\times \left[ \begin{array}{c} \alpha_p V_g \left( \sin \left( 2\pi f T + \phi \right) + e^{-\alpha_p L} \sin \left( 2\pi f T - \phi \right) \right) \\ -4\pi f \left( \cos \left( 2\pi f T + \phi \right) - e^{-\alpha_p L} \cos \left( 2\pi f T - \phi \right) \right) \end{array} \right].$$

(5.1.68)

To calculate the signal RIN, the time-averaged gain fluctuation must now be derived. This equation can be expressed as

$$\left\langle \left( \delta G \right)^2 \right\rangle = 2 \ln^2 G_A \mathrm{RIN}_p \left( \frac{V_g / L_{\mathrm{eff}}}{\left( \alpha_p V_g \right)^2 + (4\pi f)^2} \right)^2$$

$$\times \left\langle \left[ \begin{array}{c} \alpha_p V_g \left( \sin \left( 2\pi f T + \phi \right) + e^{-\alpha_p L} \sin \left( 2\pi f T - \phi \right) \right) \\ -4\pi f \left( \cos \left( 2\pi f T + \phi \right) - e^{-\alpha_p L} \cos \left( 2\pi f T - \phi \right) \right) \end{array} \right]^2 \right\rangle.$$

(5.1.69)

This equation can be simplified using the properties of time averages of trigonometric functions as previously described in Eqs. (5.1.57)–(5.1.60). Using these equations as well as (5.1.55), the time-averaged gain fluctuation and thus the RIN of the signal can now be expressed:

$$\mathrm{RIN}_s^{\mathrm{out}} = \mathrm{RIN}_s^{in} + \mathrm{RIN}_p \ln^2 G_A \frac{\left( V_g / L_{\mathrm{eff}} \right)^2}{\left( \alpha_p V_g \right)^2 + (4\pi f)^2}$$

$$\times \left( 1 - 2 e^{-\alpha_p L} \cos \left( 4\pi f T \right) + e^{-2\alpha_p L} \right).$$

(5.1.70)

For long lengths of fiber ($e^{-\alpha_p L} \ll 1$) that are typical for Raman amplifiers, this simplifies to

$$\text{RIN}_s^{\text{out}} = \text{RIN}_s^{\text{in}} + \text{RIN}_p \ln^2 G_A \frac{1/L_{\text{eff}}^2}{\alpha_p^2 + \left(4\pi f/V_g\right)^2}. \qquad (5.1.71)$$

Similar to the forward-pumped case, the RIN transfer function behaves like a low-pass filter, rejecting high-frequency RIN. The corner frequency of this low-pass filter function for long Raman amplifiers is:

$$f_c = \frac{\alpha_p V_g}{4\pi}. \qquad (5.1.72)$$

Contrary to the forward-pumped case, the corner frequency only depends on the fiber attenuation. Only slow gain fluctuations, compared to the time it takes to traverse the length where most of the gain is affected (function of pump attenuation), have an effect on the signal fluctuation. Figure 5.5 illustrates this correspondence. It can be seen from the figure that for reasonable attenuation values ($\sim$0.2 dB/km) corner frequencies of $\sim$1 kHz result. This shows a significant reduction in RIN transfer compared

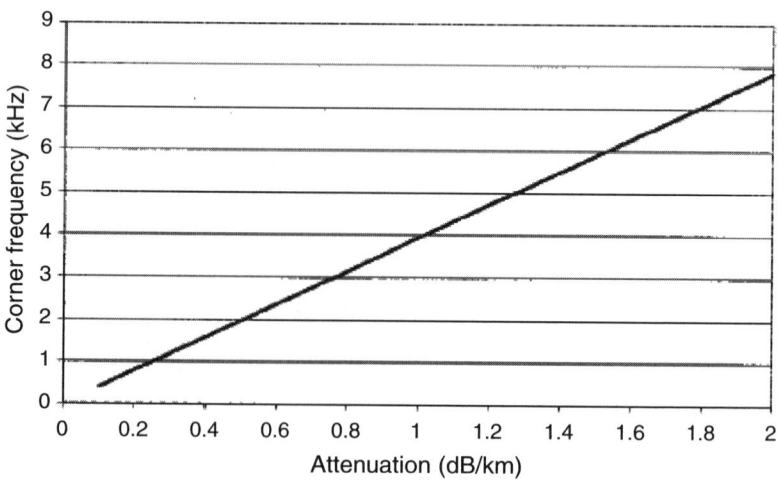

**Figure 5.5:** Corner frequency for a backward-pumped Raman amplifier as a function of pump attenuation [9, Fig. 3]. © 2004 IEEE.

to the forward-pumping case in which the typical corner frequencies are
1–100 MHz.

## 5.2.3 Performance Degradation Due to RIN

To understand the implications of the RIN transfer from pumps to signals
it is important to be able to relate the signal RIN to a performance penalty.
In general, performance penalties are calculated using numerical system
simulators including extensive receiver models taking all significant effects
into account. However, for simple evaluations of performance penalties
caused by RIN, simple models are appropriate. Also, the simpler models
provide an insight that cannot be easily gained by the more complete, and
therefore complex, models.

In Ref. [9] a very simple model for the $Q$-factor penalty associated with
signal RIN is presented. The $Q$-factor is defined as

$$Q = \frac{\langle I_1 \rangle - \langle I_0 \rangle}{\sigma_1 + \sigma_0}, \tag{5.1.73}$$

where $\langle I_1 \rangle$ and $\langle I_0 \rangle$ are the mean photocurrents for ones and zeros, respectively. Similarly, $\sigma_1$ is the standard deviation of the photocurrents for the
ones and $\sigma_0$ is the standard deviation for the zeros.

Assuming infinite extinction ratio (no photocurrent in the zeros) and that
performance is limited mainly by signal-spontaneous beat noise (standard
deviation on ones much larger than standard deviation on zeros) the $Q$-
factor in absence of signal RIN simplifies to

$$Q_s = \frac{\langle I_1 \rangle}{\sigma_1}. \tag{5.1.74}$$

These assumptions are reasonable for most optically amplified systems
using an on–off keyed modulation format. Now assuming that the signal
RIN can be treated as a Gaussian random variable, the $Q$-factor with RIN-
induced penalty included can be written as

$$Q_r = \frac{\langle I_1 \rangle}{\sqrt{\sigma_1^2 + \sigma_r^2}} = \frac{Q_s}{\sqrt{1 + \left(\frac{\sigma_r}{\sigma_1}\right)^2}}, \tag{5.1.75}$$

where $\sigma_r$ is the standard deviation of the RIN noise from the signal. This standard deviation can be found by integrating the RIN of the signal over the receiver bandwidth (from $f_1$ to $f_2$).

$$\sigma_r^2 = \langle I_1 \rangle^2 \int_{f_1}^{f_2} \text{RIN}_s(f)df \qquad (5.1.76)$$

The $Q$-factor penalty can therefore be expressed in terms of the $Q$-factor in the absence of RIN and the integrated RIN:

$$\Delta Q = \frac{Q_s}{Q_r} = \sqrt{1 + \left(\frac{\sigma_r}{\sigma_1}\right)^2} = \sqrt{1 + Q_s^2 \int_{f_1}^{f_2} \text{RIN}_s(f)\,df}, \quad (5.1.77)$$

where $Q_s$ is the $Q$-factor without RIN and $f_1$ and $f_2$ are the low and high frequency limits of the receiver bandwidth, respectively.

Using this simple formula, the penalties of the transferred RIN can be investigated. The penalties for a forward- and backward-pumped Raman amplifier based on standard single-mode fiber with a dispersion of 15 ps/nm/km were calculated assuming a receiver bandwidth from 10 kHz to 20 GHz, a Raman gain of 10 dB, pump attenuation of 0.25 dB, and a baseline $Q$-factor of 10. The results of the calculation are shown in Figure 5.6. It shows that forward-pumped amplifiers are significantly ($\sim$50 dB) more sensitive to pump RIN than backward-pumped amplifiers. To avoid any penalties from transferred pump laser RIN, a specification of $< -120$ dB/Hz is required for forward pumping and $< -70$ dB/Hz for backward pumping, if the system only consists of a single Raman amplifier. If the system consists of several amplifiers, it should be noted that RIN accumulates linearly with the number of amplifiers [9].

The impact of dispersion on the forward-pumped amplifier is further investigated in Figure 5.7 for both standard single-mode fiber and NZDSF. Low dispersion significantly increases the sensitivity to RIN and the pump RIN requirement when using such a fiber would be a pump RIN value of $< -130$ dB/Hz even for a single amplifier. If the zero-dispersion wavelength of the fiber was to lie close to mid-way between the signal and the pump the situation would be even worse as previously discussed.

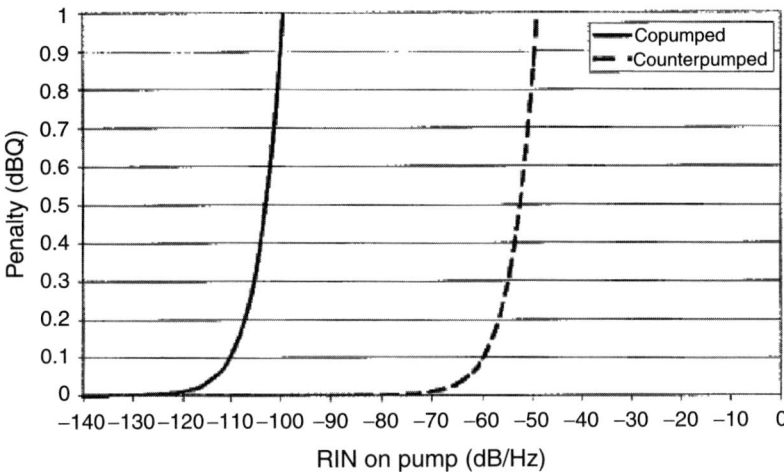

**Figure 5.6:** Estimated $Q$-factor penalty as a function of pump RIN for forward- and backward-pumped Raman amplifiers [9, Fig. 11 (corrected in [15])]. © 2004 IEEE.

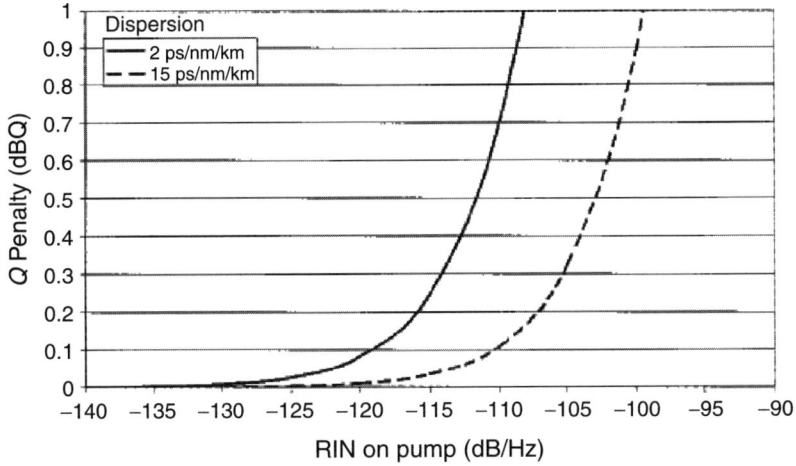

**Figure 5.7:** Estimated $Q$-factor penalty as a function of pump RIN for different values of dispersion for a forward-pumped Raman amplifier [9, Fig. 12]. © 2004 IEEE.

## 5.2.4   Impact of Pump Depletion

Most practical Raman amplifiers operate in partial saturation resulting in
some pump depletion. Whereas the gain of erbium-doped fiber amplifiers
(EDFAs) can be compressed by more than 10 dB, the gain of most practical
Raman amplifiers is only compressed by a few decibels. Even this low level
of compression can have an effect on Raman amplifier operation. In this
section, the effect of pump depletion on RIN transfer between signals and
pumps will be discussed.

Including pump depletion in the partial differential equations (5.1.35)
and (5.1.36) results in the following set of coupled linear partial differential
equations [16]:

$$\left(\frac{\partial}{\partial z} + \frac{1}{V_s} \cdot \frac{\partial}{\partial t}\right) P_s(z, t) = -\alpha_s P_s(z, t) + g_R P_p(z, t) P_s(z, t) \quad (5.1.78)$$

$$\left(\frac{\partial}{\partial z} \pm \frac{1}{V_p} \cdot \frac{\partial}{\partial t}\right) P_p(z, t) = \mp \alpha_p P_p(z, t) \mp g_R P_p(z, t) P_s(z, t).$$

$$(5.1.79)$$

The upper sign in Eq. (5.1.79) is used for forward-propagating pumps and
the lower sign for backward-propagating pumps.

This set of equations cannot, in general, be solved analytically (except
for the previously treated undepleted case) and must therefore be solved
numerically. In Ref. [16] this set of coupled partial differential equations
was simplified to a set of four linear coupled differential equations, which
were subsequently solved numerically.

As an example, a 100-km Raman amplifier with 10 dB of gain at a signal
wavelength of 1550 nm was studied in Ref. [16] for both forward-pumping
(copump) and backward-pumping (counterpump) conditions. The results
are shown in Figure 5.8, which shows the RIN transfer functions (ratio of
signal RIN after amplification to pump RIN). In the figure, curves are shown
for different input signal powers ranging from −20 dBm (undepleted)
to 10 dBm (strongly depleted). Next to each curve is also indicated the
required pump power and the corner frequency where the RIN transfer
function has been reduced by 3 dB relative to the peak.

The figure shows that as the signal power (and thus the pump depletion)
is increased, the peak RIN transfer can be reduced by as much as ∼8–15 dB,

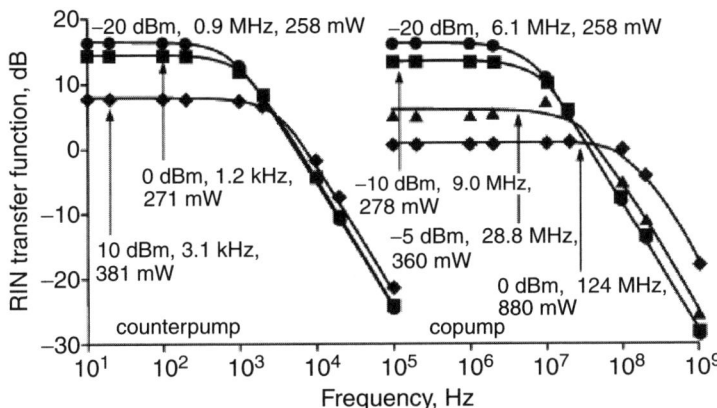

**Figure 5.8:** Calculated RIN transfer function against frequency for forward- and backward-propagating pumps [14, Fig. 1]. Parameters: $\alpha_p = 0.235$ dB/km, $\alpha_s = 0.189$ dB/km, $g_R = 1.4 \cdot 10^{-3} (\text{WM})^{-1}$, $D = 15$ ps/km/nm. © 2004 IEEE.

whereas the slope remains mostly unchanged. The integrated RIN transfer is however almost independent of pump depletion [16]. The effect of the pump depletion is to reduce the effective length where most of the amplification takes place. This is equivalent to increasing the pump attenuation rate, which was shown in the previous two sections to increase the corner frequency. The figure also demonstrates, consistent with the previous section, that the corner frequency for backward pumping is four orders of magnitude lower than for forward pumping. Pump depletion therefore significantly affects the RIN transfer function.

Of even greater importance is the impact on channel performance. Using the simple model for $Q$-factor penalty from the previous section, the impact of pump depletion on $Q$-factor penalty can be studied. In Ref. [16] the impact of pump depletion on the RIN-induced $Q$-factor penalty was studied for a 10-Gb/s receiver at a nonimpaired $Q$-factor of 10. The result is shown in Figure 5.9. The figure shows curves both with (solid line) and without (dashed) pump depletion and for both forward and backward pumping. Next to each curve is indicated the signal power and pump RIN level that would result in a 0.1-dB penalty. Even though the RIN transfer functions were significantly affected by the presence of pump depletion the

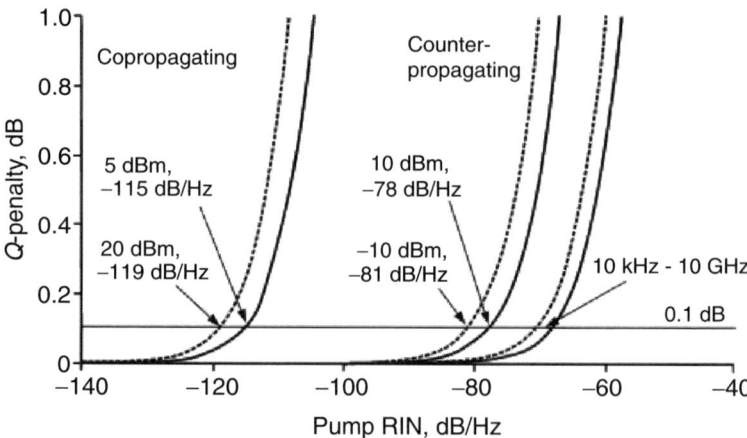

**Figure 5.9:** $Q$-penalty in decibels against pump laser RIN level, in decibels per Hertz [16, Fig. 2]. Dashed lines, without pump depletion; solid lines, with pump depletion included.

penalty is not. Even with strong pump depletion, only ∼3–4 dB/Hz of additional pump RIN can be tolerated. The combined effect of increased corner frequency and reduced low-frequency transfer results in a nearly constant level of integrated noise over the bandwidth of the signal. Thus, pump depletion only has a small impact on the transferred integrated noise [16] and does not strongly affect the $Q$-factor penalty. RIN transfer can therefore advantageously be studied using the simpler analytical models (described in the previous sections) that neglect pump depletion. This would lead to only slightly conservative specifications for tolerable levels of pump RIN.

### 5.2.5   Measurements of RIN Transfer

The RIN transfer functions discussed theoretically in the previous sections have been confirmed experimentally through direct measurements. In Ref. [9] a pump-probe measurement technique was demonstrated using an electrical network analyzer. Figure 5.10 shows the experimental setup. An electrical network analyzer modulated the drive current of the two polarization multiplexed pump lasers (each with 300-mW output) with a

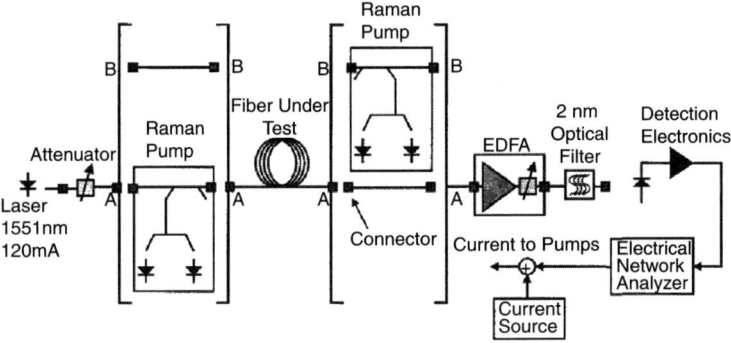

**Figure 5.10:** Experimental setup to measure the RIN transfer function for both forward-pumped (A) and backward-pumped (B) Raman amplifiers [9, Fig. 7]. © 2004 IEEE.

low modulation index. The frequency responses of the laser diodes were characterized using a photo diode and the network analyzer. The pump beam was then injected into either the input of the Raman amplifier (to study forward pumping) or the output (to study backward pumping). The signal from a single laser was used as a probe to measure the RIN transfer function. An optical attenuator adjusted the signal power to be sufficiently low so that the Raman amplifier was operating in the undepleted regime. This allowed the simple model from Section 5.2.2.2 to be used for the analysis of the data. An EDFA on the output of the Raman amplifier ensured sufficient signal into the detection system and an optical filter bandwidth limited and reduced the total amount of spontaneous scattering reaching the detector. The detected signal was then connected to the input of the network analyzer, which measured the transfer function as a function of modulation frequency.

This setup was used to study the RIN transfer function of three of the main fiber types used for Raman amplifiers: NZDSF, dispersion compensating fiber (DCF), and reverse dispersion fiber (RDF). The data for these fibers are shown in Table 5.1.

Each of these three fibers was studied experimentally with both forward and backward pumping. The RIN transfer was also calculated using the theory presented in 5.2.2.2. The results of both the experiments and the calculations are shown in Figure 5.11. The figure demonstrates excellent

**Table 5.1 Fiber data for the fibers used in the RIN measurements [9, Table 1]**

| Fiber Type | Length (km) | Added Gain (dB) | Dispersion @ 1500 nm (ps/km/nm) | Attenuation @ 1455 nm (dB/km) |
|---|---|---|---|---|
| TrueWave RS NZ-DSF | 81.1 | 12.6 | 2.3 | 0.25 |
| Coming DCM | 6.8 | 7.4 | −97.6 | 1.52 |
| RDF | 9.6 | 7.0 | −15.7 | 0.29 |

agreement between the theory and the experiments. The RIN transfer for forward pumping was least using the DCF and RDF despite their shorter length compared to the NZDSF. The low local dispersion of the NZDSF reduced the walk-off of the pump and signal and this resulted in a substantially larger RIN transfer. For backward-pumped Raman amplifiers the difference between the fiber types was not significant, since counterpropagation dominates over walk-off.

## 5.2.6   Low RIN Pump Laser Technologies

In the previous sections, coupling of RIN from the pump laser to the signal was discussed. To avoid $Q$-factor penalties associated with RIN, total system RIN levels of $< -120$ dB/Hz for forward pumping and $< -70$ dB/Hz for backward pumping, respectively, were shown to be required. Such systems can consist of more than 200 amplifiers for the longest ultra-long-haul terrestrial and submarine systems. Since RIN accumulates linearly through chains of amplifiers [9], a per-amplifier pump laser RIN requirement of $>20$ dB better than the system RIN requirement can result. The commonly used FBG-FP pump lasers exhibit typical RIN values of $-120$ dB/Hz and the cladding pumped fiber lasers used for many of the initial Raman demonstrations [5] have RIN values as high as $-110$ dB/Hz (see Section 7.2.4). Both these RIN levels are sufficient to support backward pumping even for ultra-long systems, but not for forward pumping. For example, for a 10-amplifier, forward-pumped system, the cladding pumped fiber laser would degrade the performance by ~1 dB and significantly more if additional amplifiers were added. DFB lasers have narrow linewidths of typically ~10 MHz resulting in superior RIN values of $-160$ dB/Hz

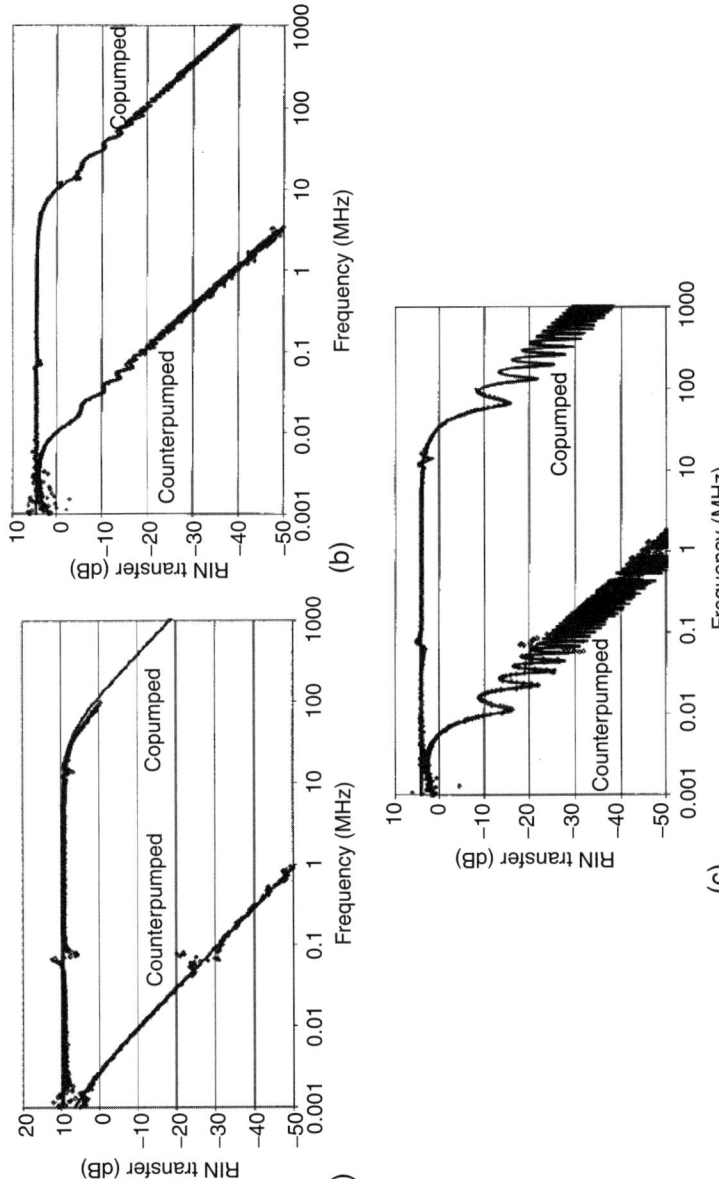

**Figure 5.11:** Measured (markers) and calculated (lines) RIN transfer functions for NZDSF (a), DCF (b), and RDF (c) for both forward and backward pumping [9, Figs. 8, 9, 10]. © 2004 IEEE.

(the narrower the linewidth the stabler the output power is). Therefore, from a RIN point of view, they would be perfect for forward pumping. Unfortunately, due to their narrow linewidth, such lasers are also very susceptible to stimulated Brillouin scattering (SBS), which limits the useful pump power to only a few milliwatts. DFB lasers are therefore not suitable as pump lasers. Initial Raman experiments all used backward pumping, even though the advantage of bidirectional pumping was well understood for long spans.

The ideal solution would be lasers with the wider linewidth of FBG-FP lasers, but with the low RIN of DFB lasers. In 2001, the solution in the form of low RIN multimode pump lasers was presented by two independent research groups [5, 6]. This type of laser was called an SB-DFB in Ref. [5] and used a modified DFB grating that allowed multiple modes to lase. In Ref. [6] the laser was called an IGM laser and the laser structure seems (from public descriptions) to work in a similar manner. Both types of lasers had RIN levels close to that of the DFBs ($-157$ dB/Hz) with linewidths that would sufficiently suppress the SBS generation.

Figure 5.12 shows the output spectra for these two types of low RIN pump lasers. Both output spectra look similar to FBG-FP lasers with multimode lasing. The IGM laser in this example was lasing in more modes than the SB-DFB.

Figure 5.13 shows the RIN spectra for both types of lasers. As can be seen on the graphs, both lasers have superior RIN performance compared to the FBG-FP lasers with RIN values of $-157$ dB/Hz, which is 20–30 dB better.

Multiwavelength backward-pumped Raman amplifiers often suffer from degraded noise performance at shorter wavelengths due to thermally enhanced spontaneous scattering close to the pump wavelength. In Ref. [6], the low RIN pump lasers were used to flatten the noise figure by using bidirectional pumping. All six pump wavelengths were injected in the backward direction. To remove the noise figure tilt, pump beams using low RIN lasers were also injected in the forward direction at the three shortest pump wavelengths. The resultant gain and noise figures are shown in Figure 5.14 and the pump configuration in Table 5.2. The figure also demonstrates the noise figure tilt that would be present if only backward pumping had been used.

To confirm the improved performance of the SB-DFB for forward-pumped Raman amplifiers, a single-channel single-span 10-Gb/s

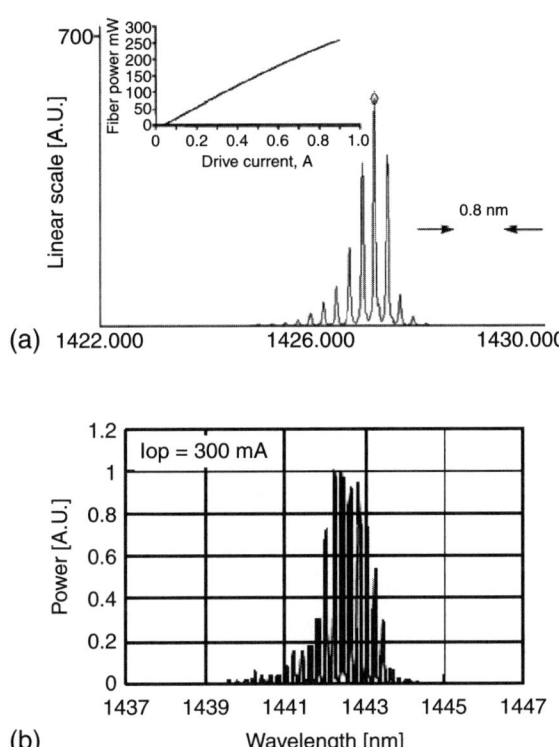

**Figure 5.12:** Output spectra for two types of low RIN multimode pump lasers: (a) SB-DFBs [4, Fig. 1a] and (b) IGM pump laser [6, Fig. 2].

transmission experiment was performed [4]. The fiber used was an NZDSF fiber with zero-dispersion wavelength between the pump and the signal. This resulted in a worst-case RIN transfer since very little walk-off took place. Also the pump and signal were polarization aligned to obtain maximum Raman efficiency. Sensitivity was measured using both an FBG-FP and the SB-DFB. Figure 5.15 shows the result of the comparison. The sensitivity of the Raman amplifier employing a grating stabilized pump degraded as the Raman on–off gain was increased. For the amplifier with the low RIN pump laser, sensitivity was almost unchanged over the range of gains investigated. This figure demonstrates that, with this type of pump laser, forward

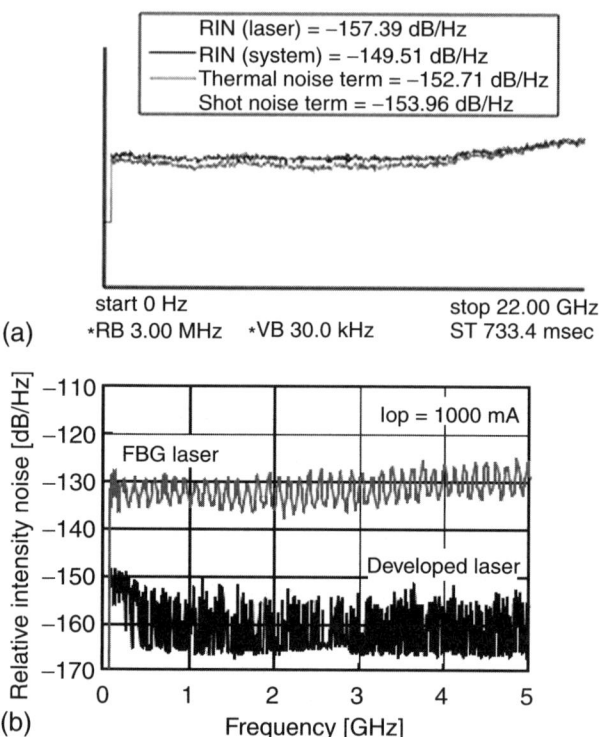

**Figure 5.13:** Relative intensity noise for both (a) the SB-DFB laser [4, Fig. 1b] and (b) the IGM laser [6, Fig. 3]. (a) also shows a comparison of laser RIN, system RIN, and the thermal and shot noise. On (b) is shown, for comparison, the RIN of a standard FBG-FP laser.

Raman pumping can be used beneficially. This allows the Raman amplifier designer additional flexibility in choosing an optimal configuration.

## 5.2.7   Summary

Coupling of RIN from pumps to signals was analyzed. The RIN transfer function is strongly affected by walk-off between the signal and the pump either due to counterdirectional propagation or due to different group velocities caused by nonzero dispersion. The worst-case RIN transfer happens

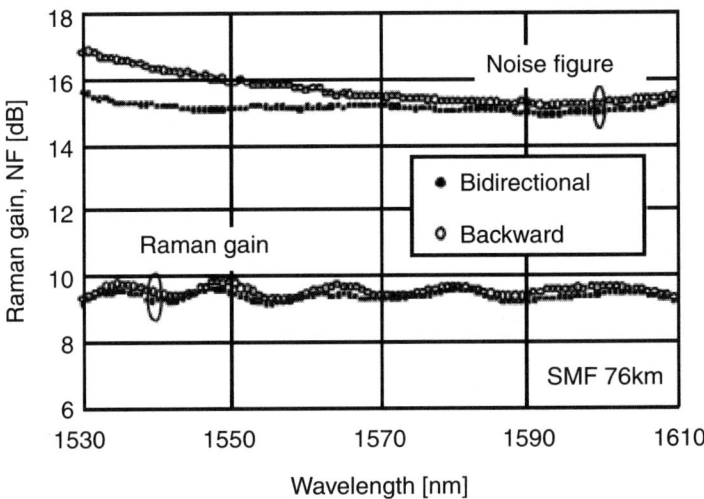

**Figure 5.14:** Gain and noise figure for six-wavelength bidirectionally pumped Raman amplifier [6, Fig. 6]. Also shown on the figure is the noise figure in the case of only backward pumping.

**Table 5.2 Power launched at each wavelength [6, Table 1]**

|  | Wavelength | Backward | Bidirectional | |
|---|---|---|---|---|
|  |  |  | Forward | Backward |
| $\lambda_1$ | 1426.2 nm | 149 mW | 31 mW | 96 mW |
| $\lambda_2$ | 1438.5 nm | 161 mW | 36 mW | 108 mW |
| $\lambda_3$ | 1451.8 nm | 91 mW | 22 mW | 65 mW |
| $\lambda_4$ | 1466.0 nm | 83 mW |  | 105 mW |
| $\lambda_5$ | 1495.2 nm | 184 mW |  | 206 mW |
| Total power |  | 668 mW | 668 mW | |

when there is no walk-off between the signal and the pump (such as when the zero-dispersion wavelength lies in the middle of the pump and signal wavelengths). An analytical model was derived for the case in which pump depletion can be ignored and it was shown that this model agrees very well with a direct numerical solution of the coupled equations taking

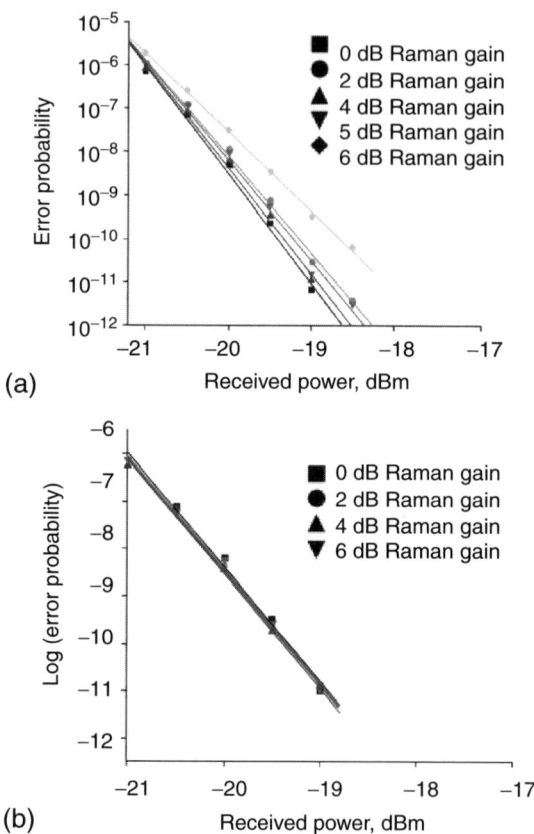

**Figure 5.15:** Sensitivity comparison for forward-pumped Raman amplifiers using 25 km of NZDSF fiber and either 1443-nm FBG-FP laser (a) or 1425-nm FB-DFB (b). The signal wavelengths were 1544 nm and 1529 nm, respectively, for the two cases [4, Figs. 3, 4].

pump depletion into account. The usage of the analytical model only leads to slightly conservative specifications on tolerable pump laser RIN. Experimental verification of the RIN transfer was presented as well as a type of pump laser, which at the same time achieves low RIN with sufficiently broad linewidths suitable for use as pump lasers. With these lasers bidirectional or forward pumping can be fully exploited by the Raman amplifier designer.

## 5.3 Multipath Interference Penalties

MPI is present in all optical transmission systems. Multiple signal paths are introduced into the transmission path through many different sources including discrete reflection sources (e.g., connectors, components within amplifiers), multimode transmission either introduced accidentally or intentionally (e.g., a higher order mode fiber for dispersion compensation) [17], and reflections caused by the distributed process of double Rayleigh scattering (DRS).

Transmission penalty is caused by the interference of a multiple (doubly) reflected (and hence time delayed) signal interfering with the intended signal. See Figure 5.16.

In a Raman-amplified system, especially one with a chain of Raman amplifiers, MPI can easily result in performance-limiting interference. The severity of the interference is due to the scattered light having passed two additional times through the Raman amplifier where it experiences the distributed gain. The presence or absence of optical isolators (which eliminate the backward traveling signal) in amplifiers has a great impact on the analysis of MPI. In optical systems with discrete amplifiers (e.g., erbium-doped fiber amplifiers) the sources of scattering can be separated from the amplifier by optical isolators. These isolators eliminate the backward traveling reflection and hence greatly reduce the MPI. Systems with lumped amplifiers will not be addressed in this section, though. In distributed Raman-amplified systems the use of isolators to completely separate the sources of reflection and amplification is not possible as both the amplification and the scattering sources are present simultaneously in the fiber. (Though impractical, isolators embedded in the transmission span have been shown to yield MPI improvements. See Section 5.3.3.)

DRS interference has been analyzed through many different techniques [18]. Various analyses of DRS MPI are performed through two basic

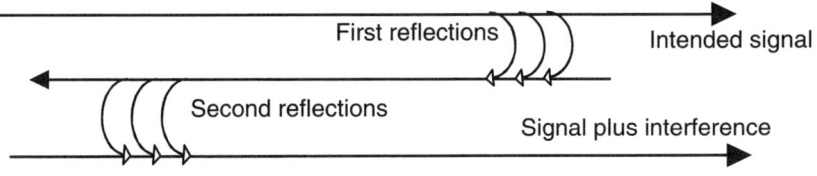

**Figure 5.16:** Double reflection causing interference with the intended signal.

techniques. The first and more complicated is through the study of interferometric conversion of laser phase noise into amplitude noise through the interference of the scattered electromagnetic field with the original [19, 20], in both the time and frequency domains. The second and simpler, but more useful, is the study of the power evolution caused by the distributed reflections [21–24].

To understand the importance of the interferometric approach, consider monochromatic light encountering a Fabry-Perot etalon (i.e., a pair of partial reflectors in the path of the light beam) or a Mach-Zehnder interferometer (i.e., recombining a light beam that has been previously split into two paths). These are the simplest cases of MPI. Clearly, an analysis of the phase relationship of the incoming light is essential to understand the properties of the system as it varies between constructive interference and destructive interference. This is the simplest example of phase to intensity conversion due to MPI. If the incoming light was not monochromatic (i.e., has a nonzero linewidth and time varying, random phase fluctuations) or the phase relationship between the interfering beams was varying, the result is a time- and frequency-dependent noise source.

The study of MPI though interferometric means is complicated [19, 20] though the analysis is easier in cases in which the optical linewidth is known and small or has a small number of discrete reflectors. Because this is rarely the case with practical, high-capacity digital transmission systems, this treatment is omitted.

MPI penalties due to multimode transmission are also not treated here. Through proper architectural design and proper design, manufacture, selection, installation, and maintenance of connectors and components, multimode effects should be insignificant in modern transmission systems.

Due to the dominance of DRS penalties in Raman-amplified systems and the relative ease of obtaining useful results through the study of the power evolution caused by the distributed reflections, this analysis will dominate the discussion.

## 5.3.1   Analysis

Double Rayleigh scattering of signals can be the most limiting factor in the design of Raman-amplified systems. The noise generated by the DRS degrades the signal by subtracting from the optical signal-to-noise ratio

(OSNR). As we will show below, the OSNR in the presence of multipath interference (MPI), $OSNR_{MPI}$, decreases with the gain of the amplifier. This $OSNR_{MPI}$ reduction places an upper limit on the useful Raman gain of an amplifier.

### 5.3.1.1 OSNR Due to Double Rayleigh Scattering

As stated earlier, the use of isolators and reasonable specifications for components removes most sources of MPI (i.e., reflections) in practical Raman amplifier systems except that caused by DRS. Rayleigh scattering is caused by small-scale inhomogeneities (compared to the wavelength of light) of the refractive index of the transmission fiber. Fiber loss due to Rayleigh reflections is characterized by a loss coefficient, $\alpha_r$, such that for power traveling in the forward direction, $P^+(z)$, the following holds:

$$\frac{dP_s^+(z)}{dz} = -\alpha_r P_s^+(z). \qquad (5.2.1)$$

Rayleigh scattering occurs in all directions though only the fraction captured in the guided mode of the fiber, $f_r$, is reflected backward. (Note that the capture ratio is dependent on the mode profile of the fiber.) For Gaussian modes, the ratio is proportional to $\lambda^2$ and inversely proportional to the effective area of the fiber [25]. The product of the loss coefficient and the capture ratio results in the Rayleigh scattering coefficient, $r_s = f_r \alpha_r$, such that the backward-propagating (i.e., reflected) signal, $P^-(z)$, is given by

$$\frac{dP_r^-(z)}{dz} = r_s P_s^+(z). \qquad (5.2.2)$$

Our concern is for the power that is doubly reflected. For the case of an unsaturated amplifier (i.e., there is no pump depletion), the equation for the signal-to-DRS noise ratio can be derived [18, 26]. A simple analysis concerns itself with only the power evolution of three signals: $P_s^+(z)$ is the forward traveling signal, $P_r^-(z)$ is the backward traveling single-reflected light, and $P_r^+(z)$ is the forward traveling double-reflected light. This derivation starts with the unsaturated Raman model [27], assumes the reflected light can be treated as a perturbation (i.e., the DRS energy is much smaller than the signal power; otherwise, the amplifier would not be a

very useful amplifier), and neglects additional reflections. The propagation
of the three signals can be described by the following three equations:

$$\frac{dP_s^+(z)}{dz} = \vec{g}(z)P_s^+(z),$$                                  (5.2.3)

$$-\frac{dP_r^-(z)}{dz} = \overleftarrow{g}(z)P_r^-(z) + r_s P_s^+(z),$$        (5.2.4)

$$\frac{dP_r^+(z)}{dz} = \vec{g}(z)P_r^+(z) + r_s P_r^-(z).$$                   (5.2.5)

Equations (5.2.3), (5.2.4), and (5.2.5) represent the forward propagation
of the signal, the backward propagation of the signal after a single reflection, and the forward propagation of the signal after two reflections. The
net forward and reverse gain per unit length of the fiber are $\vec{g}(z)$ and $\overleftarrow{g}$,
respectively. Both are given by

$$\vec{g}(z) = \overleftarrow{g} = g_R P_p(z) - \alpha_s.$$                    (5.2.6)

To help solve these linear, first-order differential equations, the following
integrating factors that physically represent the gain for forward traveling
signal from $z_1$ to $z_2$, $\vec{G}(z_1, z_2)$, and the gain for a reverse traveling signal from $z_2$ to $z_1$, $\overleftarrow{G}(z_1, z_2)$, are introduced. Though this gain distribution
is usually identical for forward and backward traveling waves, this nota-
tion allows for a nonreciprocal gain distribution such as occurs with the
placement of isolators in the transmission line

$$\vec{G}(z_1, z_2) = e^{\int_{z_1}^{z_2} \vec{g}(\zeta)d\zeta}$$              (5.2.7)

$$\overleftarrow{G}(z_1, z_2) = e^{-\int_{z_2}^{z_1} \overleftarrow{g}(\zeta)d\zeta}.$$  (5.2.8)

By applying the boundary conditions $P_r^-(L) = 0$ and $P_r^+(0) = 0$, we can
solve for $P_r^+(L)$ in terms of $P_s^+(L)$:

$$f_{DRS} = \frac{P_r^+(L)}{P_s^+(L)} = r_s^2 \int_0^L dz_2 \int_0^{z_2} \vec{G}(z_1, z_2)\overleftarrow{G}(z_1, z_2)dz_1.$$  (5.2.9)

In general, this equation must be solved numerically. However, one simple
example that can be solved analytically is the case of an ideal distributed

amplifier [28]. For this amplifier, the gain is unity (i.e., $G(z) = 1$) at all positions along the fiber. Hence, Eq. 5.2.9 reduces to

$$f_{DRS} = \frac{1}{2}r_s^2 L^2. \qquad (5.2.10)$$

As this equation demonstrates, DRS increases as the square of the fiber length (stated another way, it grows as the square of the linear gain of the amplifier). Note that this is only true for an ideal distributed amplifier. For Raman amplifiers with gain distributions that result from fiber-end pumping, the DRS cross-talk evaluation is more complicated [26] and requires numerical evaluation.

Numerical evaluation of Eq. (5.2.9) is shown in Figure 5.17 for sample amplifiers 45 and 90 km long. As expected, Rayleigh cross talk increases rapidly with fiber length and excess gain. Furthermore, it is clear that bidirectional pumping has a positive effect on double Rayleigh scattering. Intuitively, this is because DRS scales with the square of the gain and thus better distributed gain is beneficial.

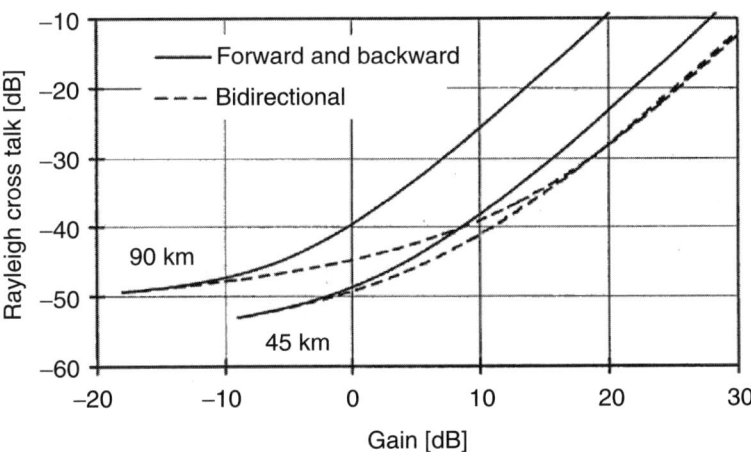

**Figure 5.17:** Rayleigh cross talk vs. on–off gain for a 45-km and 90-km unsaturated Raman amplifier pumped unidirectionally and bidirectionally. $\alpha_{pump} = 0.26$ dB/km, $\alpha_{signal} = 0.20$ dB/km.

### 5.3.1.2    Transmission Impairment Due to DRS

Though a quantitative analysis of DRS cross talk is useful, its impact on transmission performance is most important. A simple, though useful, analysis can be used that describes the general case of $N$ in-band interfering sources with Gaussian probability density distributions [29]. The following result is obtained:

$$\text{Penalty(dB)} = 10 \cdot \log 10 \left[ 1 - \frac{1}{2} f_{\text{DRS}} N Q^2 \right], \qquad (5.2.11)$$

where $N$ is the number of concatenated amplifiers in a chain and $Q^2$ is the $Q$-factor [30] of the system without cross talk.

Transmission impairment due to DRS, in general, should be studied in conjunction with other penalties present in transmission systems. In beat-noise limited receivers found in most optically amplified transmission systems (so called because the dominant noise source is not from the detector, but from amplifier noise either beating with itself or with the signal), the combined influence of both the ASE-induced beat noise and the MPI-induced beat noise must be understood. Many authors have studied the impact of both sources of noise with different approaches [31, 32]. These analyses show that the impact of DRS is dependent on the $Q$-factor of the system without MPI. In systems with poor $Q$-factors due to a large amount of ASE noise, the impact of MPI is smaller. Therefore, systems with forward error correction (and hence lower required $Q$-factors) are more robust against MPI.

The transmission impairment caused by DRS for an undepleted 45-km backward-pumped Raman amplifier and a 90-km bidirectionally pumped amplifier has been calculated according to Eq. (5.2.11) and is plotted in Figure 5.18. As shown in the figure, 7000 to 9000 km transmission distance can be achieved with near-unity gain amplifiers with either backward-pumped 45 km amplifiers or 90 km bidirectionally pumped amplifiers with only ~0.2 dB of penalty. Gain in excess of that necessary to yield the span transparent can cause significant transmission penalty, though. For example, for an excess gain of 5 dB (e.g., to compensate for component losses) a penalty of up to ~0.6 dB is seen (for 45 km backward-pumped amplifiers over 9000 km of distance).

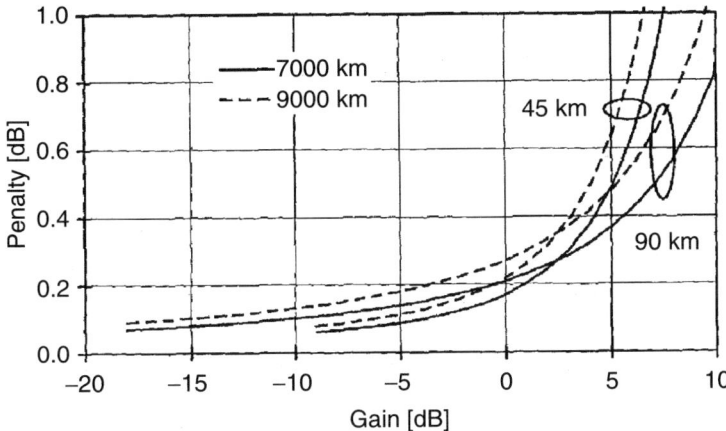

**Figure 5.18:** Transmission impairment (calculated at BER=$10^{-9}$) caused by DRS for undepleted 45-km backward-pumped and 90-km bidirectionally pumped amplifiers after 7000 km and 9000 km.

### 5.3.1.3   Polarization Properties of DRS

In a long fiber with low birefringence, Rayleigh scattering preserves the state of polarization and reduces the degree of polarization (DOP) to one third its original value. This result has been calculated and confirmed experimentally [33].

After double scattering, the DOP has been calculated to be reduced by one ninth [18]. To understand this impact, let us temporarily hypothesize that the DOP of the DRS power was reduced to zero. The DRS light could then be considered to be unpolarized. In this case, the amount of DRS light that will contribute to degrading the signal would be one half. However, since the DOP is actually one ninth, half of the remaining eight ninths is available to interfere with the signal in addition to the one ninth that is polarized. As expected this depolarization has an effect on the impact of DRS. Therefore the correction factor for the DRS noise is five ninths [33]:

$$\text{OSNR}_{\text{MPI}}^{P} = \frac{5}{9}\text{OSNR}_{\text{MPI}}. \qquad (5.2.12)$$

This correction is critical as only MPI that is in the same polarization as the signal produces beat noise in the receiver.

## 5.3.2    Measurement of DRS Noise

Measurement of DRS (and, more generally, MPI) is inherently difficult. Unfortunately use of a simple optical spectrum analyzer as would be used for measuring ASE is not adequate.[1] This is because the MPI noise occupies the same optical spectrum and state of polarization as the signal. Two measurement techniques have been widely used: an electrical beat-noise method and a time-domain extinction method. Both of these measurements attempt to quantify the optical SNR due to MPI or Rayleigh double scattering:

$$\text{OSNR}_{\text{MPI}} = \frac{1}{f_{\text{DRS}}} = \frac{P_s}{P_{\text{MPI}}} \approx \frac{P_s}{P_r}, \qquad (5.2.13)$$

where $f_{\text{DRS}}$ is the cross-talk ratio. A similar expression can be written for the polarization corrected OSNR, $\text{OSNR}_{\text{MPI}}^{P}$. Both of these methods will now be described.

### 5.3.2.1    Electrical Beat-Noise Measurement Technique

A typical apparatus to make electrical beat-noise measurements of MPI is shown in Figure 5.19. A RIN measurement is made with the electrical spectrum analyzer (ESA). Much of the difficulty of this technique is to identify and measure the sources of inaccuracy and to properly calibrate the apparatus. The RIN that is measured comes from a variety of sources besides the MPI that we wish to measure. These other contributions must be measured and subtracted from the total. The RIN contributions are due to signal-ASE beating, excess noise from the DFB, shot noise, and thermal noise from the ESA components. The optical spectrum analyzer in the test apparatus is available in the setup to measure the RIN contribution from signal-ASE beating. Spontaneous–spontaneous beat noise is made to be

---

[1]However, in principle, one could use a polarization nulling technique like that used to measure the ASE of EDFAs. By using polarization-selective optics to eliminate the signal, the remaining DRS would be measured through which the interfering DRS would be inferred [34].

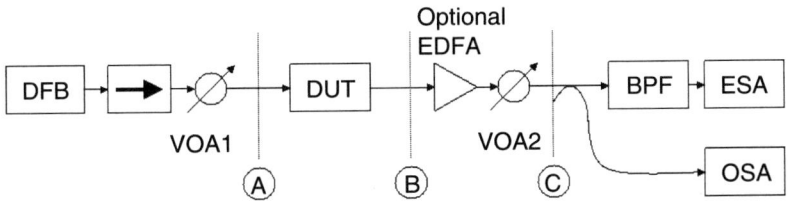

**Figure 5.19:** Test apparatus for electrical measurement of MPI.

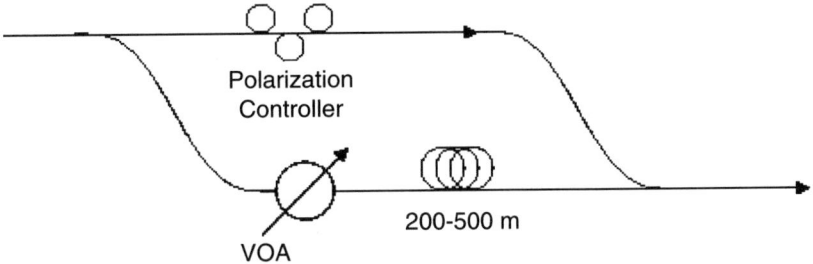

**Figure 5.20:** Mach-Zehnder interferometer for simulating MPI used to calibrate electrical beat-noise measurements of MPI.

negligible with a narrowband band-pass fiber before the photodetector in the ESA [34].

The amplifier to be tested is provided a low-noise input. The spectral properties of this signal are important. The linewidth must be large enough to ensure that its coherence length is shorter than the round-trip paths in the amplifier to ensure that the DRS light is incoherent to the straight-through light. For a typical Raman amplifier with more than a few kilometers of fiber, a laser with a linewidth of greater than 1 MHz is appropriate (e.g., a DFB laser).

An MPI simulator can be constructed to calibrate and test the measurement technique [35]. See Figure 5.20. The incoming light is split in two. One path contains a variable attenuator to vary the amount of interference provided. A suitable length of fiber is provided in the interferometer to decorrelate the interference. Suitable couplers are chosen so that greater than 98% transmission is obtained on the nonattenuated leg of the interferometer.

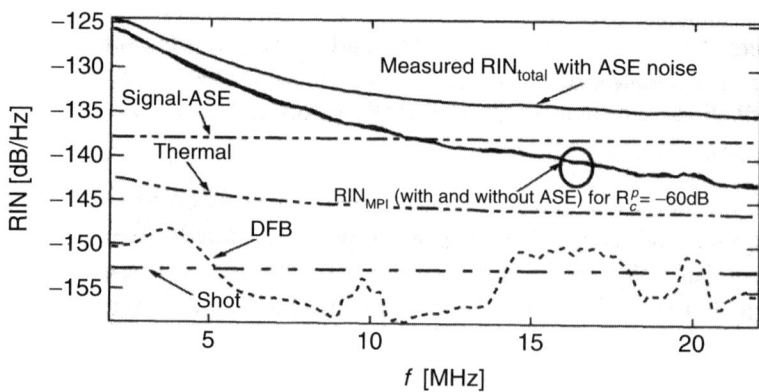

**Figure 5.21:** MPI measurement of a test interferometer. [18; Figure 15.12b].

With the simulator, RIN on the ESA can be measured and simple power measurement made on the two internal arms of the interferometer. By varying the variable optical attenuator (VOA) in the simulator, a calibration curve of various levels of cross talk and measured RIN can be made.

Figure 5.21 shows the result of measuring an MPI simulator with interference equivalent to $f_{DRS} = -60$ dB (a very low value). This figure demonstrates the subtraction of the extraneous RIN sources. RIN from the DFB was measured.

Figure 5.22 shows the result of $RIN_{MPI}$ measurements on a typical Raman amplifier (100 km of non-DSF fiber backward pumped by semiconductor pumps at 1435- and 1450-nm wavelengths). Part (a) of the figure shows the $RIN_{MPI}$ measured with the pumps adjusted from zero (minimum gain and minimum RIN) to a setting so that 24.3-dB gain was obtained. Each of the RIN curves was used to calculate a single cross-talk value in part (b) of the figure. Note the exponential increase in cross talk predicted by Eq. (5.2.10). Though not shown in these figures, this technique is capable of measuring cross talk to about −60 dB.

Note that, for low gain, the MPI cross talk is dominated by Rayleigh scattering that originates throughout the whole length of the fiber. Therefore, the addition of 1 dB of gain contributes an insignificant amount of additional cross talk. With a large Raman gain, the cross talk increases by more than 1 dB for each dB increase of Raman gain.

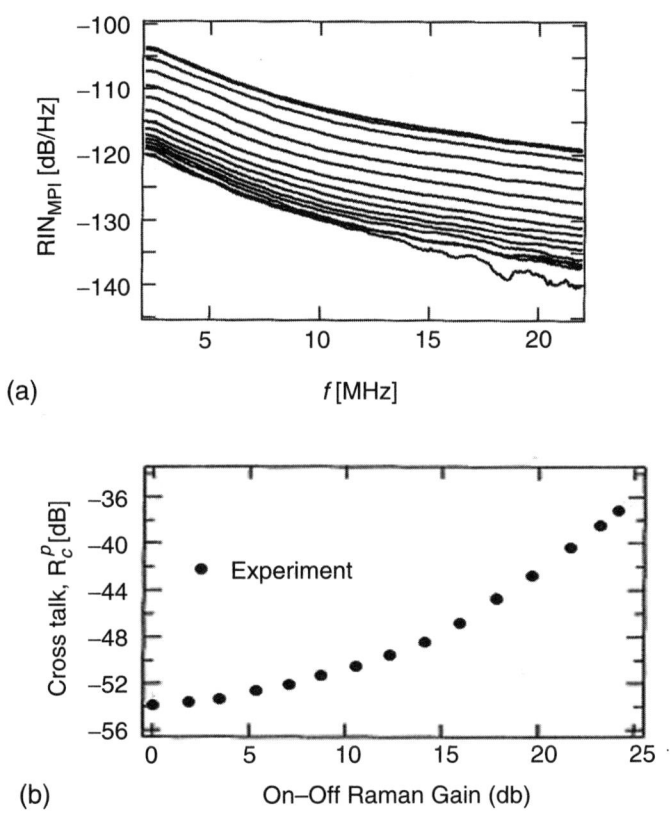

**Figure 5.22:** (a) $RIN_{MPI}$ curves for on–off gain from 0 dB (bottom curve with pumps off) to 24.3 dB (top curve). (b) Cross-talk ratio corresponding to (a). [18; Figure 15.13].

### 5.3.2.2 Time-Domain Extinction Measurement Technique

Time-domain extinction methods have previously been used to characterize noise generation by erbium-doped fiber amplifiers. Lewis et al. first established the extension of this technique to Raman amplifiers [36].

The measurement system consists of a Raman amplifier with a modulated signal at its input and an optically gated receiver at its output as shown in Figure 5.23. Acousto-optic modulators are used to provide both of these functions as they both have a high extinction ratio (~90 dB if two

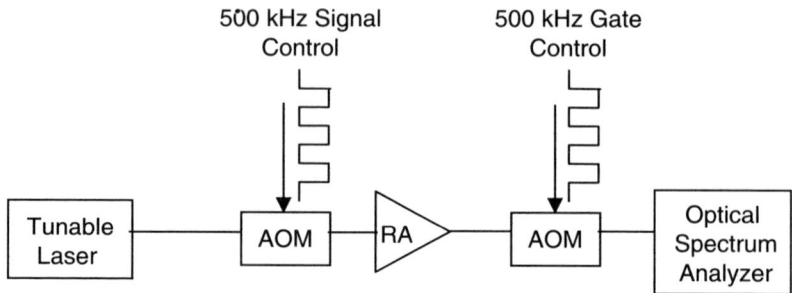

**Figure 5.23:** Time-domain extinction measurement configuration for measuring double Rayleigh scattering.

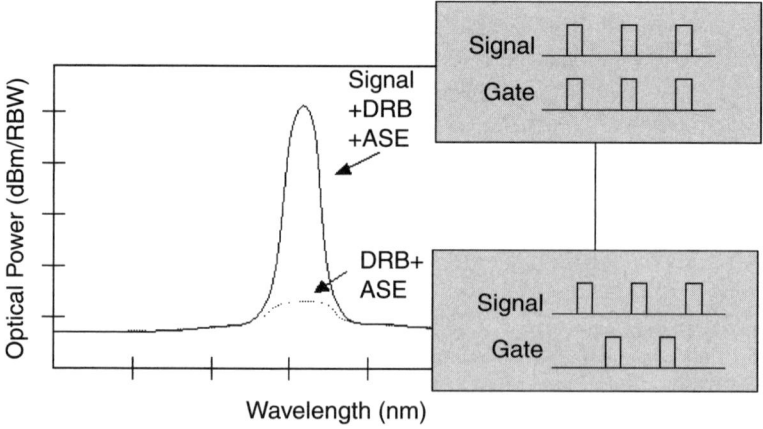

**Figure 5.24:** OSA spectra obtained from the optical time domain extinction technique and the timing required to obtain them.

modulators are paired to function as one) and an appropriate modulation frequency (hundreds of kHz) for this application.

In this technique, two traces on the optical spectrum analyzer (OSA) are required. The signal is first measured with the signal control and the gate control adjusted in phase. The Rayleigh scattered power is then measured with the receiver's sampling window out of phase with the signal. In the absence of Rayleigh scattering, only the leakage through the input acousto-optic modulator should appear on the OSA. In the presence of Rayleigh scattering, the DRS noise appears as a narrow spectral peak that stands out from the background amplified ASE. The background ASE can simply be subtracted from the total noise power to result in DRS noise power. This measurement requires a correction for the receiver insertion loss and the fact that the receiver samples only a fraction of the DRS power.

In addition to proper calibration, care must be taken to make an accurate measurement. The input to the amplifier should be small enough so that it does not affect the gain of the amplifier or transient effects will complicate the measurement [37].

Typically, this technique is adequate to measure $R_c$ to $-50$ dBm, which is not as sensitive as with the electrical methods, but the technique is simpler. Also, only amplifiers with a large delay can be measured. Short Raman amplifiers such as those made with short highly nonlinear fibers present difficulty to this measurement technique. One benefit to this technique, though, is this technique does not require a source with very low noise or specific linewidth. Therefore, any tunable laser can be used to measure Raman amplifiers over their entire gain spectrum.

### 5.3.3   MPI Suppression

It may seem that for a given fiber loss in each span, the Rayleigh scattering and its penalty would be preordained. But many researchers have discovered rules and techniques to limit or suppress the amount of MPI.

As shown in the preceding, the configuration of the amplifier has a large impact on DRS. Especially for longer amplifiers, bidirectional pumping can have a large advantage over unidirectional pumping (either forward or backward). As shown in Figure 5.17, a bidirectionally pumped 90-km amplifier has more than a 5-dB advantage. Also, any loss that is added that requires gain (i.e., in excess of that required for transparency) greatly increases the amount of DRS. An obvious technique to minimize DRS is simply to minimize the amount of gain required for a given span. Losses at the pump site (pump combiner, isolator, monitoring taps, etc.) should be minimized.

Another architectural change that can have a tremendous impact on DRS generation is the use of optical isolators. In addition to an isolator that is customarily placed adjacent to the Raman pump source, consideration should be given to placing an isolator (with a pump bypass) [38] midway in the span to break up an amplifier with gain $G$ into two amplifiers with gain $\sqrt{G}$. For example, if $G = 100$ (20 dB), then each of the two sections will have a gain of 10 (10 dB). Due to the $G^2$ dependence of DRS, the use of an isolator reduces the MPI remarkably. This technique is not practical for distributed Raman amplifiers in which amplification is performed in the transmission span, due to the difficulty in placing the isolators remotely. For discrete Raman amplifiers, mid-amplifier isolators are desirable and often necessary to limit DRS.

Another more subtle way to reduce DRS starts with the observation that Raman amplifiers are inherently not very spectrally flat. Gain flattening is usually accomplished with a gain-flattening filter. Placement of this filter within the transmission span reduces the round-trip gain experienced by the high gain signals. (Unfortunately, it also increases the noise figure of the Raman amplifier.) Lewis et al. have shown that it is possible to optimize the location of the filter mid-span [39]. In their 25-dB amplifier with 4 dB of gain ripple, the improvement in DRS OSNR was greater than 7 dB. This technique is appealing for both discrete and distributed amplifiers.

In systems where bidirectional pumping is not a feasible method to reduce MPI (e.g., because of pump-to-signal noise transfer as discussed previously in this chapter), architectural changes to the transmission topology can be useful [40]. In the configuration shown in Figure 5.25, gain is

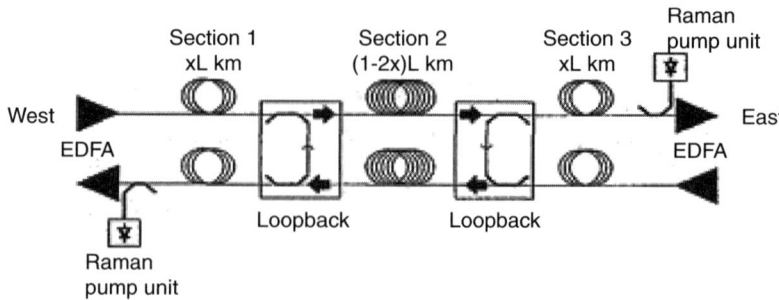

**Figure 5.25:** Passive loopbacks in a counterpumped amplifier system. [40; Figure 1]. © 2004 IEEE.

reduced in the backward-pumped transmission fiber by looping back some of its pump power onto a fiber carrying a signal in the opposite direction. This pump power is used to provide backward pumping in the middle of the opposite span. This technique provides a more distributed Raman gain while preserving the benefits of counterpropagating pumps.

In this demonstration, Fludger et al. [40] showed an improvement in noise figure of between 1.0 and 1.7 dB, a reduction in DRBS that allowed longer fiber spans and higher Raman gain to be used.

### 5.3.4 Summary

The dominant impairment caused by the transmission fiber in Raman-amplified systems was studied. Though there are many sources of reflections in a transmission system, the most important one that limits long distance (through either long spans, many concatenated amplifiers, or both) is multiple path interference caused by double Rayleigh scattering. We have seen that when there is enough gain in a span and little enough ASE noise in an amplifier, that MPI is the next important limit to long distance transmission.

## 5.4  Appendix

In support of the derivations in this chapter, the definition of the Fourier transform, autocorrelation, and properties of both are provided in this appendix. The Fourier transform $H(f)$ of a time-dependent function $h(t)$ at a frequency of $f$ is defined by

$$\Im\{f; h(t)\} = H(f) = \frac{1}{2\pi} \int_{-\infty}^{\infty} h(t)e^{-i2\pi f t} dt \qquad (5.A.1)$$

and the autocorrelation for a time-dependent function $h(t)$ is defined by

$$C_h(\tau) = \langle h(t)h(t - \tau) \rangle. \qquad (5.A.2)$$

A few useful mathematical properties of the Fourier transform and auto-correlation function will now be listed for later use. From Eq. (5.A.2) it can be seen that the autocorrelation of a constant times a time varying function $kh(t)$ is

$$C_{kh} = k^2 C_h. \tag{5.A.3}$$

Similarly, the Fourier transform of the product of a constant and a time-varying function is

$$\Im \{f; k \cdot h(t)\} = k \cdot \Im \{f; h(t)\}. \tag{5.A.4}$$

Finally, the autocorrelation of a sum of two independent random fluctuations is the sum of the autocorrelations.

$$C_{h_1 + h_2} = C_{h_1} + C_{h_2} \tag{5.A.5}$$

# References

[1] M. Nissov, C. R. Davidson, K. Rottwitt, R. Menges, P. C. Corbett, D. Innis, and N. S. Bergano, *Integrated Optics and Optical Fibre Communications, 11th International Conference on, and 23rd European Conference, on Optical Communications* (Sept. 22–25, 1997), Conf. Publ. No. 448, Vol. 5, pp. 9–12.

[2] M. X. Ma, H. D. Kidorf, K. Rottwitt, F. W. Kerfoot, III, and C. R. Davidson, *IEEE Photon. Technol. Lett.* **10,** 893 (1998).

[3] P. G. Agrawal, P. Govind, *Nonlinear Fiber Optics*, 2nd ed. (Academic Press, San Diego, CA, 1995), Chap. 9.

[4] R. P. Espindola, K. L. Bacher, K. Kojima, N. Chand, S. Srinivasan, G. C. Cho, F. Jin, C. Fuchs, V. Milner, and W. Dautremont-Smith, *Electron Lett.* **38,** 113 (2002).

[5] R. P. Espindola, K. L. Bacher, K. Kojima, N. Chand, S. Srinivasan, G. C. Cho, F. Jin, C. Fuchs, V. Milner, and W. C. Dautremont-Smith, *ECOC '01, 27th European Conference on Optical Communication* (2001), Vol. 6, pp. 36–37.

[6] S. Kado, Y. Emori, S. Namiki, N. Tsukiji, J. Yoshida, and T. Kimura, *Optical Communication, 2001. ECOC '01, 27th European Conference on Optical Communication (2001)*, Vol. 6, pp. 38–39.

[7] B. Zhu, L. E. Nelson, S. Stulz, A. H. Gnauck, C. Doerr, J. Leuthold, L. Gruner-Nielsen, M. O. Pedersen, J. Kim, and R. L. Lingle, *J. Lightwave Technol.* **22,** 208 (2004).

[8] A. Yariv, *Optical Electronics in Modern Communications*, 5th ed. (Oxford University Press, New York, 1997), Chap. 16.

[9] C. R. S. Fludger, V. Handerek, and R. J. Mears, *Lightwave Technol.* **19,** 1140, (2001).

[10] M. R. Spiegel, *Mathematical Handbook of Formulas and Tables* (McGraw-Hill, New York, 1968).

[11] L. Raade and B. Westergreen, *Beta Mathematics Handbook*, 2nd ed. (Studentlitteratur, Lund, 1990), Chap. 10.

[12] C. R. S. Fludger, V. Handerek, and R. J. Mears, *Electron. Lett.* **37,** 15 (2001).

[13] J. S. Wey, D. L. Butler, M. F. Van Leeuwen, L. G. Joneckis, and J. Goldhar, *IEEE Photon. Technol. Lett.* **11,** 1417 (1999).

[14] K. Song and S. D. Dods, *IEEE Photon. Technol. Lett.* **13,** 1173 (2001).

[15] C. R. S. Fludger, V. Handerek, and R. J. Mears, *Lightwave Technol.* **20,** 316 (2002).

[16] M. D. Mermelstein, C. Headley, and J.-C. Bouteiller, *Electron. Lett.* **38,** 403 (2002).

[17] S. Ramachandran, S. Ghalmi, I. Ryazansky, M. F. Yan, F. V. Dimarcello, W. A. Reed, and P. Wisk, *IEEE Photon. Technol. Lett.* **15,** 727 (2003).

[18] M. Islam (ed.), *Raman Amplifiers for Telecommunications 2: Sub-systems and Systems* (Springer-Verlag, New York, 2004), p. 491.

[19] P. Wan and N. Conradi, Impact of double Rayleigh backscatter noise on digital and analog fiber systems, *J. Lightwave Technol.* **14,** No. 3, 288–297 (1996).

[20] J. L. Gimlett and N. K. Cheung, *J. Lightwave Technol.* **7,** 888 (1989).

[21] P. B. Hansen, L. Eskildsen, A. J. Stentz, T. A. Strasser, J. Judkins, J. J. De Marco, R. Pedrazzani, and D. J. DiGiovanni, *IEEE Photon. Technol. Lett.* **10,** 159 (1998).

[22] C. H. Kim, J. Bromage, and R. M. Jopson, *IEEE Photon. Technol. Lett.* **14,** 573 (2002).

[23] S. H. Chang, S. K. Kim, M. J. Chu, and J. H. Lee, Limitations in fibre Raman amplifiers imposed by Rayleigh scattering of signals, *Electron. Lett.* **38,** No. 16, 856–867 (2002).

[24] W. Zhang, J. Chen, J. Peng, X. Liu, and C. Fan, *An analytical method for determining comprehensive impact of Rayleigh backscattering on performance of systems with distributed Raman amplifiers.* Lasers and Electro-Optics Society, 2001. LEOS 2001. 14th Annual Meeting of the IEEE, **1,** 350–351 (2001).

[25] E. Brinkmeyer, *J. Opt. Soci. Amer.* **70,** 1010 (1980).

[26] M. Nissov, K. Rottwitt, H. D. Kidorf, and M. X. Ma, *Electron Lett.* **35,** 997 (1999).

[27] E. Desurvire, M. J. F. Digonnet, and H. J. Shaw, Theory and implementation of a Raman active fiber delay line, *J. Lightwave Technol.* **4,** No. 4, 426–443 (1986).

[28] M. Nissov, *Long Haul Optical Transmission Using Distributed Raman Amplification* (Ph.D., Technical University of Denmark, Department of Electromagnetic Systems, December 1997).

[29] H. Takahashi, K. Oda, and H. Toba, *J. Lightwave Technol.* **14,** 1097(1996).

[30] S. D. Personik, *Bell System Technical J.* **52,** 843 (1973).

[31] C. R. S. Fludger and R. J. Mears, *J. Lightwave Technol.* **19,** 536 (2001).

[32] J. Bromage, C. H. Kim, R. M. Jopson, K. Rottwitt, and A. J. Stentz, *Proceedings of Optical Amplifiers and Their Applications* (2001), paper OTuA1.

[33] M. O. van Deventer, *J. Lightwave Technol.* **11,** 1895 (1993).

[34] D. Derickson, ed., *Fiber Optic Test and Measurement* (Prentice Hall, Upper Saddle River, NJ, 1998).

[35] C. R. S. Fludger and R. J. Mears, *J. Lightwave Technol.* **19,** (2001).

[36] S. A. E. Lewis, S. V. Chernikov, and J. R. Taylor, *IEEE Photon Technol. Lett.* **12,** 528 (2000).

[37] C.-J. Chen and W. S. Wong, *Electron. Lett.* **37,** 371 (2001).

[38] A. J. Stentz, S. G. Grubb, C. E. Headley, J. R. Simpson, T. Strasser, and N. Park, *Proceedings of the Optical Communications Conference* (1996), paper TuD3, p. 16.

[39] S. A. E. Lewis, S. V. Chernikov, and J. R. Taylor, *Proceedings of the Optical Fiber Communications Conference* (2000), paper FF5-1, p. 109.

[40] C. R. S. Fludger, V. Handerek, and N. Jolly, Inline loopbacks for improved OSNR and reduced double Rayleigh scattering in distributed Raman amplifiers, *Optical Fiber Communication Conference and Exhibit* (2001), **1** pp. Mi1-1–MI1-3.

# Chapter 6

# Semiconductor Pump Lasers

Naoki Tsukiji and Junji Yoshida

This chapter describes semiconductor pump lasers used for pumping broadband Raman amplifiers and referred to as 14xx-nm pump lasers because they operate in the wavelength region extending from 1400 to 1500 nm. In Section 6.1 we introduce the fundamental concepts behind high-power pump lasers. Section 6.2 then provides details from the standpoint of requirements for counter- and co-propagating Raman amplifiers.

## 6.1 Technology Basis of High-Power Semiconductor Lasers

After the initial use of a high-power, semiconductor, 14xx-nm laser as a pump laser for Raman amplification in telecommunication systems [1], such lasers were developed as a 1480-nm pump laser for erbium-doped fiber amplifiers (EDFA) [2–6]. The increase in channel counts in wavelength-division-multiplexed (WDM) systems using EDFA has been a driving force behind the development of high-power pump lasers. Later, high-power 14xx-nm pump lasers triggered their actual application in Raman amplifiers employed in commercial WDM systems [7, 8]. Since the commercial use of over 200-mW-output 14xx-nm laser modules [9–11] in 1998, the development of high-power 14xx-nm diode lasers has accelerated.

Currently, up to 400 mW of fiber-coupled power has became practical [12]. In this section we discuss the fundamentals of high-power semiconductor pump lasers with high reliability.

## 6.1.1   High-Power Semiconductor Laser Module

Figure 6.1 shows the appearance of the standard 14-pin butterfly laser module, most widely used for practical systems [12, 13]. Such a module contains a package of dimensions 15 mm × 22 mm × 8 mm together with a single-mode fiber pigtail of about 1-m length.

Figure 6.2 shows a plane view of the laser module package. The laser module consists of the laser chip, an optical coupling system to couple the laser beam to the optical fiber, a monitor photodiode (PD), a thermistor, and a thermoelectric cooler (TEC). The laser chip is mounted on the TEC via a submount (heat sink) with high thermal conductivity to dissipate the heat generated from the laser chip as effectively as possible. The temperature of the laser chip is measured by a thermistor, mounted near the laser chip and stabilized at constant temperature by the TEC with an

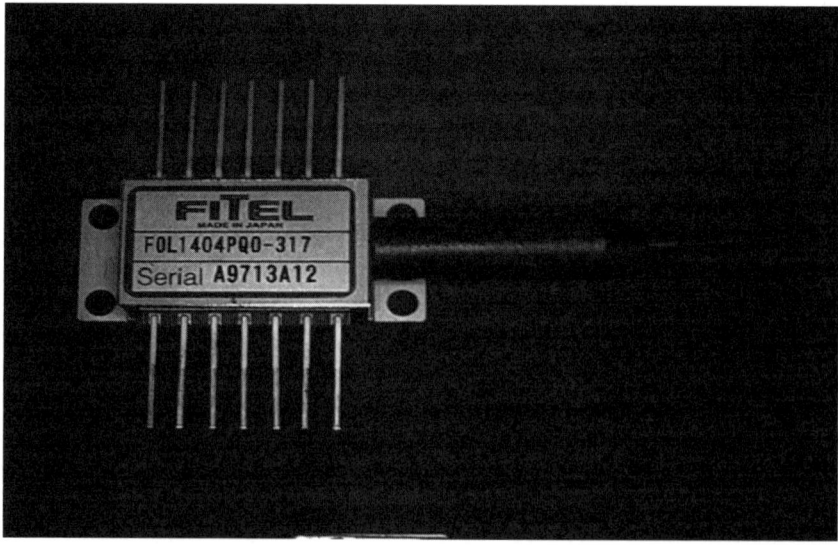

**Figure 6.1:** Pump laser module.

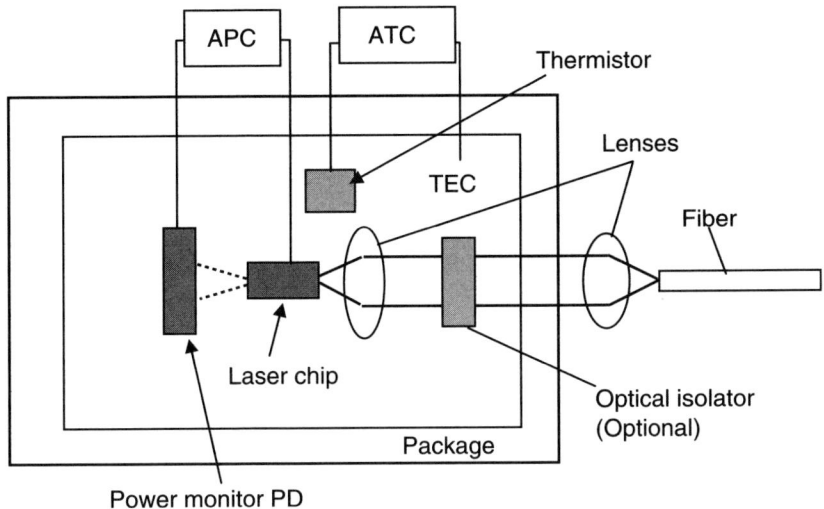

**Figure 6.2:** A schematic plane view of the laser module package.

automatic temperature-control circuit (ATC). Optical output power exiting from the back facet of the laser (approximately proportional to the optical power from the front facet) is detected by monitor PD, whose current is then used to control the laser power through an automatic power-control circuit (APC).

To realize a high-power laser module, the following three items are important:

- Design of semiconductor laser chip
- High capability of heat exhaustion
- High coupling efficiency

## 6.1.2   Fundamentals of High-Power Semiconductor Laser Chip

Figure 6.3 shows the structure of the 14xx-nm semiconductor laser chip most widely used in practice. The chip employs the GRIN-SCH-CS-MQW-BH-laser (graded-index-separate-confinement-compressive-strained-multi-quantum-well-buried-heterostructure-laser) design made of a GaInAsP/InP

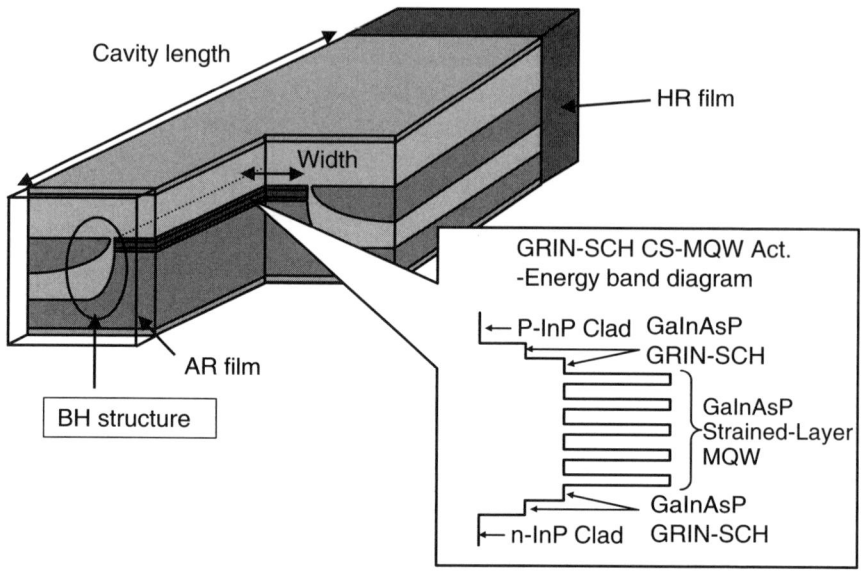

**Figure 6.3:** Structure of 14xx-nm laser chip.

compound semiconductor grown by an epitaxial growth technique such as MOCVD (metal-organic chemical vapor deposition) or MBE (molecular beam epitaxy). A GRIN-SCH-CS-MQW-active region is suitable for high-power-output operation. The BH structure is useful to minimize leakage current and also to get stable single lateral mode operation. Asymmetrical coatings comprising AR (anti-reflective) and HR (high-reflective) coatings are fabricated on the front and rear facets, respectively, to get higher power output from the front facet of the laser. Cavity length and width of the active region are important parameters for high-power performance. Details of each item are described in the following sections.

### 6.1.2.1  Active Region

To enhance the stimulated emission for laser operation, population inversion by current injection in semiconductor lasers is an important phenomenon. To achieve effective population inversion, the double-heterostructure (DH) laser [14] has been widely used as a basic structure

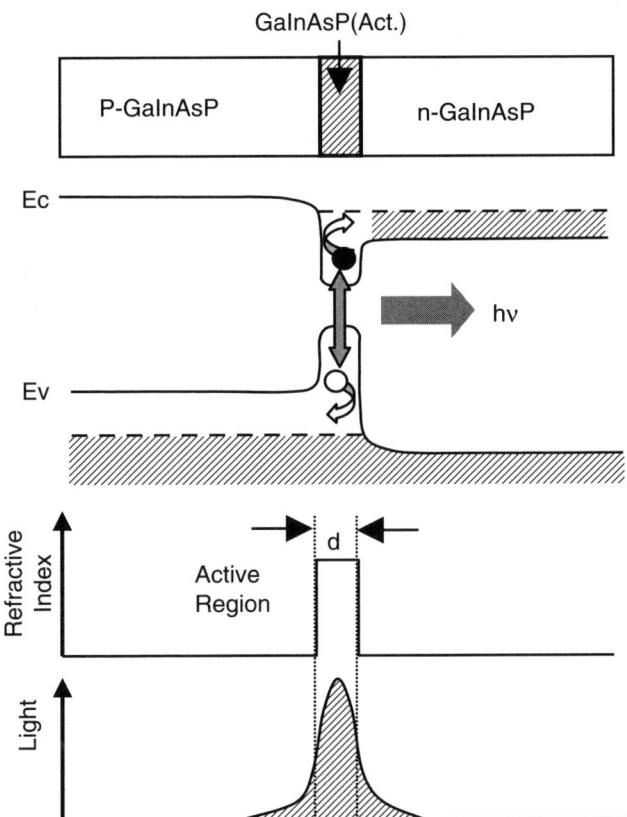

**Figure 6.4:** Schematic representations of the band diagram under a forward-bias condition, refractive index profile, and optical-field distribution of light generated at the junction of a DH laser.

for highly efficient laser diodes. The DH structure consists of a thin semiconductor active layer sandwiched between two layers of different semiconductor materials. Figure 6.4 represents the schematic of the band diagram under a forward-bias condition, the refractive index profile, and the optical-field distribution of the light generated at the junction of a DH laser.

Because the refractive index of the active layer is larger than the index of its surrounding layers, a three-layer optical waveguide is formed in parallel to the layer interfaces. Thus, the laser light is confined inside the active region and propagates along the cavity direction. As the bandgap energy of the active layer is lower than other layers, the carriers are also confined inside the active region, where radiative recombination occurs by the heterojunction barriers to prevent the carriers from moving out of the active region without recombination.

Using the so-called quantum-well (QW) structure for the active region in a DH laser is one of the most popular techniques to improve the laser performance such as low threshold current and high power operation. The QW structure consists of several well layers separated by barrier layers. The band gap energy of well layers is lower than that of barrier layers. In practice, the well thickness is set to be below 20 nm to take advantage of the quantum-size effect [15] of the carriers. When a single QW is employed, the structure is called a single-quantum-well (SQW) structure. However, when the number of QWs is two or more, we call it a multi-quantum-well (MQW) structure.

Further improvement can be realized by introducing strain in the QW layer. Strain occurs when the lattice constant of the layer is different from that of the substrate by a few percentage points. Generally, each epitaxial layer should be grown under the lattice-matched condition so as not to produce dislocations in the crystalline. Here, the lattice-matched condition is defined such that the lattice constant of the layer in the epitaxial plane coincides with that of the substrate. However, one can grow a lattice-strained layer without any dislocations when the layer is so thin that its thickness is only a few nanometers. Such strained QW layers can improve the laser characteristics considerably because one can arbitrarily control the valence band structure of the QW structure through the strain. More specifically, the strain is expected to reduce the intervalence-band absorption (IVBA) and Auger recombination in the GaInAsP/InP material [16–18].

Compared to a lattice-matched MQW active layer, a compressive-strained MQW layer provides a larger optical gain at the same injection current. Therefore, such a laser can be operated even with a small optical confinement factor because of a large optical gain. Here, the optical confinement factor represents the ratio of the optical intensity distributed in the active layer to the total optical intensity. The resulting small internal loss of the laser cavity leads to a high slope efficiency, defined as the ratio of the optical output change to the change of injected current.

A GRIN-SCH structure has the advantage of higher carrier confinement into the active region in comparison with a single-step SCH structure or the structure without SCH. This enables both low-threshold and high-power operation. Thus, compressive-strained MQW with a GRIN-SCH structure is suitable to realize high-power semiconductor lasers.

### 6.1.2.2   BH Structure

The BH structure shown in Figure 6.3 consists of the p–n–p InP current blocking layers, formed around the active region. We note that a heterojunction is formed at the interface between the active region and the current-blocking layer. The fundamental transverse mode operation is stabilized by the large refractive index difference between the active region and the current-blocking region. Further, since the current-blocking region also acts as the p–n–p–n thyristor, injection current is prevented from flowing into this region. Thus, we can control the leakage current and make it relatively small. When the BH structure is applied for fabricating a high-power laser, one should carefully optimize the other design parameters such as impurity concentration and layer thickness to obtain the high threshold of thyristor turn-on voltage.

It is very important to control the oscillation of the higher order transverse mode even at high power levels. When the laser operates in a higher order of a transverse mode, a kink appears in the light-current characteristics, as seen in Figure 6.5. Such kinks drastically affect the coupling efficiency between the laser chip and the fiber pigtail within the pump laser module and should be avoided. In practice, unstable operation is caused by changes in the refractive index due to the carrier-induced plasma effect [19], temperature variations induced by the injection current, and the spatial hole burning [20] under high photon-density operation. To suppress the influence of such effects, an index-guided structure such as the BH structure is useful. In an index-guided structure, the refractive index difference is large in the lateral direction perpendicular to the laser cavity direction. To reduce the spatial hole burning, it is necessary to design the waveguide width below the diffusion length of the injection carrier (about 2 to 3 $\mu$m) in order to produce uniform optical gain distribution. Moreover, a BH structure can be designed to produce a nearly circular beam pattern, leading to a high coupling efficiency between the laser chip and the single-mode fiber.

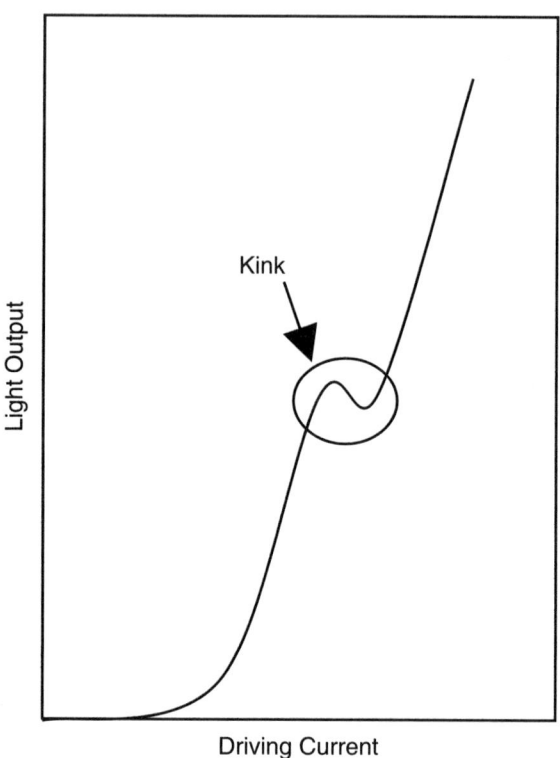

**Figure 6.5:** Kink in the L-I curve.

### 6.1.2.3  Asymmetric Coating

Generally, to obtain higher output power from the front facet, the AR and HR films are coated on the front and rear facets, respectively. When the reflectivity of the front and rear facets is denoted by $R_f$ and $R_r$, respectively, the ratio of front-facet power to the rear-facet power is given by

$$\frac{P_f}{P_r} = \sqrt{\frac{R_r}{R_f}} \cdot \frac{1 - R_f}{1 - R_r}. \tag{6.1.1}$$

In order to obtain larger front-facet power, the rear-facet reflectivity $R_r$ should be as high as possible, as is evident from Eq. (6.1.1). Typically,

$R_f$ exceeds 90%. However, a small value of $R_f$ improves $P_f$ in general. However, a low $R_f$ causes an increase in the threshold carrier density because of increased mirror losses. This change may increase the nonradiative component of the injection current, the carrier leakage from the active region, and the spatial hole burning along the laser cavity direction. Therefore, we have to optimize the reflectivity $R_f$ to get the maximum output power while considering these trade-off effects. In general, $R_f$ is in the range of 1–10%.

In practice, $\lambda/4$ coatings on the front facet consist of dielectric thin films of materials such as $SiO_2$, $SiN_x$, and $Al_2O_3$. In contrast, multiple layers of materials such as $SiO_2$ and amorphous Si ($\alpha$-Si) are applied to realize the high reflectivity of the rear facet.

### 6.1.2.4   Cavity Length

An adoption of a laser with longer cavity length gives smaller series resistance and small thermal impedance. Those result in lower power consumption and suppression of optical power saturation at a given operating current. On the other hand, slope efficiency of the laser decreases with increasing current due to internal loss of the laser cavity. However, by using a GRIN-SCH compressive-strained MQW active region, which shows small internal loss of the cavity, as described in Section 6.1.2.1, we can keep the slope efficiency sufficiently high even for a long-cavity laser.

Figure 6.6 shows the cavity length dependence of the laser driving current using light output power as a parameter. The driving current at 300-mW output power was minimized for a cavity length of 1.3 mm. When the cavity length is shorter than 1.3 mm, the driving current increases due to the effect of thermal saturation. In contrast, when the cavity length is longer than 1.3 mm, the driving current increases due to the reduction of slope efficiency.

Figure 6.7 shows the cavity length dependence of the laser driving power using light output power as a parameter. From the viewpoint of laser driving power, the advantageous feature of the laser chip with 0.8-mm cavity length (an 0.8-mm chip) does not appear even in the low output power range. It was found, however, that the laser driving power of a 1.5-mm chip was comparable to that of a 1.3-mm chip at 300-mW output power. Further, over 300-mW optical output power, the 1.5-mm chip has

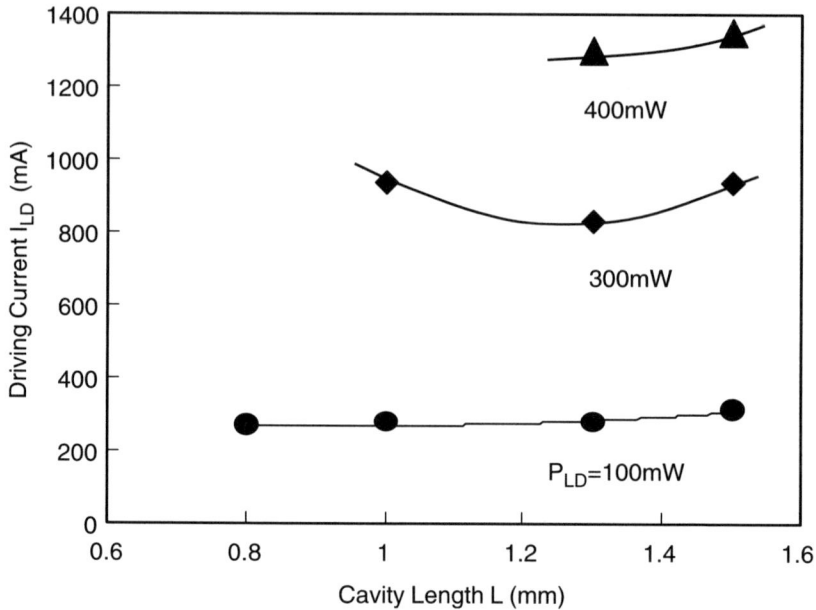

**Figure 6.6:** Driving current versus cavity length.

an advantage in the laser driving power. Thus, the cavity length of the laser for minimum driving power is longer in comparison with the case that minimizes the driving current due to the reduction of series resistance and thermal impedance.

Figure 6.8 shows the Arrhenius relationship of the median lifetime for four types of laser cavities. The median lifetime was estimated through accelerated aging tests for various conditions [12]. The 1.5-mm chip has a median lifetime longer than other lasers with shorter cavity length at the same junction temperature and also has the same activation energy of 0.62 eV. More important, the 1.5-mm chip shows higher power output at the same junction temperature due to reduced power consumption and low thermal impedance. This proves that a longer cavity laser is very effective in highly reliable operations. Clearly, consideration of the driving electrical power consumption is very important in designing the optimum cavity length.

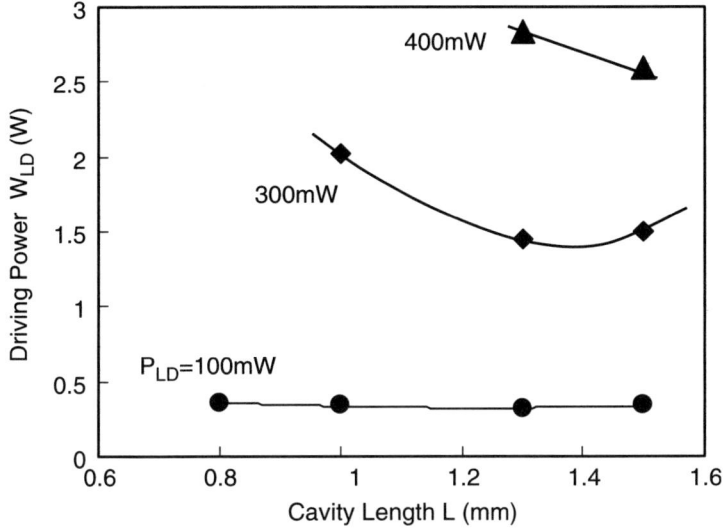

**Figure 6.7:** Driving power versus cavity length.

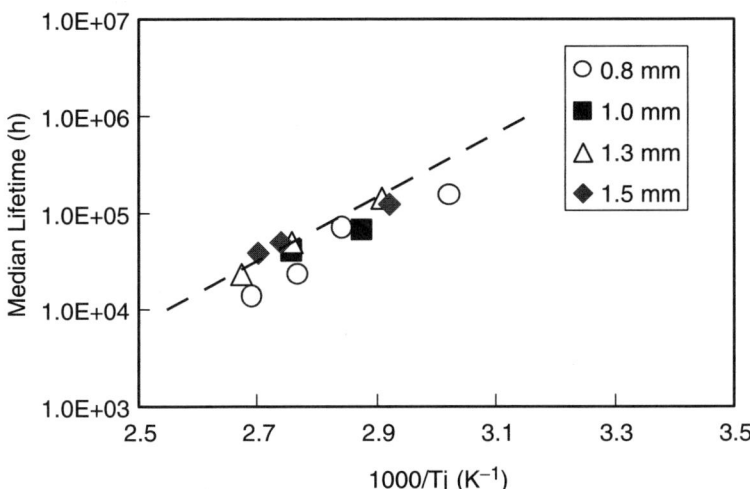

**Figure 6.8:** Arrhenius relationship of the median lifetime.

### 6.1.2.5    Width of Active Region

The small optical confinement in the active region allows a design using a lower equivalent refractive index difference between the active region and the InP cladding layers. We can set the cut-off width of the active optical waveguide wider, due to this small index difference. Here, the cut-off width is defined as the maximum width for which only one fundamental transverse mode exists. As a result, we can also reduce both series resistance and thermal impedance and hence achieve high power operation of the laser.

## 6.1.3    Other Approaches to High-Power Pump Laser

### 6.1.3.1    Ridge Waveguide Laser

A schematic of the ridge waveguide structure is shown in Figure 6.9. Ridge waveguide lasers with a refractive index difference as small as $5 \times 10^{-3}$ [21]

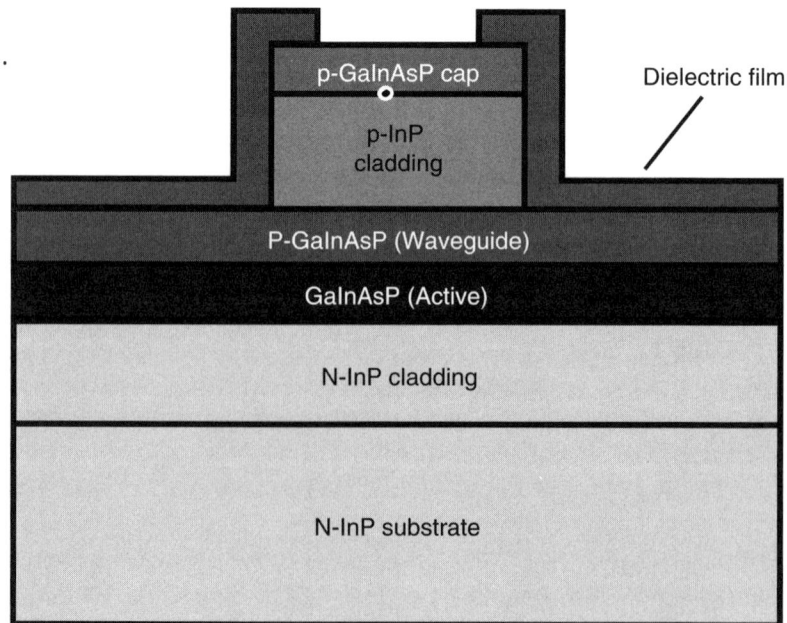

**Figure 6.9:** Cross-sectional schematic of ridge waveguide structure.

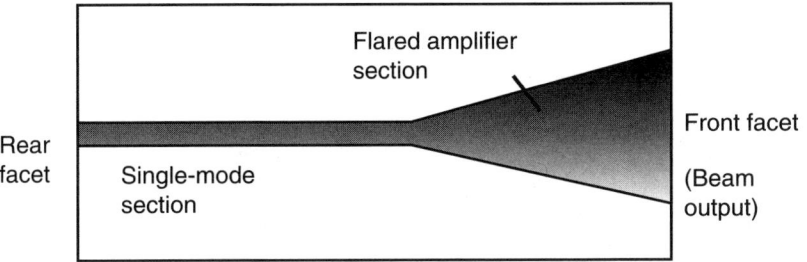

**Figure 6.10:** Flared laser structure.

operate in the index-guided regime. The small refractive index difference allows an adoption of wider active stripes, as described earlier.

As the injection current spreads inside the active region, the optical gain is distributed inhomogeneously in the lateral direction under the high injection current condition. Then the spatial hole burning arises at high power operation, resulting in unstable transverse mode oscillation. We often observe kinks in the L-I characteristics. To get stable high-power operation from a ridge-waveguide laser, it is necessary to optimize the laser structure such that kinks appear at power levels much higher than those at which the device is intended to operate. Furthermore, it is difficult to realize a high coupling efficiency, when compared to the BH structure, because of an elliptical beam pattern. The advantage is, however, the simple fabrication process.

The following is the state of the art of the ridge waveguide lasers in high power operation. By using a very long cavity length of 3 mm and a wide ridge width, a maximum fiber coupled power of 710 mW has been achieved [22]. The flared laser with ridge waveguide structure, which consisted of a single-mode section and a flared amplifier section, is shown in Figure 6.10 [23]. The flared amplifier section has a wider area than the single mode section. By using this structure, electrical resistance is remarkably reduced due to the large path area of the current flow and heat dissipation and over 525 mW of fiber-coupled power was reported [23].

### 6.1.3.2  Other Novel Structures

Other approaches for widening the active stripe of the laser are application of an active MMI (multimode-interference) laser [24] as shown in

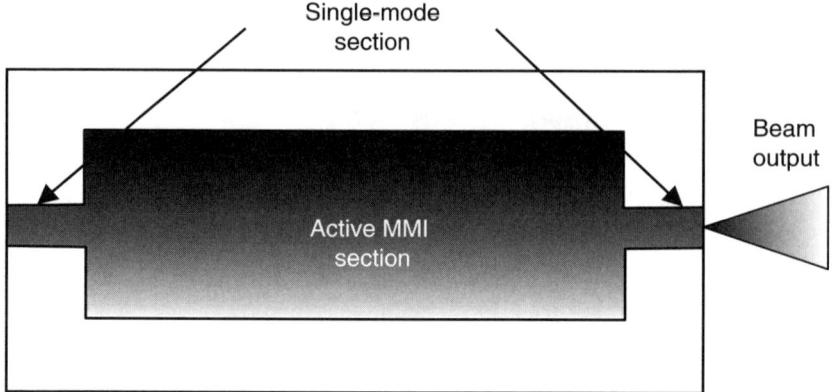

**Figure 6.11:** Active MMI laser structure.

Figure 6.11 and a BH laser with asymmetric cladding structure using different materials in each cladding layer [25].

The active MMI laser consists of a laser cavity comprising a single-mode section and a multimode section. The regular single-mode input and output waveguides are connected to a $1 \times 1$-MMI coupler (i.e., a multimode section). The optical mode field of the multimode section is coupled into the single-mode section with a very small coupling loss by the MMI effect. Also, the active multimode section is much wider than the single-mode section, resulting in a small power consumption by the laser. It is reported that the electric power consumption was reduced by 50% compared to a conventional single-stripe laser [24].

An asymmetric cladding laser [25] has an n-GaInAsP cladding layer with a 7.5-$\mu$m layer thickness, of which the bandgap wavelength is 0.95 $\mu$m and a p-InP cladding layer. As the refractive index of the GaInAsP cladding layer is larger than that of InP cladding, the intensity of the electric field of the light distributes toward the n-InP substrate to the active region. This asymmetric structure enables the reduction of the optical loss caused by intervalence-band absorption in the p-cladding layer. Further, the refractive index difference between the region involving the active region and the region involving the current blocking region becomes smaller in comparison with a conventional structure having an InP cladding layer. Thus the active region width could be expanded with maintaining the single transverse mode operation. It is reported that the laser chip provides an output of over 1 W at 3.5 A with a laser cavity of 3 mm and an active region width of 3.5 μm.

### 6.1.3.3   New Material

The other approach would be changing basic materials. Recently, by use of the AlGaInAs/InP-based ridge-waveguide structure, over 500 mW of fiber-coupled power was realized [26]. This high-power operation was achieved by improvement in the temperature dependence of the light-current characteristics. AlGaInAs/InP materials have a larger conduction band offset compared to that in InGaAsP/InP, providing strong carrier confinement that leads to lower carrier leakage at high operating junction temperature.

## 6.1.4   Heat Exhaustion of the Laser Chip

Larger power consumption and higher junction temperature of the laser chip worsen the reliability. Thus, it is very important to effectively dissipate the heat generated in the active region of a high-power semiconductor laser.

For the effective heat dissipation, the laser chip was bonded onto the heat sink under the epi-side down (the so-called junction down) condition. The desirable materials for a heat sink is that the thermal conductivity is large in terms of high-power operation and that the thermal expansion coefficient is close to that of semiconductor material in terms of reliability. Generally, diamond and AlN are very popular materials for the heat sink. Table 6.1 represents the material parameters of thermal conductivity and thermal expansion coefficients. The maximum operating ambient temperature is determined by the following three factors:

   (i)  the power consumption of the laser chip,
  (ii)  the thermal impedance throughout the heat dissipation path inside the laser module,
 (iii)  the heat absorption performance of TEC.

As the power consumption of the laser chip becomes high with increasing operating current, the power consumption of TEC increases drastically. Then it is difficult to operate at the high ambient temperature.

**Table 6.1 Material properties**

| Materials | Thermal Expansion Coefficient $(K^{-1})$ | Thermal Conductivity (W/m K) |
|-----------|------------------------------------------|------------------------------|
| Diamond | $2.1 \times 10^{-6}$ | 1000 |
| AlN | $4.5 \times 10^{-6}$ | 200 |
| AuSn | $1.75 \times 10^{-6}$ | 44 |
| Au | $1.42 \times 10^{-6}$ | 315 |
| InP | $4.5 \times 10^{-6}$ | 70 |

Thus, TEC capability should be optimized according to the laser-driving power consumption [12, 13] and the required ambient temperature.

## 6.1.5 Optical Coupling System

Launching optical power from a laser into a single-mode fiber entails considerations such as the numerical aperture, core sizes, refractive index profiles, and core-cladding index difference of the fiber, plus the size, radiance, and angular power distribution of the optical source. Fiber coupled power is given by

$$Pf = \eta Po, \qquad (6.1.2)$$

where, $Pf$ and $Po$ indicate the power coupled into the fiber and the power emitted from the pump source, respectively, and $\eta$ is the coupling efficiency. The coupling efficiency depends on the radiation patterns of the light source and on the coupling system.

Figure 6.12 shows examples of the coupling system. In the single discrete lens system as shown in Figure 6.12a, the optical beam is focused by the lens onto the core facet of the optical fiber. In the two discrete lenses system as shown in Figure 6.12b, the optical beam is collimated by the first lens and then is focused by the second lens on the core facet of the optical fiber. The laser chip is spatially coupled to the optical fiber; therefore, it is easy to insert optical bulk components such as an optical isolator between the lenses and optical fiber, which prevents the reflection of the light at the end of the fiber. In the lensed-fiber system, the launching power is directly

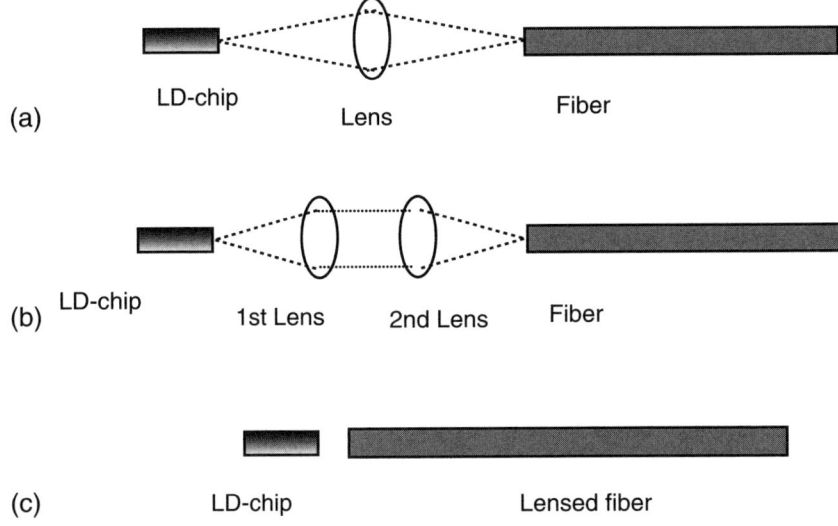

(a)  LD-chip          Lens          Fiber

(b)  LD-chip     1st Lens     2nd Lens     Fiber

(c)  LD-chip               Lensed fiber

**Figure 6.12:** Optical coupling systems: (a) single discrete lens system, (b) two discrete lenses system, and (c) lensed fiber system.

coupled to the optical fiber as shown in Figure 6.12c. The position accuracy of these optical components should be within 1 to 10 μm. In particular, the core position of optical fiber should be fixed with submicron accuracy.

The BH laser launches a light beam in a cone angle of approximately 20 degrees. Discrete lens systems as shown in Figures 6.12a and 6.12b are widely used. However, the ridge waveguide laser launches a light beam in an elliptical radiation pattern. Thus, it is difficult to use a discrete spherical lens system because of a drastic decrease of coupling efficiency. A widely used coupling system is the lensed-fiber system.

## 6.1.6    Performance of 14xx-nm Pump Lasers

### 6.1.6.1    Characteristics of Pump Laser Module

Figure 6.13 shows the L-I characteristics of the module. The laser has a 1.5-mm-long cavity and is operated at a laser chip temperature of 25°C. Fiber-coupled power of 400 mW was obtained at an operating current

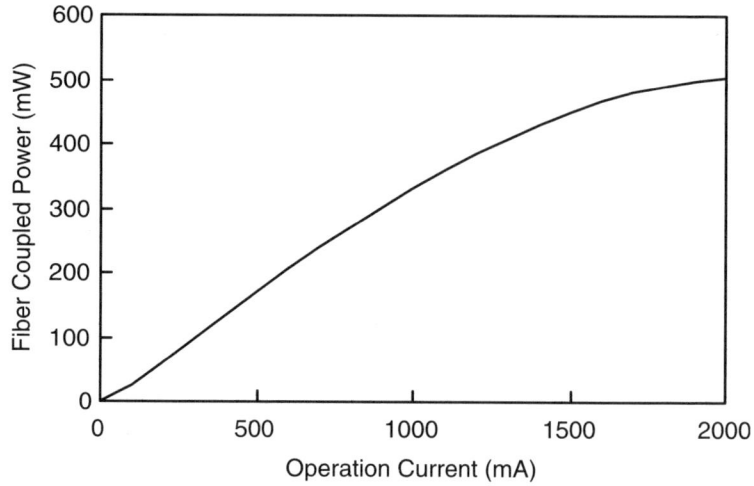

**Figure 6.13:** L-I characteristics of pump laser module.

of 1270 mA. We cannot observe kinks in the L-I curve and the maximum output power of half-watt was limited by thermal roll-off. Coupling efficiency between the laser chip and the single-mode fiber was 80%.

The relation between the fiber-coupled power and laser-driving power is shown in Figure 6.14. Figure 6.15 shows the cooling capacity of the TEC of this pump-laser module. It is remarkable that this pump-laser module can be operated at up to 75°C of case temperature. The pump-laser module consumes electrical power of 2.8 W at a fiber-coupled power of 400 mW as shown in Figure 6.15. At this time, as Figure 6.15 demonstrates, TEC draws an electric power of 7.2 W when case temperature should sustain at a high temperature of 75°C.

### 6.1.6.2    Reliability

Figure 6.16 shows the long-term life test results by plotting changes in operating current of 18-laser chips with 1.5 mm of cavity length. The test condition is 80% of maximum output power at 60°C [12]. All lasers exhibited stable operation, and any random sudden failures have never been observed for over 5000 h.

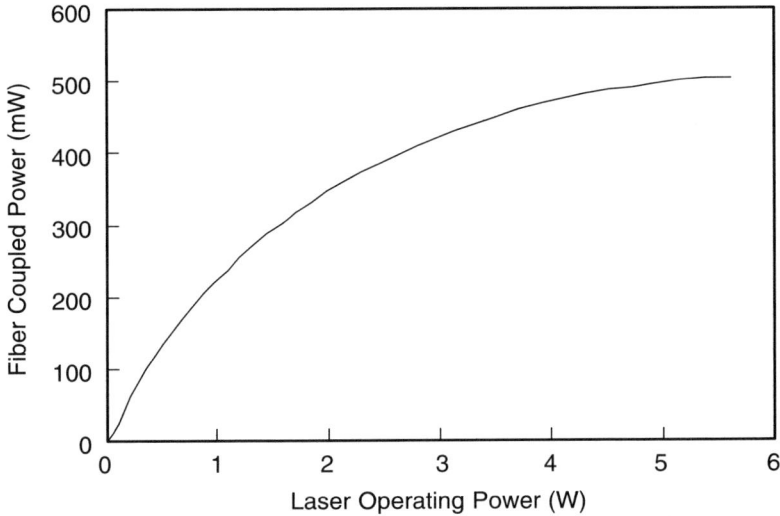

**Figure 6.14:** L-W characteristics.

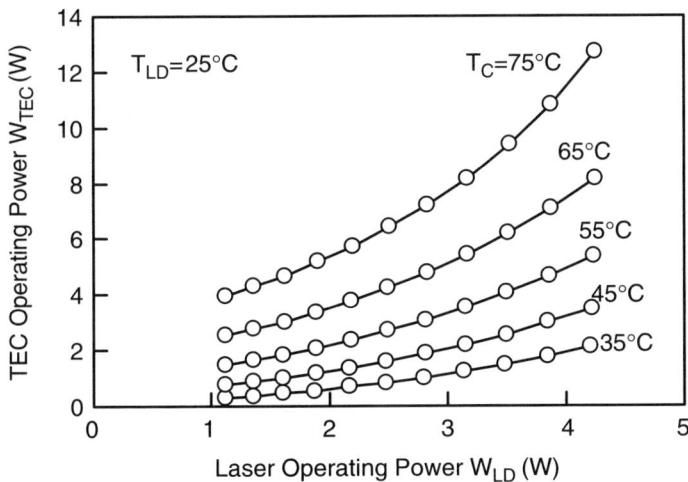

**Figure 6.15:** TEC operating power.

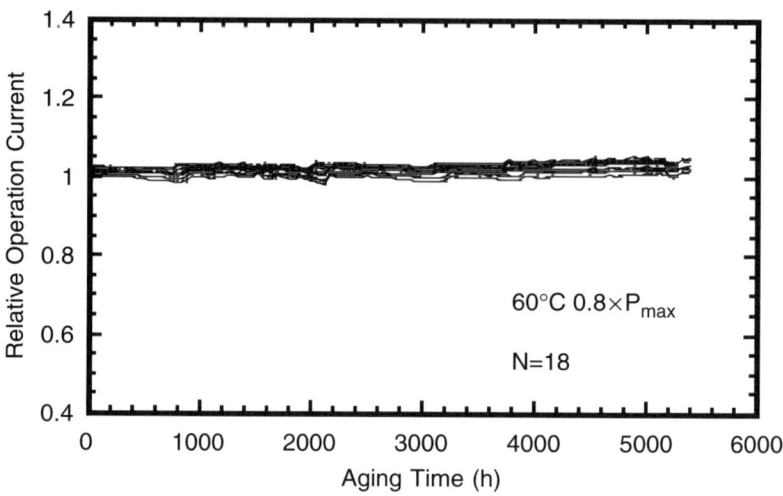

**Figure 6.16:** Life test drift plots.

## 6.2  Semiconductor Pump Lasers for Raman Amplifiers

In this section, several structures used for semiconductor pump lasers designed for Raman amplifiers are described. Table 6.2 shows a summary of the requirements for such pump sources needed for Raman amplification systems. An appropriate choice of pump lasers is necessary in accordance

**Table 6.2 Brief summary of requirements for semiconductor pump laser sources**

| Characteristics | Counterpumps | Copropagating pumps |
|---|---|---|
| Optical output power | 100~300 mW | 100~300 mW |
| Reliability | Telecommunication grade less than 1000 FIT @ 25 years | Telecommunication grade less than 1000 FIT @ 25 years |
| Wavelength stability | Important for WDM pumping | Important for WDM pumping |
| RIN | Not so important | Important |
| SBS | SBS free | SBS free |
| DOP (availability for depolarizing) | Not so important | Important |

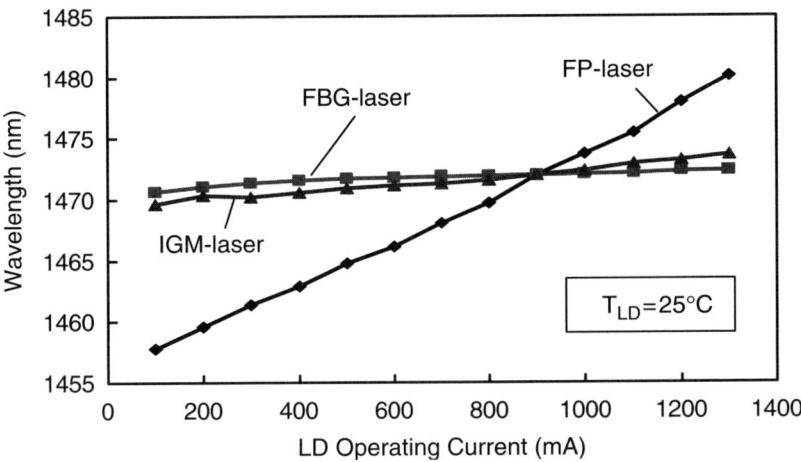

**Figure 6.17:** Operating current dependencies of oscillation wavelength of laser modules at 25°C of $T_{LD}$.

with the performance requirements for the system to operate reliably. Currently, semiconductor pump lasers used commonly for counterpumping and copumping are the Fiber Bragg grating (FBG) laser and the Fabry-Perot (FP) laser, respectively. Figures 6.17 to 6.20 show the comparison of important characteristics for Raman amplification systems designed with various pump lasers [27, 28] described in the following sections.

## 6.2.1   Fiber Bragg Grating Lasers

The FBG laser [9, 29] is a laser module whose wavelength has been stabilized by using an FBG. The module contains an FP semiconductor laser chip and an external FBG that is fabricated in the fiber pigtail as shown in Figure 6.21. It exhibits excellent wavelength stability, sufficiently narrow lasing spectral width, very high stimulated Brillouin scattering (SBS) threshold power, and comparatively large relative intensity noise (RIN), the features shown in Figures 6.17 to 6.20, respectively. Because of these features, it is currently the most widely used pump laser for counterpumping a Raman amplifier. The FBG laser has coupled-cavity design, comprising

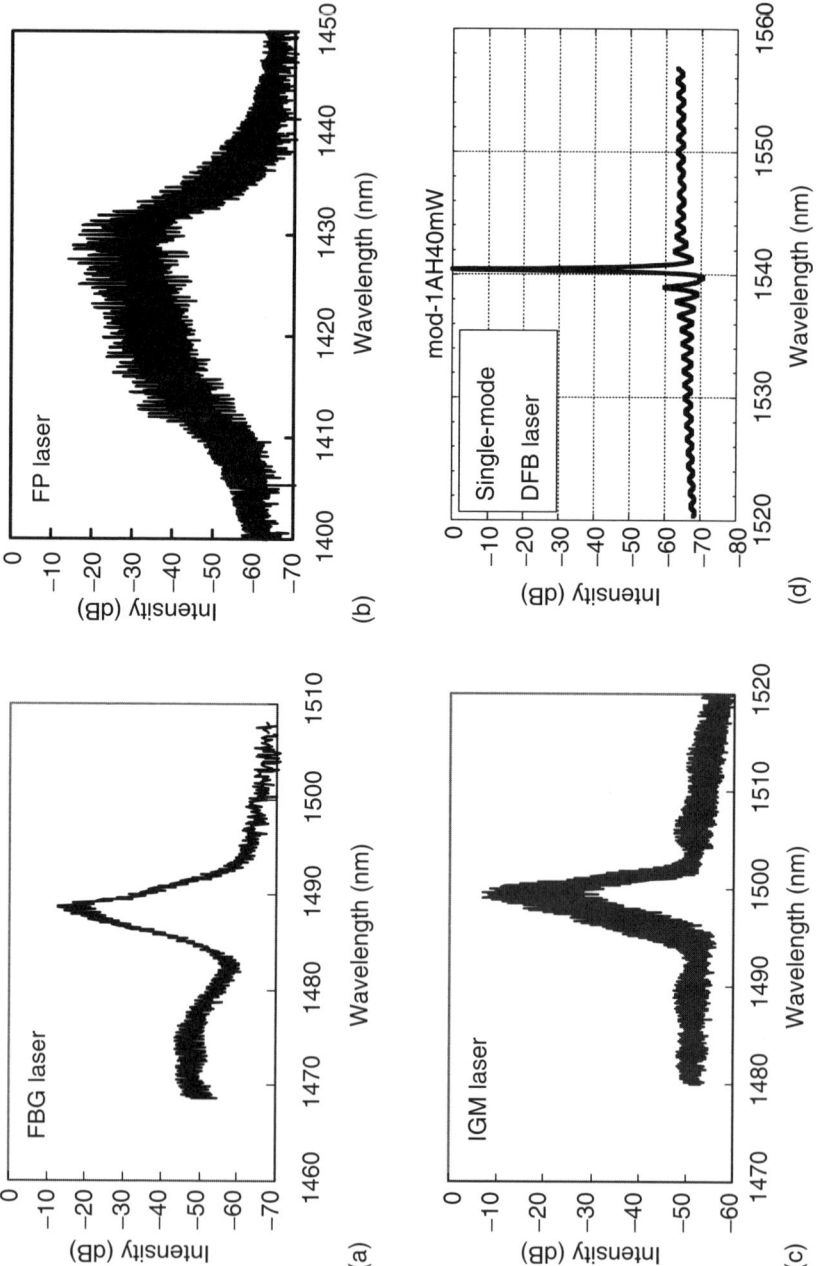

**Figure 6.18:** Optical output spectra of different types of pump lasers.

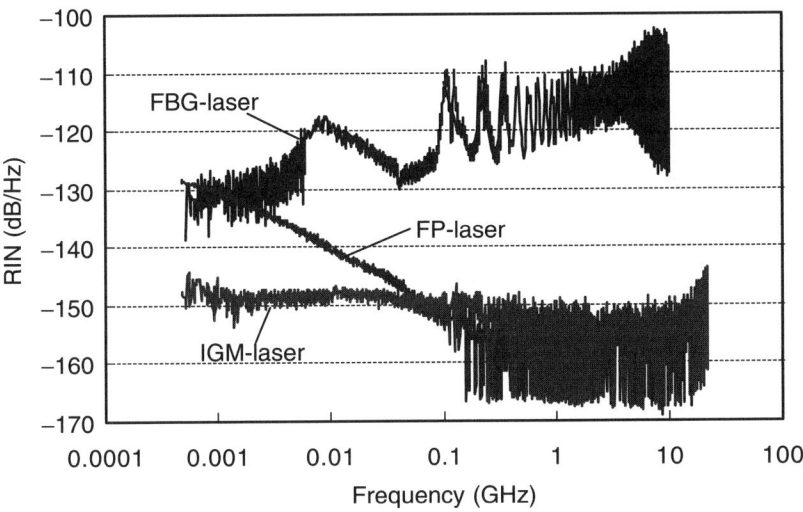

**Figure 6.19:** Comparison of RIN characteristics.

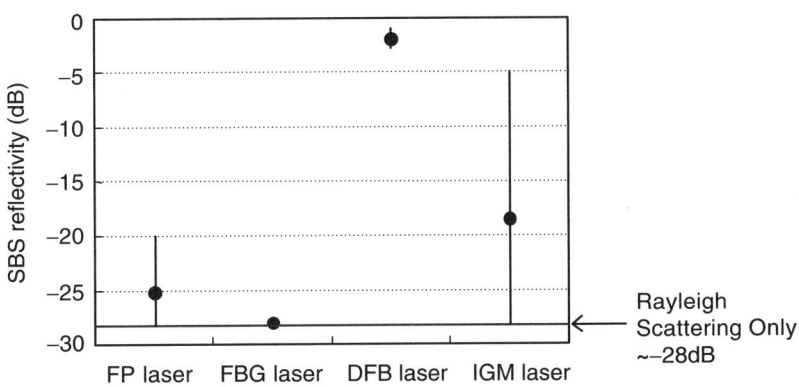

**Figure 6.20:** Comparison of SBS performances for a 55-km DSF for transmission line with the transmission loss of 0.21 dB/km at 1550 nm, the dispersion of $-0.07$ ps/km/nm at 1550 nm, the $A_{\text{eff}}$ of 47 $\mu$m$^2$ at 1550 nm, and the zero-dispersion wavelength of 1551 nm.

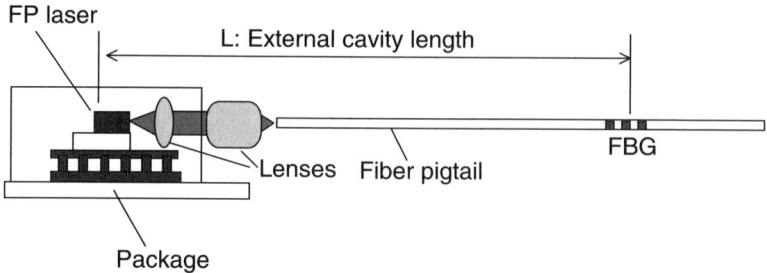

**Figure 6.21:** Schematic of the FBG laser.

the cavity of the FP laser chip and the external cavity between the FBG and the FP laser chip. The FP laser has an anti-reflection coating on its front facet to reduce the influence of the laser's own FP modes. A narrow oscillation spectrum according to the bandwidth of FBG is obtained. The spectrum of the FBG laser shown in Figure 6.18a consists of FP modes and external cavity modes, and its envelope is dictated by the reflection spectrum of the FBG.

The FBG laser for Raman pumping is generally designed to show multiple longitudinal modes in the oscillation spectrum [29]. One can see the difference between single and multilongitudinal mode spectra in Figure 6.18a and 6.18d. The module has advantages such as good linearity in L-I characteristics, high SBS threshold power, and fitness for depolarizing. In the case of a single-mode FBG laser, a nonlinearity (kink) in the L-I characteristics, resulting from longitudinal mode hopping, is observed, as shown in Figure 6.22. However, if a large number of longitudinal modes exist within the oscillation bandwidth of the FBG laser, the impact of optical power fluctuations from longitudinal mode hopping is averaged by power distribution in various modes, and a smooth L-I characteristic is obtained.

The optical beam exiting from the semiconductor laser chip is strongly polarized. So, for using such a laser as a Raman pump source, a technique to get a low degree of polarization (DOP), such as depolarizing by using a depolarizer or polarization combining two laser beams by a polarization beam combiner (PBC), is used. The minimum required length of polarization maintaining fiber (PMF) used as a depolarizer is inversely proportional

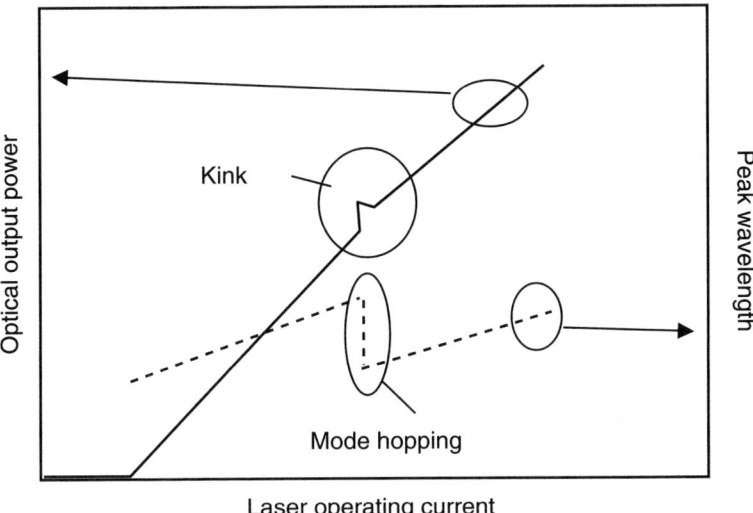

**Figure 6.22:** L-I and wavelength-current characteristics of single-longitudinal-mode FBG laser.

to the oscillation bandwidth of the laser [30, 31]. For this reason, a laser with multiple longitudinal modes has the advantage for depolarizing.

Generally, the RIN of an FBG laser is relatively large, as shown in Figure 6.19. The periodic peaks in the RIN spectrum of the FBG laser originate from resonances between the FBG and the laser chip. Such peaks appear about every 100 MHz, which corresponds to the distance between the laser chip and the FBG of around 1 m. The resonance frequency interval is given by the following equation:

$$\Delta f = \frac{c}{2 \times n_{\text{eff}} \times L}. \tag{6.2.1}$$

Here, $n_{\text{eff}}$ and $L$ are an effective refractive index and the length of the external cavity between the laser chip and the FBG, respectively (see Figure 6.21). Further, $c$ is the velocity of light in a vacuum.

A relatively long cavity of FBG lasers is beneficial for suppressing SBS, as shown in Figure 6.20. The FBG laser is operated in the coherence collapse state [32, 33]. In other words, the distance between the FBG and

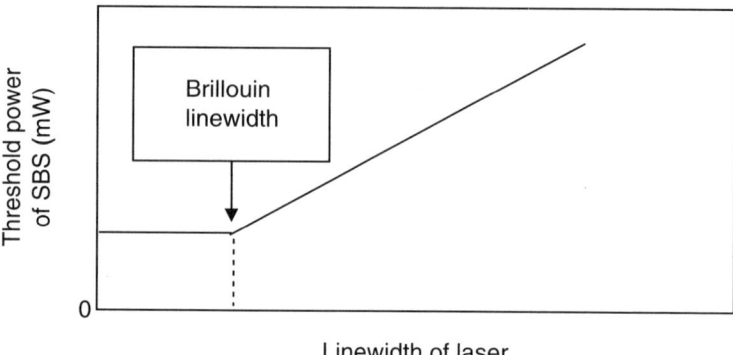

**Figure 6.23:** Theoretical relationship between the linewidth of laser and SBS threshold power.

the laser chip is longer than the coherence length of the FP laser beam. This condition makes the phase of optical feedback from the FBG to the laser chip random, and consequently, the linewidth in each longitudinal mode becomes very wide because the phase in the laser beam is washed out by random phase variations. As a result, an FBG laser has a much wider linewidth compared with other types of lasers, and this feature leads to a very high SBS threshold power.

The threshold power of SBS is approximated by the following simple expression [35],

$$P_{\text{th}} = 21 \frac{A_{\text{eff}}}{G_b L_{\text{eff}}} \left\{ 1 \ (F_l < F_b), \frac{F_l}{F_b} \ (F_l > F_b) \right\}, \tag{6.2.2}$$

where $F_l$ is the spectral linewidth of the laser, $F_b$ is the Brillouin linewidth, $A_{\text{eff}}$ is the effective area of fiber, $L_{\text{eff}}$ is the effective length of fiber, and $G_b$ is the material parameter representing Brillouin gain. As shown in Figure 6.23, when the linewidth is narrower than the Brillouin linewidth, SBS threshold power is constant. And when the linewidth is wider than the Brillouin linewidth, SBS threshold power increases proportionally to the linewidth of the lasing mode.

Equation (6.2.1) implies that the resonance frequency can be increased by shortening the cavity length $L$. By attaching an FBG to a butterfly package (i.e., using shorter $L$), a 10-dB reduction of RIN up to 5 GHz has been reported [34]. However, the adoption of a shorter external cavity length often results in the nonlinear light-current characteristics. A linear L-I characteristic can be obtained by using two FBGs with slightly different reflection peak wavelengths. For practical use of the laser, one must consider both the SBS threshold and the RIN performance and design the laser with a proper trade-off between the two.

## 6.2.2 Fabry-Perot Lasers

The detailed structure of FP lasers is described in Section 6.1. Such lasers exhibit relatively good RIN and SBS performance, as shown in Figure 6.19 and 6.20, and thus they can be used as a pump source for copumped Raman amplifiers. To get excellent RIN performance, suppression of backward reflected light to the laser chip is important. For this purpose, an AR coating is fabricated onto the surface of optical components such as lenses, windows of the package, and the edge of the fiber pigtail. Also, angled fiber facets and an integrated optical isolator are often used. The mean oscillating wavelength of this laser depends strongly on the operating current. The dependency is several times larger than that of FBG lasers as shown in Figure 6.17. Variations in the wavelength of FP lasers are due to an increase in the junction temperature of the active region with increasing current, and they are dominated by temperature dependence of the bandgap of the active region. For Raman amplifiers, such wavelength variations are often compensated by decreasing laser temperature as the operating current is increased.

Recently, a wavelength-stabilized FP using a wavelength locker, containing a wavelength monitor PD and a short-wavelength-pass filter, was proposed [36]. The schematic of such a laser is shown in Figure 6.24. The transmission of the short-wavelength-pass filter changes depending on the wavelength. So, the detected power ratio of the power-monitor PD and the wavelength-monitor PD (that detects transmitted light) changes depending on the oscillation wavelength of the FP laser. Consequently, if the ratio of power-monitor current and wavelength-monitor current is kept constant by using both APC and automatic wavelength (AWC) to control the laser operating current and the TEC operating current, respectively, the oscillation wavelength of the FP laser can remain fixed.

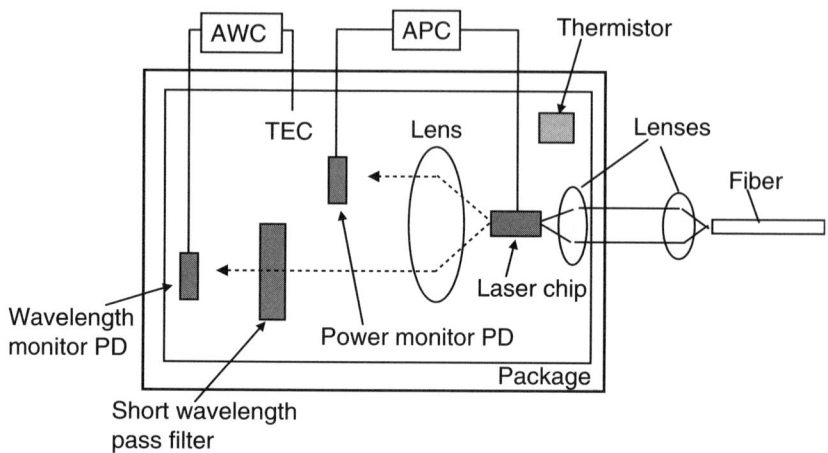

**Figure 6.24:** Schematic of the pump laser with wavelength locker.

## 6.2.3 Inner-Grating Multimode Lasers

The inner-grating-stabilized multimode (IGM) laser [27, 28, 37, 38] is characterized by a grating structure built into the laser cavity. The wavelength of the laser is stabilized by an internal grating that is optimized to select several longitudinal modes as shown in Figure 6.18c. The IGM laser has been proposed as a pump laser for the copumped Raman amplifier because of its features such as low noise and wavelength stability [39]. Another design similar to the IGM laser employs the multimode operation of a distributed feedback (DFB) laser and is called a multimode DFB laser [40].

The package structure of the IGM laser is similar to that shown in Figure 6.2. The IGM laser could be operated up to 400 mW without a kink in L-I characteristics by using an optimized inner-grating structure. By eliminating external FBG, the low-frequency resonance according to the external cavity between the laser chip and the FBG is also eliminated; low RIN characteristic could then be obtained, as shown in Figure 6.19. An optical isolator can be integrated inside of the package because the laser chip does not need optical feedback from the external resonator.

The shift in the oscillation wavelength with increasing operating current is dominated by the thermal expansion and the temperature dependence of the refractive index in the internal grating layer. The dependence is about four times smaller than that of the FP laser as shown in Figure 6.17. More important, the laser is designed to ensure multilongitudinal-mode oscillation that helps in reducing SBS. The SBS light strongly aggravates the noise performance of Raman amplifiers.

The SBS threshold power of the IGM laser is lower than that of the FBG laser, because the spectrum linewidth of each mode in the IGM laser that works coherently is much narrower than that of the FBG laser (see Section 6.2.1). It was shown that the threshold power of SBS strongly depended on the number of longitudinal modes. Figure 6.25 shows the relation between SBS reflectivity and the number of longitudinal modes in the IGM laser [38]. In this experiment, a 55-km-long dispersion shifted fiber (DSF) with a loss of 0.21 dB/km, a dispersion of $-0.07$ ps/km/nm, an effective core area $A_{eff}$ of 47 $\mu m^2$ (all at 1550 nm), and a zero-dispersion wavelength at 1551 nm was used for the transmission line. Here, the number of longitudinal modes is defined as the number of modes in the spectral width at 10 dB down from the peak. The SBS reflectivity tends to decrease with increasing number of longitudinal modes in the IGM laser, as shown in Figure 6.25. Especially, the IGM lasers, having more than 18 longitudinal modes, exhibit sufficiently suppressed SBS reflectivity (below $-29$ dB, which corresponds to Rayleigh scattering). It is confirmed that SBS free

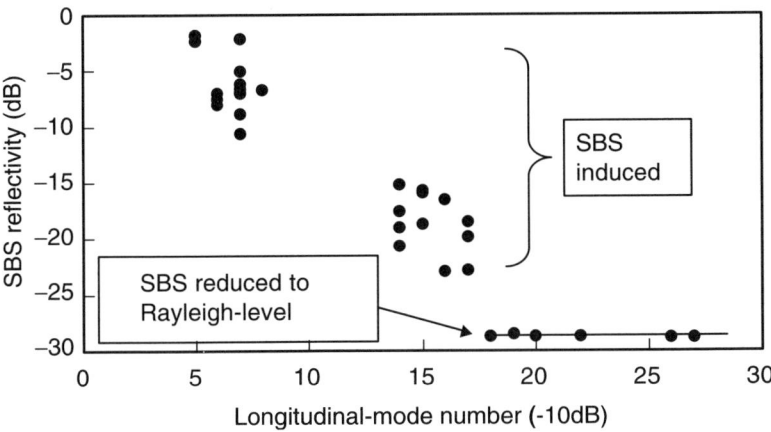

**Figure 6.25:** Dependency of SBS reflectivity on the number of longitudinal modes.

operation up to 270 mW of fiber input power could be obtained by using an IGM laser with over 18 longitudinal modes.

It was shown through detailed investigations that an increase in the number of modes results in a wider linewidth of each longitudinal mode. It is thought that mode competition is responsible for this linewidth broadening effect. Thus, SBS is reduced by the following two factors: one is power distribution in several longitudinal modes and the other is the increase of SBS threshold power in each longitudinal mode [38].

It was shown that frequency modulation (dithering) of an IGM laser is also effective for suppressing SBS [28, 41]. With frequency modulation, the spectral linewidth of the laser is broadened because optical cavity length of a laser chip fluctuates as a result of thermal variations related to injected current fluctuations. This is a well-known technique for suppressing SBS in the case of DFB lasers [42]. Figure 6.26 shows an example of experimental results for SBS suppression by frequency dithering. Here the modulation index is defined as the ratio of the modulation amplitude of optical power to the average output power. When a dithering at 1% of the modulation index was applied, the SBS was successfully suppressed down to the Rayleigh scattering level (i.e., the background level) over the dithering frequency range of 10 to 100 kHz. Further suppression could be achieved at 10%

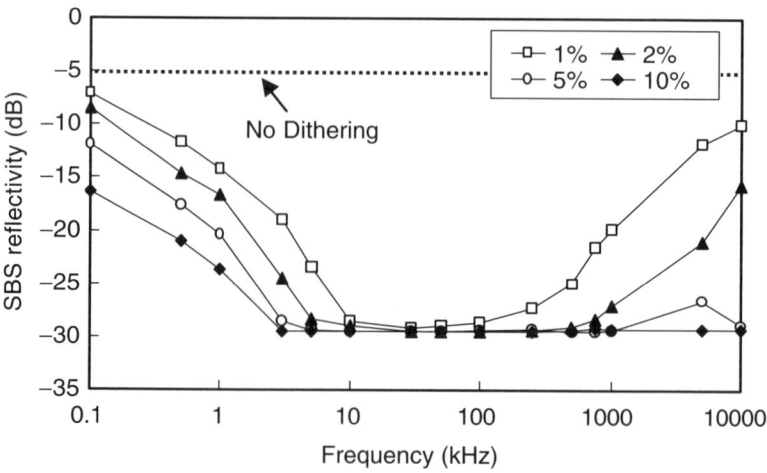

**Figure 6.26:** SBS suppression versus dithering frequency in IGM laser.

modulation index for a wider range extending from 3 kHz to 10 MHz. Frequencies less than 3 kHz are not effective because they are too slow in comparison with the response time of SBS. Also, applying frequencies higher than 10 MHz is not effective because the amplitude of temperature change in an active region caused by current injection is reduced by the delay of thermal response due to thermal impedance.

It is known that the RIN of multilongitudinal-mode lasers increases during transmission along a dispersive fiber line due to the mode partition noise (MPN) [43]. It was shown that the increase in RIN after fiber transmission for an IGM laser is smaller than that for an FP laser, which is attributable to the smaller number of longitudinal modes in the IGM laser compared with that of the FP laser [28].

### 6.2.4 Hybrid Pump

A hybrid pump [44] is a pump with a low DOP and a very high fiber-coupled power (over 1 W) realized by integrating a two-stripe laser chip with a PBC in the package, as shown in Figure 6.27. It is also advantageous that a hybrid pump needs a small assembly area in comparison with the conventional pump sources with a low DOP. The PBC block is located between the first and the second lenses. The two laser beams are combined by an integrated PBC and coupled into a common single-mode fiber pigtail through the second lens. Two laser stripes are electrically connected in parallel. The integrated PBC includes a half-wave plate at one input path in order to make the two input beams orthogonally polarized.

The coupling efficiency into a single-mode fiber is typically about 80% for hybrid pump lasers. This value is almost equal to the product of the insertion loss of PBC and the coupling efficiency of conventional laser modules. The active stripe of the two-stripes laser chip is twice as wide as that of a single-stripe laser. The active areas of both types have almost the same series resistance and thermal impedance per unit active area. The power consumption of the single-stripe laser chip with 2-mm cavity length and the two-stripes laser chip with 1.5-mm cavity length is compared in Figure 6.28. By using a hybrid pump with a two-stripes laser chip, about 20% reduction in power consumption of the laser is achieved at the output power of 500 mW. Figure 6.29 shows the spectrum of the hybrid pump at an output power of 600 mW, with and without FBG, respectively. The

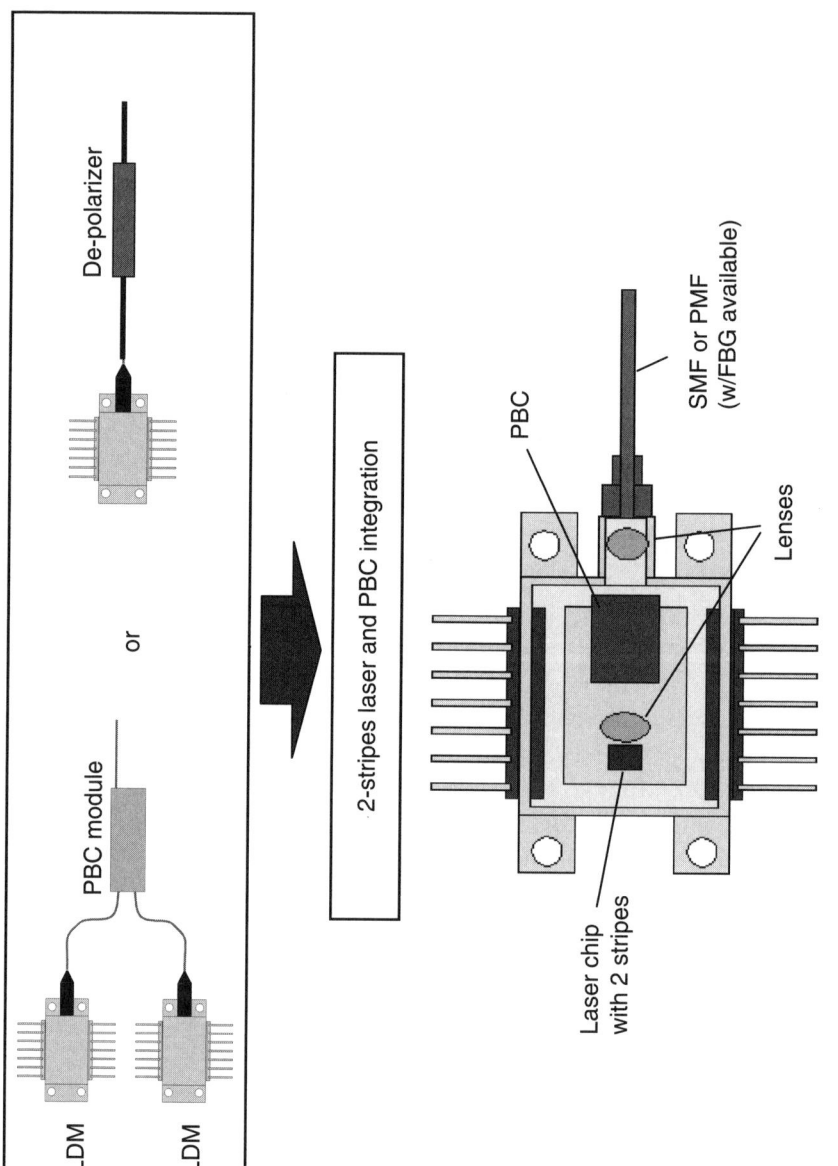

**Figure 6.27:** Schematics of the hybrid pump and conventional solutions for pump sources with low DOP.

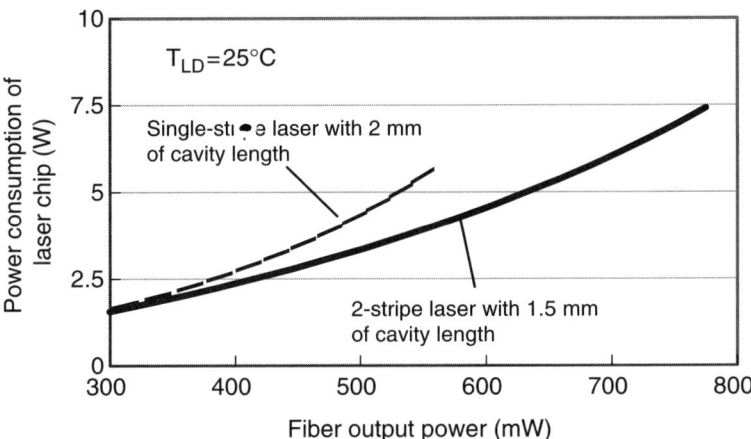

**Figure 6.28:** Power consumption of the laser chip with single stripe and two stripes.

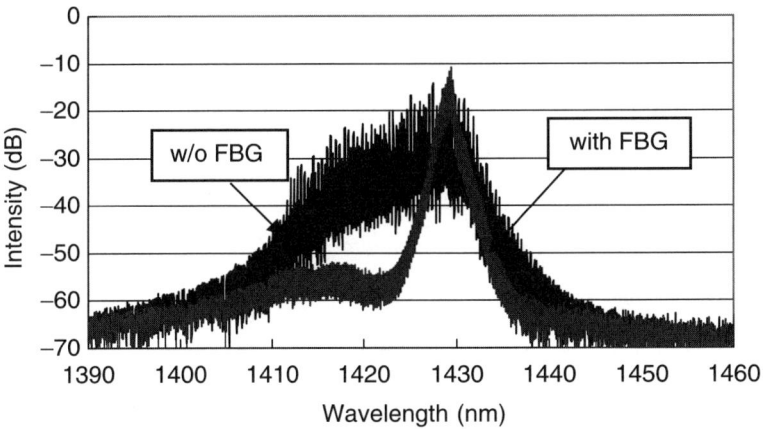

**Figure 6.29:** Spectra of hybrid pump with and without FBG.

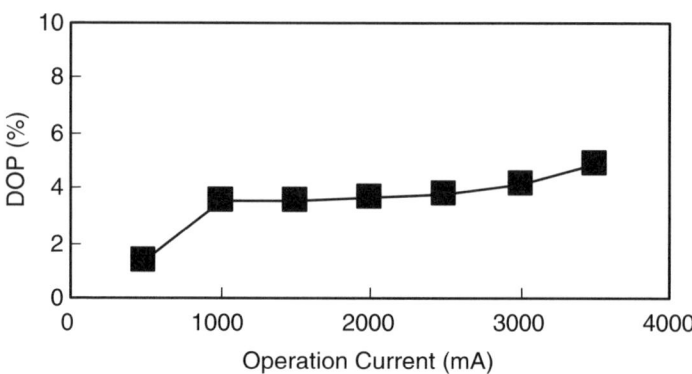

**Figure 6.30:** Operating current dependency of DOP in hybrid pump.

oscillation spectrum of the hybrid pump is stabilized by an FBG. More-over, the DOP was reduced to 5% or less in the operating current range, as shown in Figure 6.30. These characteristics are generally suitable for Raman amplifiers.

# References

[1] M. Nakazawa, Y. Kimura, and K. Suzuki, *Appl. Phys. Lett.* **54,** 295 (1989).

[2] I. Mito, H. Yamazaki, H. Yamada, T. Sasaki, S. Takano, Y. Aoki, and M. Kitamura, *Proc. 7th Int. Conf. Integrated Opt. & Opt. Fiber Commun.* Kobe, Japan (1989), 20 PDB-12.

[3] H. Asano, S. Takano, M. Kawaradani, M. Kitamura, and I. Mito, *IEEE Photon. Technol. Lett.* **3,** 415 (1991).

[4] K. Yamada, S. Oshiba, T. Kunii, Y. Ogawa, T. Kamijoh, T. Nonaka, and Y. Kawai, *Proc. Optical Amplifiers and Their Applications* (Monterey, CA, 1990), paper WA-3.

[5] H. Kamei, N. Katoh, J. Shinkai, H. Hayashi, and M. Yoshimura, *Technical Digest of the Optical Fiber Communication Conference* (San Jose, CA, 1992), p. 45.

[6] A. Kasukawa, T. Namegaya, N. Iwai, N. Nakayama, Y. Ikegami, and N. Tsukiji, *IEEE Photon. Technol. Lett.* **66,** 4 (1994).

[7] Y. Emori, Y. Akasaka, and S. Namiki, *Electron. Lett.* **34,** 2145 (1998).

[8]  Y. Emori, K. Tanaka, and S. Namiki, *Electron. Lett.* **35,** 1355 (1999).

[9]  S. Koyanagi, A. Mugino, T. Aikiyo, and Y. Ikegami, *Proc. Optical Amplifiers and Their Applications* (1998), paper MC2.

[10]  T. Kimura, N. Tsukiji, A. Iketani, N. Kimura, H. Murata, and Y. Ikegami, *Proc. Optical Amplifiers and Their Applications* (1999), paper Thd12.

[11]  N. Tsukiji, J. Yoshida, T. Kimura, S. Koyanagi, and T. Fukushima, *Proc. ITcom '01 on Active and Passive Optical Components for WDM Communication* (Denver, CO, 2001), Vol. 4532, p. 349.

[12]  J. Yoshida, N. Tsukiji, A. Nakai, T. Fukushima, and A. Kasukawa, *Proc. Photonics West '01 on Testing, Reliability, and Applications of Optoelectric Devices* (San Jose, CA), 4285, p. 146.

[13]  T. Hosoda, Y. Sasaki, H. Yamazaki, and K. Komatsu, *Proc. Optical Amplifiers and Their Applications* (1999), paper Thd11.

[14]  Y. Suematsu and A. R. Adams, *Handbook of Semiconductor Lasers and Photonic Integrated Circuits* (Chapman & Hall, London), p. 28.

[15]  G. P. Agrawal and N. K. Dutta, *Long-Wavelength Semiconductor Lasers* (Van Nostrand Reinhold, New York), Chap. 9.

[16]  A. R. Adams, *Electron. Lett.* **22,** 249 (1986).

[17]  G. Fuchs, J. Horer, A. Hangleiter, V. Harle, F. Scholz, R.W. Gelw, and L. Goldstein, *Appl. Phys. Lett.* **60,** 231 (1992).

[18]  W. S. Ring, A. R. Adams, P. J. A. Thijis, and T. van Dongan, *Electron. Lett.* **28,** 569 (1992).

[19]  F. R. Nash, *J. Appl. Phys.* **44,** 4946 (1973).

[20]  Y. Suematsu and A. R. Adams, *Handbook of Semiconductor Lasers and Photonic Integrated Circuits* (Chapman & Hall, London), p. 168.

[21]  G. P. Agrawal and N. K. Dutta, *Long-Wavelength Semiconductor Lasers* (Van Nostrand Reinhold, New York), Chap. 2.

[22]  D. Garbuzov, R. Menna, A. Komissarov, M. Maiorov, V. Khalfin, A. Tsekoun, S. Todorov, and J. Connolly, in *Optical Fiber Communication Conference* (Anaheim, CA, 2001), paper PD18.

[23]  A. Mathur, M. Ziari, M. Hagberg, in *OSA Trends in Optics and Photonics Series, OAA '99*, Vol. 30 (1999), p. 187.

[24]  K. Hamamoto, M. Ohya, K. Naniwae, in *OECC '02* (Yokohama, Japan, 2002), 10C3-7.

[25]  Y. Nagashima, S. Onuki, Y. Shimose, A. Yamada, and T. Kikugawa, *Proc. Laser & Elecro Optics Society, 2002* (Glasgow, Scotland, 2002), PD 1.4.

[26]  A. Hohl-AbiChedid, A. Rice, J. Li, X. Chen, R. Salvatore, Y. Qian, R. Bhat, M. Hu, and C. Zah, in *Proc. Optical Amplifiers and Their Applications* (Vancouver, Canada, 2002), OMB2.

[27] N. Tsukiji, J. Yoshida, S. Irino, T. Kimura, Y. Ohki, and M. Funabashi, in *OECC'02* (Yokohama, Japan, 2002), 10C3-7.

[28] Y. Ohki, N. Hayamizu, H. Shimizu, S. Irino, J. Yoshida, N. Tsukiji, S. Namiki, in *OAA'02* (Vancouver, Canada, 2002), PD7.

[29] A. Hamanaka, T. Kato, G. Sasaki, and M. Shigehara, *Proc. ECOC* (1996), p. 1.119.

[30] Y. Emori, S. Matsushita, and S. Namiki, in *Optical Fiber Communication Conference, OSA Technical Digest Series* (Optical Society of America, Washington, DC, 2000), paper FF4.

[31] S. Matsushita, J. Shinozaki, Y. Emori, and S. Namiki, in *Optical Fiber Communication Conference* (Anaheim, CA, 2002), WB3.

[32] R. W. Tkach and A. R. Chraplyvy, *J. Lightwave Technol.* **4,** 1655 (1986).

[33] C. R. Giles, T. Erdogan, and V. Mizrahi, in *OAA, Technical Digest Series* (1993), Vol.14, PD11.

[34] G. Sasaki, in *LEOS 2000* (Rio Grande, Puerto Rico, 2000), WA3.

[35] R. G. Smith, *Appl. Opt.* **11,** 2489 (1972).

[36] A. Miki, H. Yabe, J. Shinkai, H. Go, and G. Sasaki, in *OAA'02* (Vancouver, Canada, 2002), OMB3.

[37] S. Kado, Y. Emori, S. Namiki, N. Tsukiji, J. Yoshida, and T. Kimura in *ECOC'01* (Amsterdam, Netherlands, 2001), PD.F.1.8.

[38] N. Tsukiji, N. Hayamizu, H. Shimizu, Y. Ohki, T. Kimura, S. Irino, J. Yoshida, T. Fukushima, and S. Namiki, in *OAA'02* (Vancouver, Canada, 2002), OMB4.

[39] B. Zhu, L. Nelson, S. Stulz, A. Gnauck, C. Doerr, J. Leuthold, L. Gruner-Nielsen, M. Pedersen, J. Kim, R. Lingle, Y. Emori, Y. Ohki, N. Tsukiji, A. Oguri, and S. Namiki, in *OFC 2003* (Atlanta, GA, 2003), PD19.

[40] R. P. Espindola, K. L. Bacher, K. Kojima, N. Chand, S. Srinvivasan, G. C. Cho, F. Jin, C. Fuchs, V. Milner, and W. C. Dautremont-Smith, in *ECOC '01* (Amsterdam, The Netherlands, 2001) PD. F.1.7.

[41] J. Yoshida, N. Tsukiji, T. Kimura, M. Funabashi, and T. Fukushima, *Proc. ITcom '02 on Active and Passive Optical Components for WDM Communication II* (Boston, MA, 2002), Vol. 4870, p. 169.

[42] Y.-K. Park, D. A. Fishman, and J. A. Nagel, US patent no. 5329396 (1994).

[43] R. H. Wentworth, G. E. Bodeep, and T. E. Darcie, *J. Lightwave Technol.* **10,** (1992).

[44] T. Kimura, M. Nakae, J. Yoshida, S. Iizuka, A. Sato, H. Matsuura, and T. Shimizu, in *OFC'02* (Anaheim, CA, 2002), ThN5.

# Chapter 7

# Cascaded Raman Resonators

**Clifford Headley**

In the previous chapters the use of stimulated Raman scattering (SRS) to amplify signals was discussed. This chapter describes Raman fiber lasers (Rfls), which use SRS to generate light at pump wavelengths for Raman amplification. The first claim to an RFL in an optical waveguide can perhaps be made by Ippen [1], who used a liquid-filled glass tube as the waveguide. The first RFL in a conventional fiber was reported by Stolen et al. [2] in 1972. This and subsequent work in the 1970s [3–7] employed bulk optics for the cavity mirrors and to focus light into and out of the cavity. These devices were inefficient and had low output powers. It was the advent of Bragg gratings and multimode diodes that made efficient practical devices possible [8–11]. The RFL played a critical role in the initial experiments demonstrating Raman amplification in telecommunications, by initially providing higher pump powers at many more wavelengths than are available from the single-mode diodes.

In the next section an overview of an RFL is given, followed by more detailed descriptions of the constituent parts, the pump laser, the fiber, and the Bragg gratings. In Section 7.3 a numerical model for an RFL is described and the effect of various laser parameters are studied. Section 7.4 explores more advanced laser concepts such as multiple wavelengths and multiple order Rfls.

303

# 7.1   Overview

A schematic of an exemplary 1455-nm RFL pump module is shown in Figure 7.1. As with any other laser it consists of three parts: a pump source, a gain medium, and a means of providing feedback. For clarity in this chapter, the RFL refers to the complete device shown in Figure 7.1. The pump portion of the RFL consists of two parts: a set of multimode 9xx-nm diodes that are the optical pumps [12–14] and a rare-earth-doped cladding-pumped fiber laser (CPFL) [17–20], which converts the multimode diode light into single-mode light at another wavelength. The cascaded Raman resonator (CRR) refers to the portion of the device indicated in Figure 7.1 [10, 11, 20–25]. It consists of input and output grating sets separated by a fiber with an enhanced Raman gain coefficient. In reference to Figure 7.1, light from the 1100-nm laser enters the CRR. As it propagates down the fiber it is converted into light at the next Stokes shift, 1156 nm. Any 1100-nm light that is not converted will be reflected by a high reflector (HR) (∼100%) on the output grating set. The light at 1156 nm is confined in the cavity by two HR gratings on either end of the cavity and is itself converted to 1218-nm light. The process cascades through with nested pairs of HRs at all intermediate Stokes shifts forming intermediate cavities. When the desired output wavelength is reached, the output grating set contains a grating whose reflectivity is less than 100% so that light is coupled out of the cavity. This grating is called the output coupler (OC). By proper choice of the input pump wavelength and grating sets, CRR at almost any wavelength greater than 1 μm can be made.

The output power at 1455 nm versus input power at 1100 nm for a CRR is shown in Figure 7.2a. The optical-to-optical slope efficiency for this device is 52% with a threshold power, $P_{th}$, of 425 mW. If the output is measured relative to the 915-nm diodes, the slope efficiency is reduced to 40%. The output spectrum is shown for the same device in Figures 7.2c and 7.2d. The intermediate Raman orders can be seen in Figure 7.2c. The ratio between the peak power of the desired output wavelength and the intermediate Stokes order with the highest output power is called the suppression ratio. In Figure 7.2c, the largest intermediate order is at 1156 nm, with a suppression ratio of approximately 20 dB. The suppression ratio will vary with pump power, generally increasing as the pump power does. Finally, Figure 7.2d shows that at the output wavelength of the RFL, the spectrum

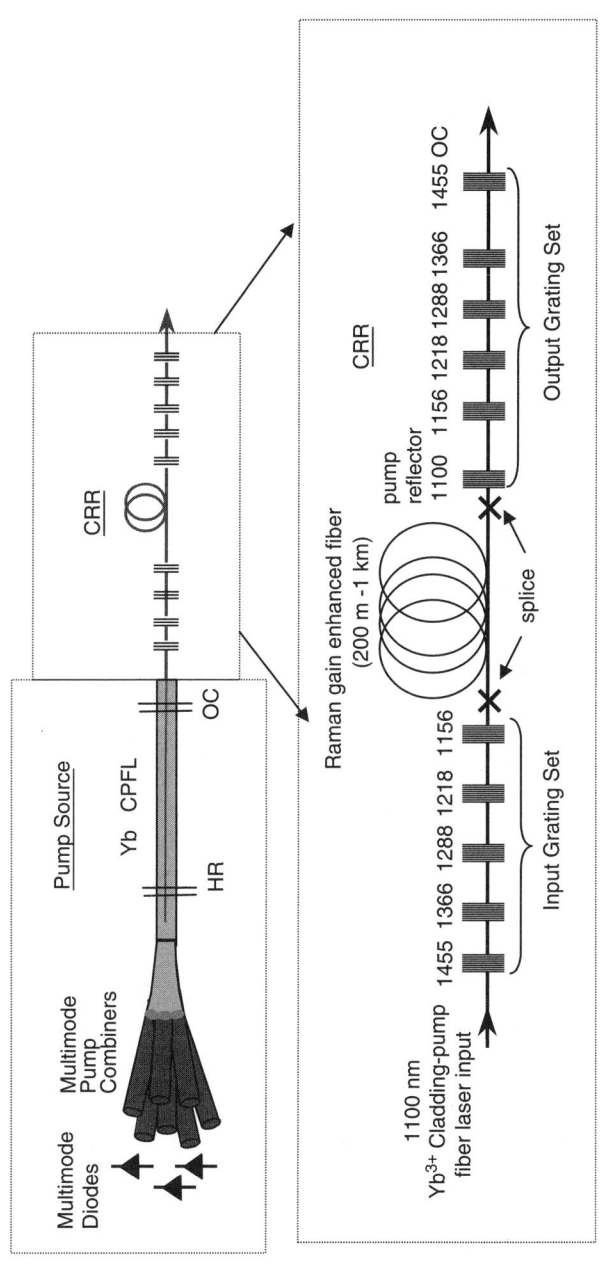

**Figure 7.1:** Schematic of an exemplary 1455-nm-output wavelength RFL consisting of a pump source and a CRR. A more detailed drawing of the CRR is provided in the lower box.

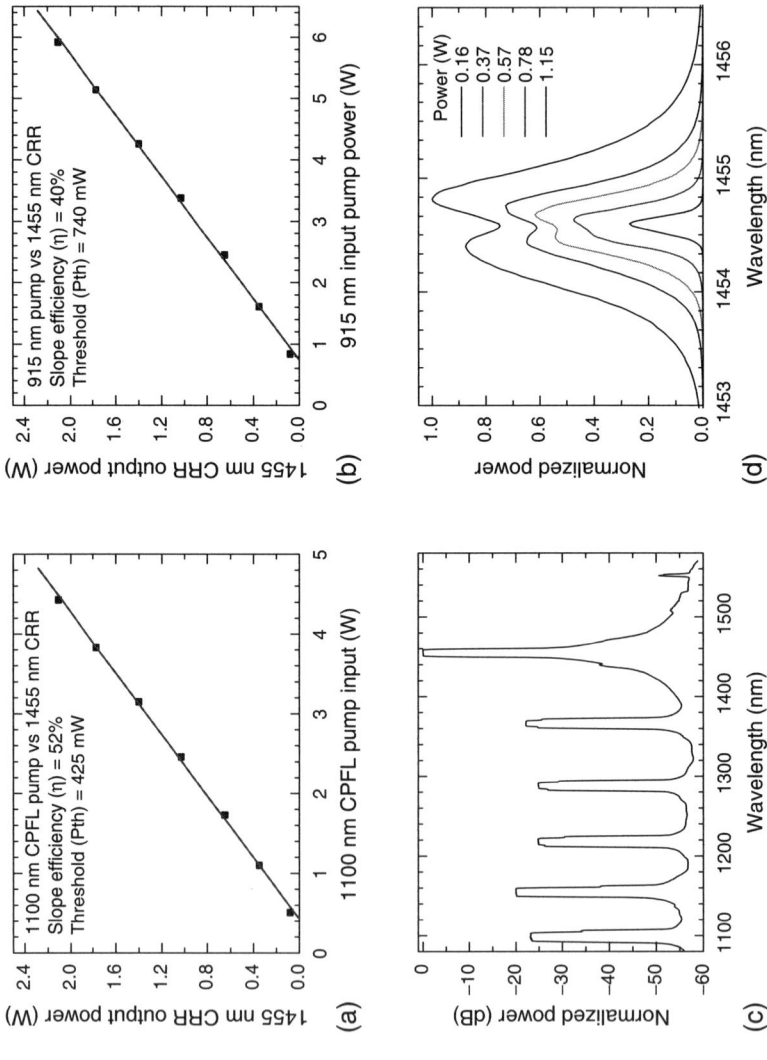

**Figure 7.2:** Plots of the CRR output power versus (a) the 1100-nm input power and (b) the 915-nm diode power. Also shown are the output spectra of the device for (c) a broad (at 1-W output power) and (d) a narrow spectral plot shown as power is increased.

broadens with increasing power, and a dip may appear in the center of the spectrum.

Raman fiber lasers have been used in several of the pioneering experiments in distributed Raman amplification. For example, the first demonstrations of (i) capacity upgrades using Raman amplification by Hansen et al. [26], (ii) multiwavelength pumping for large bandwidth by Rottwitt and Kidorf [27], and (iii) higher order pumping by Rottwitt et al. [28] all used single wavelength Raman fiber lasers. Many other systems' results have also established an RFL as a viable Raman pump source. In the following sections, individual parts of the RFL, that is, the pump source, the Raman gain fiber, and Fiber Bragg gratings, are described.

## 7.1.1 Pump Laser

Due to the power requirements for Raman amplification, high-power multimode diodes are used as pump lasers. These diodes extract more power from a semiconductor laser by increasing its size. The resultant larger beam size means it is more difficult to couple the light into a single-mode fiber; therefore, larger diameter multimode fibers are used as the output fiber pigtails. The 915- or 975-nm laser diodes used for RFL are typically GaAs-based devices. These diodes are capable of emitting 1–4 W of fiber-coupled power from a 100-$\mu$m-diameter fiber core with 0.15 NA (numerical aperture) [29]. Recently, diodes emitting as much as 15 W into a 0.15-NA fiber were reported, and it is possible to couple even more light into the fiber by increasing the fiber NA [29]. However, such high powers are not necessary for telecom applications. High-power multimode diodes have some advantages over their single-mode counterparts. The larger chip size lowers both the junction temperature and the facet intensity, increasing the yield in the manufacturing of these diodes and thereby lowering their cost. The reduced thermal stress on the diodes due to the lower intensity also means that these diodes have the potential to be run without thermal electric cooling. This reduces electrical power consumption and provides a saving in capital and operational costs.

The light from the multimode diode is next coupled into a CPFL. A schematic of the index profile of a cladding-pumped fiber (CPF) (also referred to as double-clad fiber) is shown in Figure 7.3a [15, 16]. It consists of a rare-earth-doped core surrounded by a silica glass cladding. Typical

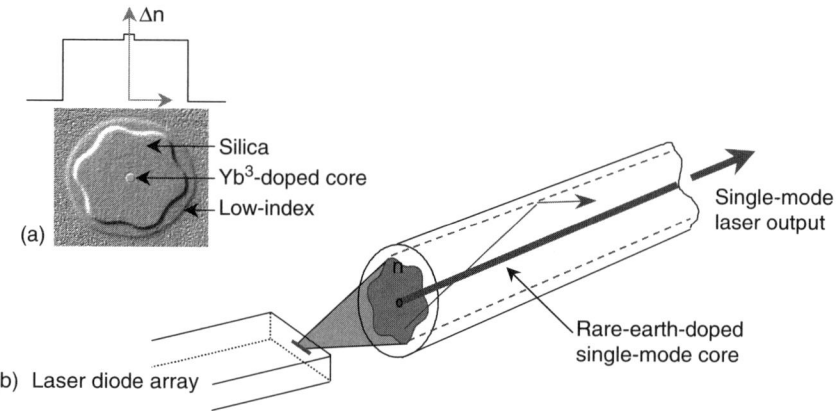

**Figure 7.3:** (a) Scanning electron microscopy picture of a CPF and the corresponding index profile. The core cladding and low-index polymer are clearly visible. (b) Propagation of pump light down the core of the fiber illustrating the absorption of the pump by the doped core. (Photo courtesy of OFS Fitel Laboratory.)

core diameter and NA are 5–6 $\mu$m and 0.1 $\mu$m, respectively, comparable to transmission fiber. What differentiates the fiber from transmission fiber is that surrounding the glass is a polymer whose index is lower than that of silica. This allows light to be guided by the silica cladding as well as the core. Typical values of the cladding diameter and NA are 125–200 $\mu$m and 0.45–0.6, respectively. This large area and NA allow efficient input coupling of multimode diodes.

Light from a multimode diode is transmitted along the cladding of the fiber as indicated in Figure 7.3b. As the light propagates through the core it is absorbed by the rare earth dopant. A significant portion of the stimulated emission generated in the core will be guided in the core. By placing a Fiber Bragg grating (FBG) with the appropriate reflectivity at either end of the single-mode core of the CPF (see Figure 7.1), a CPFL is formed. It should be noted that if the silica cladding was circular, then some modes of the pump light could propagate without crossing the single-mode core, reducing the efficiency of the device. The noncircular shape forces mode mixing, so all the modes eventually cross the core. For long enough lengths of fiber and efficient pump mode mixing, the absorption coefficient of the CPF can be approximated by the absorption coefficient of the single-mode core times the ratio of the core to cladding area [30]. In summary, the CPFL is a brightness and wavelength converter, coupling the large-area high-NA multimode light from the diodes into a small-area low-NA fiber.

Ideally, the rare earth dopant used in the single-mode core would allow lasing of 14xx nm for Raman amplification. These wavelengths allow Raman amplification over a wide range of signal bands (e.g., S-, C-, or L-band) used for telecommunications. However, no such dopant exists and a CRR is used. The rare earth dopant most often used for the pump stage of the CRR is ytterbium ($Yb^{+3}$), and its absorption and emission spectra are shown in Figure 7.4 [31]. Among the several advantages of Yb are a very broad absorption spectrum from $\sim$800 to $\sim$1064 nm. This means a wide range of pump laser sources can be used. In addition the broad emission spectrum allows lasing from $\sim$970 nm to 1200 nm. Finally the simple $Yb^{3+}$ energy level consists of two energy levels each with some Stark splitting. This means there is no excited state absorption, and the large energy gap between the two levels does not allow multiphoton absorption or concentration quenching. As a result these devices are very efficient.

The choice of diode pump wavelengths for a Yb-doped CPFL is based on a trade-off, which can be seen from examining the Yb absorption spectrum shown in Figure 7.4. The use of 975-nm pumps allows for a much higher absorption coefficient and hence a more efficient device. However, due to the narrow width of the absorption spectrum around this wavelength, the diodes would have to be thermally stabilized to prevent a drift in wavelength with temperature. Alternatively, pumping around 920 nm reduces the efficiency of the device, but alleviates concerns about the wavelength stability of the multimode diodes.

The discussion of the pump source is now completed by describing the coupling of the multimode diode light into the CPF. Though several approaches have been demonstrated, a few are more widely used. The most obvious is to free space couple the light from the multimode diodes into the ends of the fiber. This approach is shown in Figure 7.5. It is simple to use in a laboratory, but it does not have a high coupling efficiency. In addition the ends of the fiber are blocked, and if counterpumping is to be used, then the signal will have to pass through a wavelength-dependent element, introducing further loss. Finally, this setup is not very rugged.

The second approach shown in Figure 7.1 is a multimode pump combiner. In this approach, a number of fibers, $N_P$, are fused together and their diameter adiabatically tapered down into one multimode core [32].

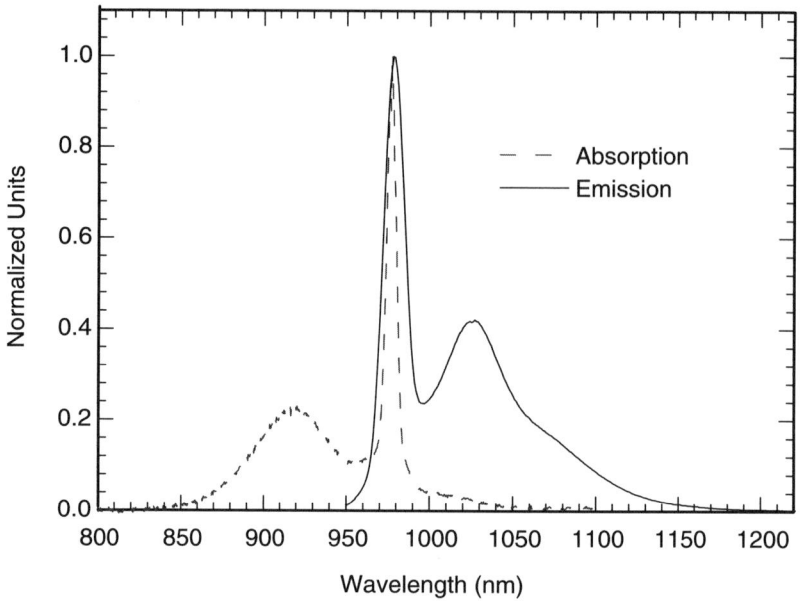

**Figure 7.4:** A plot of the normalized absorption and emission coefficient of a Yb-doped CPF.

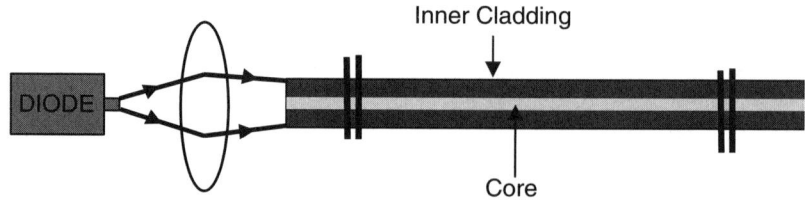

**Figure 7.5:** A simplified diagram of a diode pump laser imaged onto the end of a CPFL.

The fiber at the center may contain a single-mode core for transmitting the laser output or signal light in an amplifier configuration. A cross-section of the tapered-down fibers is shown in Figure 7.6 for both 7 and 19 input arms. The tapered-down bundle of fibers is spliced to an output pigtail fiber, and this assembly is recoated with the same low-index polymer used for

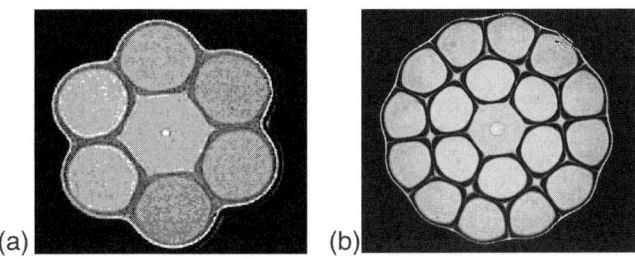

**Figure 7.6:** Microscope pictures of cross-sections of (a) a 7-arm and (b) a 19-arm multimode pump combiner, before the splice to the output pigtail fiber. (Photo courtesy of OFS Fitel Laboratory.)

the CPF. Efficient coupling into the output pigtail is obtained if

$$\sqrt{N} D_i (\text{NA})_i \leq D_o (\text{NA})_o, \tag{7.1}$$

where $D_{i/o}$ and $(\text{NA})_{i/o}$ are the input/output fiber diameters and numerical apertures, respectively. Coupling efficiencies greater than 95% for the pump lasers have been obtained. In order to maintain circular symmetry when fusing the bundles together, the number of input pigtails cannot be increased in increments of one. An expression for $N_p$ based on adding additional layers of fibers is given by the expression [33]

$$N_p = 1 + 6 \sum_{i=1}^{N_L} i, \tag{7.2}$$

where $N_L$ is the number of layers. Hence the number of input pigtails increases from 7 to 19 to 37 and so on. The limitation of adding new layers is the increased diameter and NA required of the output pigtail fiber as described by Eq. (7.1).

A third approach shown in Figure 7.7 is a side pumping technique [34–37], in particular v-groove side pumping [36, 37]. First the low-index polymer coating is stripped off the fiber. A 90° v-groove is cut into the glass inner cladding of the fiber, without crossing into the core. A low-index glass substrate is bonded to the inner glass cladding below the v-groove to provide mechanical strength. Light from broad-stripe diodes is imaged through the glass using a micro-lens assembly to focus the light onto the v-groove.

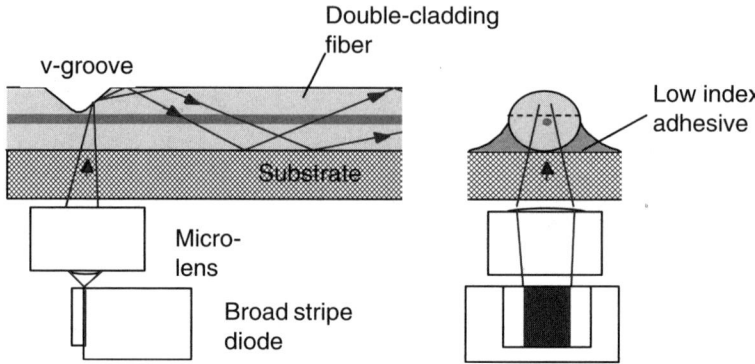

**Figure 7.7:** Schematic showing the coupling of the pump light into a CPF using the v-groove side-pumping technique. (Diagram courtesy of Lew Goldberg Keopysy.)

The pump light is reflected at the v-groove such that it propagates along the fiber. Coupling efficiencies of up to 90% have been reported using this technique. It has the advantages of (i) leaving the fiber ends unobstructed; (ii) being a very scalable approach, that is, in principle a large number of pumps can be added [38]; and (iii) no wavelength selective couplers are required. In a variation of this technique called embedded-mirror side pumping [39], a notch is cut in place of the v-groove and a right triangular–isosceles prism is placed in the groove. Light is then coupled into this mirror, which then reflects it down the inner glass cladding. The reported coupling efficiency for this device was 91%.

Finally in another approach two or more fibers are drawn such that their claddings are in optical contact with each other [40]. These fibers are all then recoated with the same low-index polymer. This is conceptually similar to the previously described multimode bundle. One of these fibers contains the dope core, while pump light is coupled into the others. The pump light will couple back and forth between fibers, but when the light couples into the fiber containing the doped core it will be absorbed. The advantage of this approach is that the individual fibers can be independently addressed so that the ends of the fiber containing the single-mode light is unobstructed, and in theory more and more pump fibers can be added, meaning it is a very scalable approach.

## 7.1.2 Raman Fiber

The fiber choice for a CRR is based on five considerations: increasing the Raman gain coefficient, reducing the effective area, decreasing the fiber loss, reducing splice losses, and the ability to write Bragg gratings in the fiber.

The largest influence on the shape and wavelength at which the peak Raman gain is located is the dopant used and the quantity of the dopant [41–43]. Figure 7.8 shows the relative strength and shape of the Raman gain spectrum for different glass formers: $SiO_2$, $GeO_2$, $B_2O_3$, and $P_2O_5$. The results are summarized in Table 7.1, with correction with the data shown in Figure 7.8 to account for Fresnel reflections. It can be correctly inferred from these data that adding these dopants to the silica will increase the Raman gain coefficient and change the shape of the gain spectrum in a fiber.

Cascaded Raman resonators have been built using either Ge- or P-doped fibers. Figure 7.9a is a diagram of the normalized gain spectrum of P-doped and Ge-doped fiber. Germanium has been the traditional dopant used in optical fibers, and the techniques for incorporating Ge into fibers are well understood. On the other hand, P-doped fibers with a larger Stokes shift (39.9 THz) can reduce the number of Raman shifts needed (e.g., two shifts from 1060 nm to 1480 nm) as compared to Ge-doped fiber (six shifts from 1060 nm to 1480 nm). Further flexibility is gained in P-doped fiber because lasing can be achieved from the silica peak at 13.2 THz. Therefore, the number of Stokes shifts needed to arrive at a particular wavelength can be minimized by utilizing the P and Si Raman gain peaks [25].

The next goal of the fiber design is to reduce $A_{\text{eff}}$. This can be accomplished by reducing the core area. Although the reduction in the core

**Table 7.1 A list of glass formers, their relative intensities, and frequency shifts (From Ref. [41])**

| Dopant | Relative Intensity | Frequency Shift (THz) |
|--------|--------------------|-----------------------|
| $SiO_2$ | 1 | 13.2 |
| $GeO_2$ | 7.4 | 12.6 |
| $B_2O_3$ | 4.6 | 24.2 |
| $P_2O_5$ | 4.9 | 19.2 |
| | 3.0 | 41.7 |

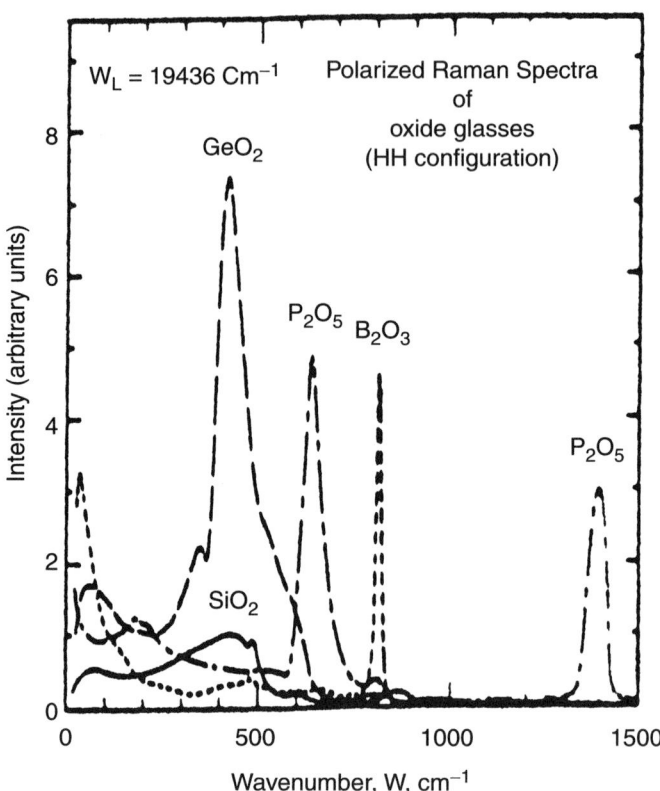

**Figure 7.8:** The relative strength, shape, and frequency locations for various glass formers. (From Ref. [41]).

radius, $r$, does not have a one-to-one correlation with the mode diameter, it is an effective means of controlling mode diameter. There is a lower limit on $r$, however, as eventually the mode field diameter will begin to increase and the bend loss becomes excessive. This loss can be reduced by increasing the index difference, $\Delta n$, between the core and the cladding, which also more tightly confines the mode, reducing $A_{eff}$.

However, there is another design constraint and that is the cutoff wavelength, $\lambda_c$. It is intuitive that the most efficient Raman pumping will take place if the pump and signal wavelengths spatially overlap. Therefore, the light at all the intermediate Raman wavelengths in the fiber should be

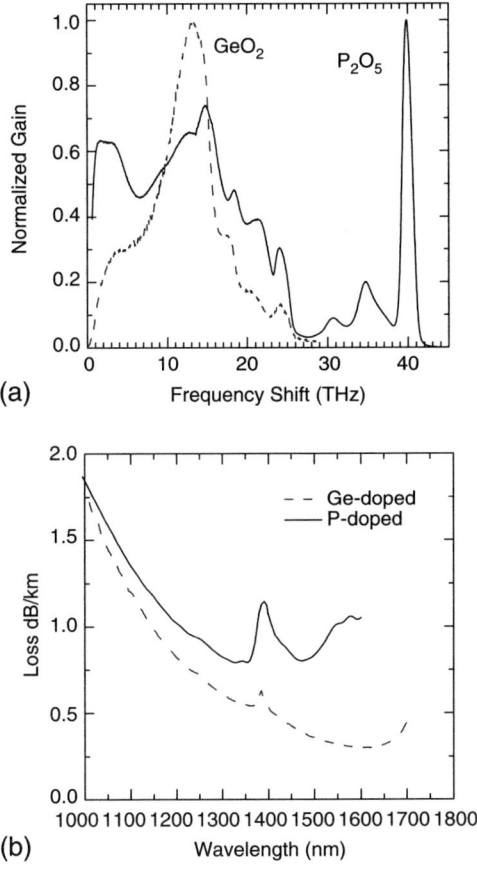

**Figure 7.9:** Normalized (a) gain spectra and (b) intrinsic fiber loss for P-doped and Ge-doped fibers.

guided in a single mode. The wavelength of the CPFL determines the $\lambda_c$ to which the fiber will be designed. An expression for $\lambda_c$ for step index fibers is given as [49]

$$\lambda_c = \frac{2\pi r}{V_c} \left[ (n_1 + n_2)\, \Delta n \right]^{1/2} , \tag{7.3}$$

where $n_1$ and $n_2$ are the refractive indices of the core and cladding, respectively, and $V_c = 2.405$ is the normalized frequency at $\lambda_c$. It is now clear that in order to reduce $A_{\text{eff}}$, $r$ should be reduced and $\Delta n$ increased with the changes constrained by Eq. (7.3). However, there are additional problems associated with using fibers with a high $\Delta n$.

The use of high-index difference fibers is both beneficial and detrimental. The dopants such as Ge and P added to increase the core index also increase the Raman gain coefficient as outlined earlier. However, an undesirable effect is that the fiber loss is also increased [45–48]. This is the third fiber design parameter. For low values of $GeO_2$ in the core it is straightforward to predict the fiber loss by extrapolating out the increased Rayleigh scattering due to the increased $GeO_2$. It is well documented that there is an excess loss in highly doped $GeO_2$ fibers that goes beyond that predicted from the Rayleigh scattering losses. This loss is independent of the Rayleigh loss and adds to it. One explanation is that this excess loss arises from draw-induced fluctuations in the core radius. Examples of the fiber loss spectra for Ge- and P-doped fibers are shown in Figure 7.9b with a standard telecom fiber shown for comparison.

The fourth design consideration for a Raman fiber is the ability to obtain low-loss splices between the Raman fiber and other transmission fibers typically used in telecommunications. The reduction in $A_{\text{eff}}$ results in a significant mode field mismatch between the Raman gain fiber and other fibers and the different melting points and diffusion rates between high-index difference fibers and standard fibers present splicing challenges. In telecommunication applications, the RFL will have to be spliced to a standard fiber and any splice losses will have to be compensated for by increased pump power.

The need to minimize the splice losses in the cavity leads to the final design parameter for a Raman fiber laser cavity, and that is the ability to write low-loss Bragg gratings in the same fiber that is used to provide gain in the cavity [23]. Without this ability, dissimilar fibers would have to be used in the cavity, which would increase the intercavity splice losses, hence reducing the efficiency of the device. The severity of this effect will be quantified in Section 7.3.1.

The trade-offs between the amount of dopant to add to increase the gain and $\Delta n$ and at the same time minimize loss is quantified by a figure of merit (FOM) defined as [47–49]

$$\text{FOM} = \frac{g_R}{A_{\text{eff}} \alpha}, \tag{7.4}$$

**Table 7.2 A list of different fiber types, their Raman gain coefficients, intrinsic loss, and FOM. Also shown are the simulation results for optimum fiber lengths and output reflectivity for optimum performance for 2 W and 5 W of pump power**

| $P_p$(W) | Fiber Type | $g_R A_{eff}$ $(km^{-1} W^{-1})$ | $\alpha_{1.55 \mu m}$ (dB/km) | FOM $(W^{-1} dB^{-1})$ | L (m) | $R_{oc}$ (%) | $P_{out}$ (W) | $\eta_T$ (%) |
|---|---|---|---|---|---|---|---|---|
| 2 | RF | 2.4 | 0.30 | 8.0 | 400 | 45 | 0.9 | 46 |
| | HSDK | 3.3 | 0.64 | 5.1 | 250 | 60 | 0.8 | 40 |
| | TWRS | 0.7 | 0.21 | 3.4 | 800 | 95 | 0.7 | 35 |
| 5 | RF | 2.4 | 0.30 | 8.0 | 300 | 15 | 2.7 | 55 |
| | HSDK | 3.3 | 0.64 | 5.1 | 200 | 25 | 2.6 | 51 |
| | TWRS | 0.7 | 0.21 | 3.4 | 600 | 55 | 2.4 | 48 |

where $g_R$ is the Raman gain coefficient, $A_{eff}$ is the effective area of the fiber, and $\alpha$ is the loss at a specified wavelength. Table 7.2 shows among other things the ratio of $g_R/A_{eff}$, $\alpha$ (1550 nm), and FOM for three different fibers. These fibers are all manufactured by OFS Fitel and are high slope dispersion compensating fiber (HSDK), TrueWave RS (TWRS), and a Raman-enhanced fiber (RF). It is seen that even though HSDK has a higher gain coefficient than RF, its higher loss makes it an inferior fiber for Raman amplification compared to RF. Likewise, though the loss of the TWRS fiber is lower than the RF fiber, its lower gain coefficient makes this fiber inferior to the RF. In comparing the FOM between fibers with different dopants, such as P, other factors such as the number of shifts needed to reach a certain wavelength must also be considered. More will be said about this in Section 7.3.1.

## 7.1.3 Fiber Bragg Gratings

With an understanding of the gain medium, the feedback mechanism used in a CRR is now described. At the interface between two dielectric materials, scattering occurs. The fraction of light reflected depends on the angle of incidence and the refractive index of the materials. For a multilayered dielectric medium such as that shown in Figure 7.10, where a small index difference has been assumed, scattering will occur at each interface. In order for there to be a significant amount of reflection from the multiple layers, the reflections from each layer should add in phase. The condition under which this is satisfied is when the path length shown as $AB + BC$ is an integer number, $m$, of wavelengths $(\lambda/n)$ and is given as

$$2\Lambda \sin \theta = m\frac{\lambda_B}{n_2}, \tag{7.5}$$

where $\Lambda$ is the layer thickness, $\theta$ is defined as in Figure 7.10, and $\lambda_B$ is called the Bragg wavelength and is the wavelength at which the phase matching is satisfied. When Eq. (7.5) is satisfied, this is known as a Bragg reflection. For multiple layers with a period $\Lambda$, a significant amount of light can be reflected backward. Due to the wave-guiding nature of optical fibers $\theta = 90°$ and the Bragg condition becomes $\Lambda = m\lambda_B/2n_2$.

A multilayered structure is reproduced in optical fibers by inducing periodic changes in the refractive index of the core of the fiber. The coupling between the forward (transmitted) and the backward (reflected) traveling waves in a Bragg grating is represented by coupled mode equations. From these equations a reflection coefficient, $R$, can be extracted and is

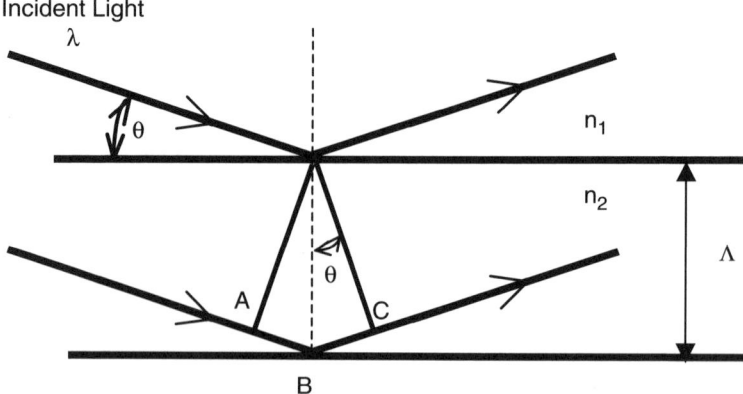

**Figure 7.10:** Schematic showing light reflected from two different layers of a dielectric medium.

given by [9, 50–53]

$$R = \frac{\kappa^2 \sin h^2(|S|L)}{\delta\beta^2 \sin h^2(|S|L) + |S|^2 \cos h^2(|S|L)} \qquad \kappa^2 > \delta\beta^2 \qquad (7.6a)$$

and

$$R = \frac{\kappa^2 \sin^2(|S|L)}{\delta\beta^2 - \kappa^2 \cos^2(|S|L)} \qquad \kappa^2 < \delta\beta^2. \qquad (7.6b)$$

In these equations, $\delta\beta = 2\pi n_{\text{eff}}/\lambda - m\pi/\Lambda$ (from Eq. (7.5)) measures the mismatch at the free-space wavelength $\lambda$ from the Bragg condition ($n_{\text{eff}}$ is the mode effective index) and $\Lambda$ is the period of the index change. $\kappa$ is the coupling coefficient between the transmitted and the reflected field. $S = [\delta\beta^2 - \kappa^2]^{1/2}$ is the dispersion relationship for the backward propagating wave determined from the boundary conditions of the coupled mode equations. Finally, $L$ is the length of the grating. For a uniform sinusoidal modulation of the refractive index throughout the core, $\kappa$ can be expressed as

$$\kappa = \frac{\pi \delta n}{\lambda_{\text{B}}} \eta, \qquad (7.7)$$

where $\delta n$ is the periodic index difference and $\eta$ is the overlap of the mode of the electric field with the core. Note that Eq. (7.6a) describes when the coupling of the grating at $\lambda$ is larger than the detuning of the grating from Bragg resonance, whereas Eq. (7.6b) is the reverse case.

Equations (7.6) show how grating properties can be manipulated in order to obtain the reflectivities and spectral widths needed to make RFLs. The peak reflectivity is obtained when the Bragg condition ($\delta\beta = 0$) is satisfied. From Eqs. (7.6),

$$R_{\text{max}} = \tan h^2(\kappa L). \qquad (7.8)$$

Hence for $\kappa L = 3$ or larger the reflectivity approaches 100%. The magnitude of $\kappa$ and hence $R_{\text{max}}$ is primarily determined by the index difference induced between consecutive sections of the grating.

Another important property of a grating is its spectral bandwidth $\Delta\lambda$. There are several measures of bandwidth. A simple one to derive is the difference between the first minima occurring on either side of the reflection peak. This occurs when $\sin h^2(|S|L) = 0$ in Eqs. (7.6), and an expression is given by

$$\Delta\lambda = \frac{\lambda_B^2}{\pi n_{\text{eff}} L}\left[(\kappa L)^2 + \pi^2\right]^{1/2}, \qquad (7.9)$$

where it has been assumed that $\lambda \approx \lambda_B$.

### 7.1.3.1  Photosensitivity

The periodic index changes necessary to produce a grating can be achieved by exposing the fiber to ultraviolet light [8, 51]. These index changes are analogous to those light induce on a photographic plate. During the fabrication of optical fibers under ideal circumstances all the bonds formed would be $SiO_2$ (silica dioxide) bonds. When dopants such as germanium (Ge) are added they would then replace the Si atoms. However, Si–Si, Si–Ge, and Ge–Ge bonds may form. These bonds are defects in the fiber, and when exposed to ultraviolet (UV) light around 244 nm they break, freeing an electron. It is believed that the increase in the number of this type of defect changes the absorption spectrum of the light, resulting in a change in the refractive index of the fiber through a Kramers–Kronig relationship. A second explanation is that there is an increase in density of the glass through the creation of these defects. The removal of the electron changes the shape of the molecule. This densification results in index changes. Although the exact mechanism by which this occurs is not fully agreed on, it is likely that both mechanisms play a role, the determination of which dominates dependent on the fiber type, the power of the laser used to write the grating, and wavelength. While Ge is the dopant most generally used, P, which has also been used to make CRR, has been shown to be photosensitive at 193 nm [21, 22]. An important aspect of UV induced index changes is that they require thermal annealing to ensure long-term stability.

The quality of a grating can be improved by increasing the photosensitivity of the fiber. This provides a larger amount of index change, and it reduces the exposure time needed to achieve the index change. The improved sensitivity is accomplished through a high-pressure exposure of the fiber to hydrogen ($H_2$) [54] or deuterium ($D_2$) [55]. It appears that these elements can produce Si–OH or Si–OD groups as well as oxygen-deficient Ge sites. These are the sites mentioned in the previous paragraph that cause index changes when exposed to UV light. The photosensitivity of fibers treated in this way can be two orders of magnitude higher than that of untreated fiber. Any unreacted $H_2$ or $D_2$ diffuses out of the fiber.

### 7.1.3.2  Grating Writing

Two techniques for grating writing are the dual-beam holographic method and the phase mask approach [9, 51–53, 56–58]. A diagram of the dual-beam holographic method is shown in Figure 7.11. The bare fiber is exposed to two beams originating from the same ultraviolet source. The interference pattern produced by the two beams leads to a periodic modulation of the refractive index of the glass core. Cylindrical lenses are used to elongate the beam along the fiber length. The grating period is given by the expression

$$\Lambda = \frac{n_{\text{eff}} \lambda_{\text{laser}}}{n_{\text{laser}} \sin \theta}, \tag{7.10}$$

where $\lambda_{\text{laser}}$ and $n_{\text{laser}}$ are the laser wavelength and index in the fiber, respectively, and $\theta$ is the angle between the two beams as shown in Figure 7.11.

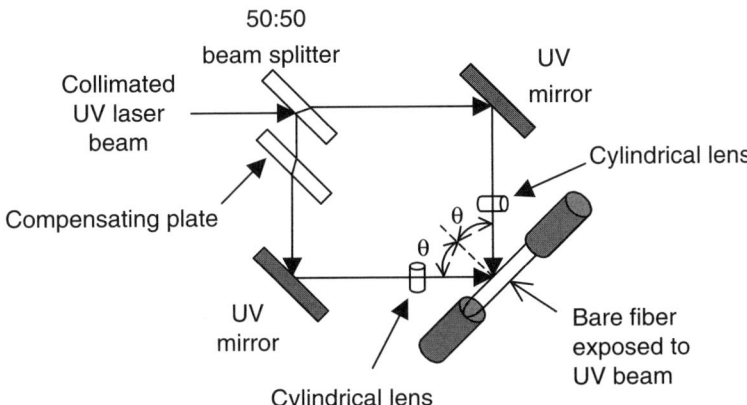

**Figure 7.11:** A schematic of the experimental setup of the dual-beam holographic method for writing Bragg gratings.

By adjusting $\theta$, and hence $\Lambda$, the wavelength at which the peak reflectivity occurs can be varied. The dual-beam holography method was one of the first approaches to remove the limitations on the wavelength of the grating obtainable imposed by the ultraviolet source. This technique, however, has one disadvantage. The source required needs a high degree of spatial and temporal coherence. This is because mechanical vibrations and air turbulence can affect the quality of the interference pattern generated in the fiber affecting grating quality. For low coherence sources, it is usual to insert a glass plate in the arm of the interferometer path that undergoes a reflection at the beam splitter in order to compensate for the path difference. This makes this method ideally suited for gratings that can be written with high-energy pulsed sources with a very short exposure time.

The second approach shown in Figure 7.12 overcomes some of the disadvantages of the dual-beam holographic approach, which is to use a phase mask. The phase mask is obtained using photolithography techniques commonly employed for fabrication of integrated electronic circuit devices.

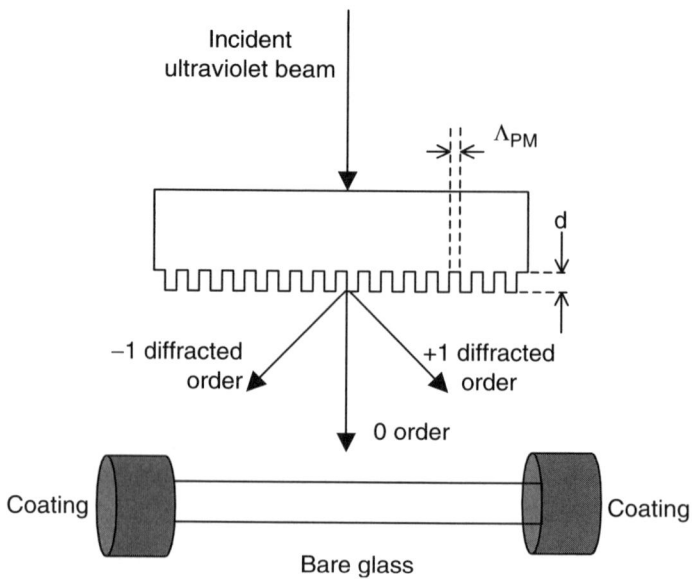

**Figure 7.12:** A schematic of an experimental setup for writing Bragg gratings using a phase mask.

The majority of the light incident on the phase mask (typically normal incidence) is diffracted into the $m = 0$ and $\pm 1$ orders. The interference between the $\pm 1$ orders sets up a standing wave pattern that then induces periodic changes in the index of a fiber placed in the field. In this approach the grating period $\Lambda$ is truly independent of the ultraviolet source and is given by

$$\Lambda = \Lambda_{PM}/2, \qquad (7.11)$$

where $\Lambda_{PM}$ is the period of the phase mask. For efficient diffraction into the first order, the phase mask should be written to have a groove-space ratio of 1:1. The depth of the grooves, $d$, affects the amount of light diffracted into the first order and is chosen to minimize the light in the 0th order with the expression for $d$ with the minimum transmission given by

$$d = \frac{\lambda_{laser}}{2(n_{laser} - 1)}. \qquad (7.12)$$

Hence the zero order minimization can only be done for a given laser wavelength. The advantages of this technique are reduced demand on the temporal and spatial coherence of the ultraviolet source, the gratings are very reproducible, gratings with a variation in period (chirped) can be made, and very long gratings can be fabricated. Some disadvantages to this approach are that the quality of the grating is almost completely mask dependent, and for a given phase mask the range of center wavelengths of gratings which can be written using it is only $\sim 5$ nm.

## 7.2 Design of a Cascaded Raman Resonator

The design of a CRR requires the optimization of the various components described in the previous sections. In this section, the behavior of a CRR is studied numerically, and the effect of fiber length, OC reflectivity, splice loss, fiber type, pump wavelength, and linewidth as well as the noise properties of the CPFL is examined.

In Chapter 2, equations describing Raman amplification were presented. These equations have to be modified in order to account for the backward and forward traveling waves in the laser cavity. The evolution of the powers, $P_1$ to $P_n$, in a CRR with $n - 1$ Raman shifts, in which $P_1 = P_p$ represents the pump light and $P_n$ the light at the desired output wavelength, can be described by a set of nonlinear ordinary differential equations [60–65]. These are

$$
\begin{aligned}
\frac{d P_j^{\mathrm{F/B}}}{dz} = {} & \mp \alpha_j P_j^{\mathrm{F/B}} \pm \sum_{k<j} \frac{g_R(k,j)}{A_{\mathrm{eff}}^k} \left( P_k^{\mathrm{F}} + P_k^{\mathrm{B}} \right) P_j^{\mathrm{F/B}} \\
& \mp \sum_{k>j} \frac{\nu_j}{\nu_k} \frac{g_R(j,k)}{A_{\mathrm{eff}}^k} \left( P_k^{\mathrm{F}} + P_k^{\mathrm{B}} \right) P_j^{\mathrm{F/B}} \\
& \pm \sum_{k<j} \frac{g_R(k,j)}{A_{\mathrm{eff}}^k} h\nu_j \Delta\nu_{\mathrm{spon}}(k) \left( P_k^{\mathrm{F}} + P_k^{\mathrm{B}} \right) \\
& \mp \sum_{k>j} \frac{\nu_j g_R(j,k)}{\nu_k A_{\mathrm{eff}}^k} h\nu_k \Delta\nu_{\mathrm{spon}}(k) P_k^{\mathrm{F/B}},
\end{aligned}
\tag{7.13}
$$

where the superscript F/B designates the power, $P$, in the forward and backward traveling waves, respectively, $g_R(k,j)$ is the gain coefficient between the $j$th (signal) wavelength and the $k$th (pump) wavelength, $A_{\mathrm{eff}}$ is the effective area of the fiber, $\alpha$ is the intrinsic fiber loss, $\Delta\nu_{\mathrm{spon}(k)}$ is the bandwidth of the spontaneous emission, and $\nu$ is the frequency of a given wave. The forward direction is from the Yb-doped CPFL to the OC as shown in Figure 7.1. The first term on the right-hand side of Eq. (7.13) is the intrinsic fiber loss of that wave. The next term represents the gain of the $j$th wave through Raman pumping by all wavelengths less than $\lambda_j$. The third term in Eq. (7.13) describes the depletion of that wave by the forward and backward orders with longer wavelengths. The fourth term represent the gain from light spontaneously scattered from shorter wavelength Raman orders, and the final term is the loss to longer wavelengths due to spontaneous scattering. The boundary conditions for these equations are given by

$$
\begin{aligned}
P_p^{\mathrm{f}}(0) &= P_{\mathrm{in}} & P_p^{\mathrm{B}}(L) &= R_p^{\mathrm{f}} \cdot P_p^{\mathrm{F}}(L) \\
P_i^{\mathrm{F}}(0) &= R_i^{\mathrm{f}} \cdot P_p^{\mathrm{B}}(0) & P_i^{\mathrm{B}}(L) &= R_i^{\mathrm{b}} \cdot P_i^{\mathrm{F}}(L) \\
P_n^{\mathrm{F}}(0) &= R_n^{\mathrm{f}} \cdot P_n^{\mathrm{B}}(0) & P_n^{\mathrm{B}}(L) &= R_{\mathrm{oc}} \cdot P_n^{\mathrm{F}}(L),
\end{aligned}
\tag{7.14}
$$

where $R$ is the reflectivity of the input/output (f/b) Bragg grating, $L$ is the length of the RFL cavity, and $R_{oc}$ is the reflectivity of the OC. For notational simplicity $P_{out} = (1 - R_{oc})P_n^F(L)$ is used.

In the simulations presented here an 1117-nm to 1480-nm CRR was modeled, using the experimentally measured parameters for RF described in Table 7.2 and Figure 7.9b. The splices between the fiber containing the gratings and the Raman enhanced fiber, and the gratings themselves were initially assumed to have no loss. The three measures of the RFL performance used are the slope efficiency, $\eta_s$, pump threshold power, $P_{th}$, and the overall (total) efficiency, $\eta_T$. These quantities are defined from the linear fit of a graph of the launched pump power $P_p$ versus output power $P_{out}$ as

$$P_{out} = \eta_s (P_{in} - P_{th}),$$

$$\eta_T = \frac{P_{out}}{P_{in}}. \tag{7.15}$$

Since the focus is on improving the performance of the cascaded RFL, $P_p$ is the power from the CPFL entering the CRR.

## 7.2.1 Optimization of Fiber Length, Output Coupler Reflectivity, and Splice Loss

Figure 7.13a is a plot of $\eta_s$ as a function of $L$ for different $R_{oc}$. As $L$ is increased, $\eta_s$ decreases almost linearly for the length range under consideration. This decrease is due to the increased intrinsic fiber loss as its length is increased. This behavior is very similar to that seen when the reflectivity of the OC is increased, as shown in Figure 7.14b. This decrease in $\eta_s$ is expected since a higher $R_{oc}$ means less of the power that is generated in the cavity is extracted from it.

The trade-offs in designing a CRR become apparent in comparing Figure 7.13a with 7.13c and Figure 7.13b with 7.13d. Figures 7.13c and 7.13d are plots of the CRR's threshold power level as a function of length and OC reflectivity. All of the changes that had a negative effect on $\eta_s$ now have a positive effect on $P_{th}$. Increasing the length of fiber reduces $P_{th}$, since the Raman threshold is decreased as the fiber length is increased. Eventually, there is no benefit to increasing the fiber length as the increase

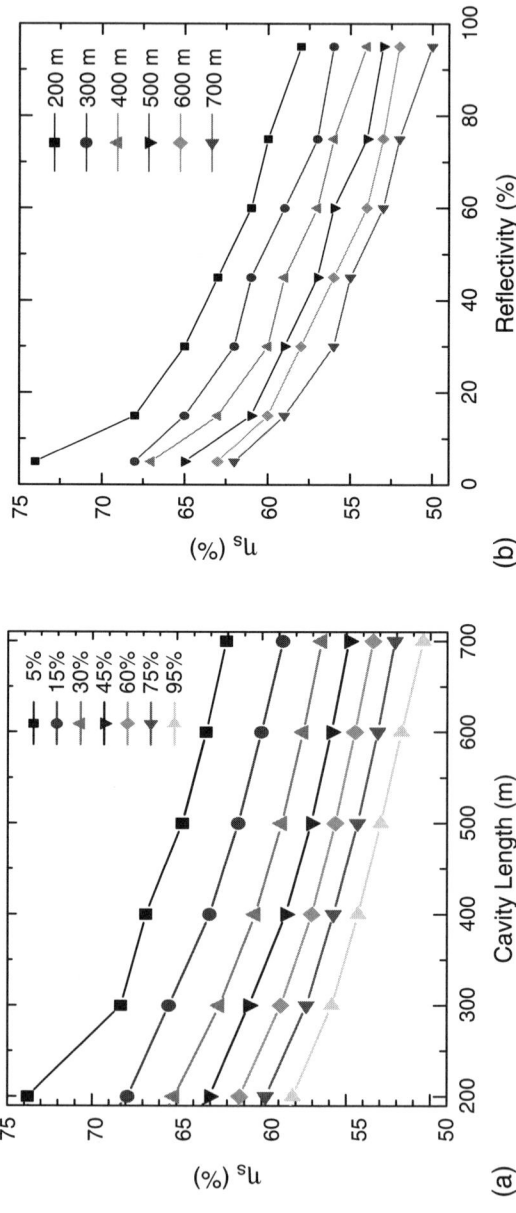

**Figure 7.13:** Simulation results showing the effect of (a) cavity length and (b) reflectivity on slope efficiency. Also shown is the dependence of threshold power on (c) cavity length and (d) reflectivity.

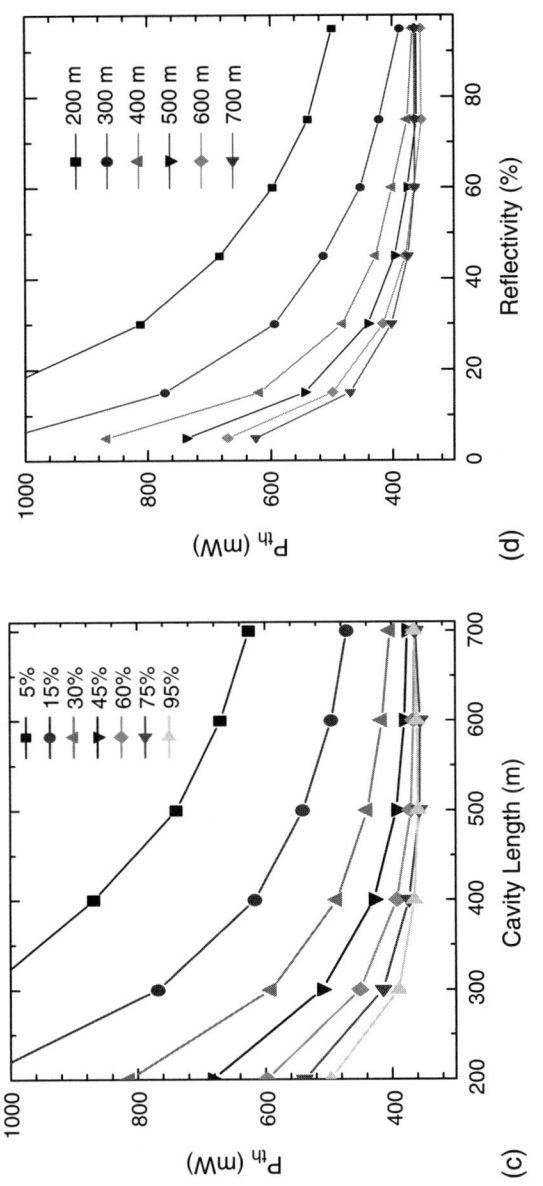

**Figure 7.13:** Continued.

loss exceeds any decrease in the Raman threshold, and $P_{th}$ can begin to increase. Figure 7.13d shows $P_{th}$ decreases nonlinearly as $R_{oc}$ is increased, eventually saturating. This behavior is expected since as $R_{oc}$ is increased the cavity losses are reduced, there is more intracavity power, and the Raman threshold is reduced. However, an overview of the results leads to the conclusion that for long lengths or high reflectivities $P_{th}$ is fairly insensitive to changes in cavity length or OC reflectivity.

The behavior described in the previous paragraphs points to the need to balance the values of $L$ and $R_{oc}$ between maximizing the slope efficiency and minimizing threshold power. The overall efficiency is a measure of this trade-off. The results of simulations to maximize $\eta_T$ by varying $L$ and $R_{oc}$ are shown in Table 7.2 for three different fiber types and pump power values of 5 W and 2 W. The fibers are all manufactured by OFS Fitel and are HSDK, TWRS, and RF. For the current discussions the focus is on the RF. The results show that the optimized parameters depend on the desired output power. For low $P_{out}$, it is more important to design the cavity to minimize $P_{th}$ because it represents a large percentage of pump power. Increasing $L$ and $R_{oc}$ does this. For higher powers, $\eta_s$ is more important since the device is operating far above threshold; hence smaller values of $L$ and $R_{oc}$ are required. It is also seen that the maximum $\eta_T$ increases with $P_p$. This is again a reflection of the decrease in the percentage of pump power light used to reach $P_{th}$.

The sensitivity of the optimum values of $L$ and $R_{oc}$ for the RF listed in Table 7.2 is shown in Figure 7.14. Unless otherwise stated, the parameters used are those in Table 7.2. Qualitatively, it is seen that $\eta_T$ is fairly insensitive to $L$ and $R_{oc}$ around the optimum parameters. It is noteworthy, however, that a laser designed to operate at 2 W (5 W) will not perform optimally at 5 W (2 W). Therefore, a laser required to operate over a wide power range will have to be a compromise between the two designs for optimum performance.

The effect of splice loss is examined in Figures 7.15a and 7.15b. For these simulations, the loss indicated is divided evenly between the splices of the input and output grating set of the CRR in Figure 7.1. As was indicated earlier an ideal splice loss of 0 dB was used in the previous simulations. Just adding 0.05-dB splice loss (considered an excellent splice loss) to each end of the cavity produces a predicted decrease in $\eta_T$ of 4% (5%) for an RFL optimized for 5 W (2 W). The drop in overall efficiency is worse at lower powers since the increased cavity loss raises $P_{th}$, which represents a

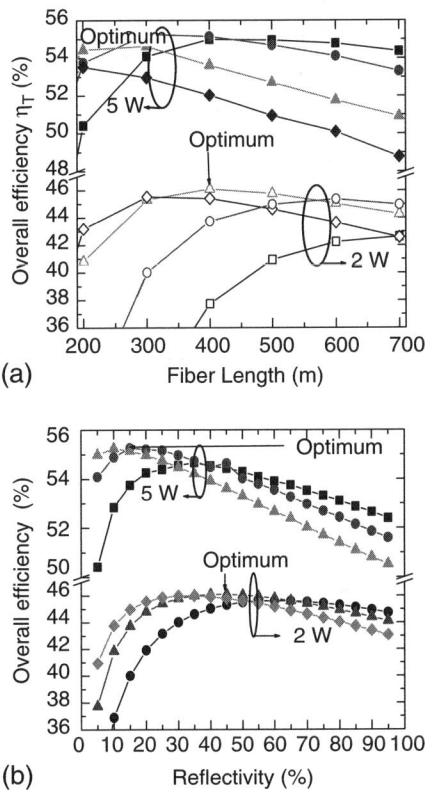

**Figure 7.14:** Overall efficiency as a function of (a) fiber length and (b) output coupler reflectivity for input pump powers of 5 W (closed symbols) and 2 W (open symbols). Plot (a) is for $R_{oc} = 5\%$ (squares), 15% (circles), 45% (triangles), and 70% (diamonds). Plot (b) is for $L = 200$ m (squares), 300 m (diamonds), 400 m (circles), and 600 m (triangles).

larger percentage of the pump power. Since in most applications a constant output power is required, Figure 7.15b shows the percentage increase in pump power required to maintain the same output power for a given splice loss compared to an ideal 0-dB splice loss. In order for a 5 W (2 W) device to maintain the same operating power, a 7% (7.5%) increase in pump power is needed. These results also highlight how relatively insensitive the CRR is to length and $R_{oc}$ compared to splice losses.

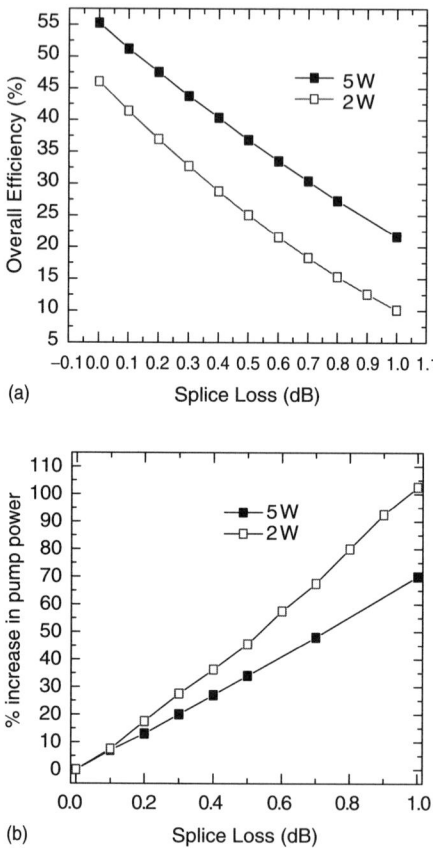

**Figure 7.15:** (a) The effect of splice loss on overall efficiency, and (b) the percentage increase in pump power required to maintain the same output power compared to a cavity with no splice losses for a 5-W and a 2-W device.

## 7.2.2    Fiber Type and Pump Wavelength Optimization

The choice of fiber to use in the Raman cavity is now considered. Referring to Table 7.2, fiber type RF was used for the prior simulations. Relative to the RF, the HSDK fiber has higher gain and loss coefficients, whereas the reverse is true for the TWRS fiber. The best performance, however, is obtained from the RF, which has the highest FOM (see Eq. (7.4)). Several other observations about the fiber type can be made. The fibers with the highest gain need the shortest length to maximize $\eta_T$. The difference is especially noticeable at the lower power level. The optimum reflectivity increases for the HSDK (TWRS) compared to the RF fiber because the higher fiber loss (lower gain); emphasizes minimizing cavity losses. Finally, the best $\eta_T$ is more sensitive to fiber design at the lower power since $P_{th}$ is a larger percentage of the total power.

The discussion so far has been focused on fibers whose gain has been enhanced using Ge-dopant. However, in optimizing the whole RFL pump module, consideration should be given to the choice of CPFL input wavelength and to the dopant used in the fiber [66]. The efficiency of the CPFL depends on the desired output wavelength. Efficient lasing in Yb-doped fibers is obtained from 1.06 to 1.13 μm. However, for very short wavelengths the signal light can be reabsorbed by the doped fiber, since the tail of the absorption spectrum for Yb extends out to these wavelengths (see Figure 7.4). At the longer wavelengths there is a reduction in the emission cross-section of the Yb-doped ions. Hence an optimum lies between these two extremes. In one experiment [66] the efficiency of the CPFL at different wavelengths was optimized with respect to reflectivity. The optimum efficiency was shown to be in the 1080- to 1100-nm region.

In Section 7.1.3 it was pointed out that P-doped fibers have a larger Raman shift (see Figure 7.9a) compared to Ge-doped fibers. This decreases the number of wavelength shifts necessary to obtain light in the 14xx-nm region from five Raman shifts to two. This reduces the complexity of the CRR and in principle increases the conversion efficiency. There are, however, several disadvantages. The larger Raman shift lowers the wavelength of the input pump light (from, for example, 1100 nm to 1049 nm). This pushes the operation of the Yb-doped fiber laser into a more inefficient regime. In addition, the increased intrinsic loss and difficulty in writing grating in P-doped fiber must also be considered [23]. In one study of this trade-off, four different devices were considered [66]. The results of this study are summarized in Table 7.3. The first device used a 1049-nm Yb-doped CPFL and two shifts of the 39.9-THz shift of the P-doped fiber; the second used a 1089-nm CPFL, one shift of the 39.9-THz, and two shifts of the 13.2-THz peaks in a P-doped fiber; the third used an 1108-nm CPFL, one shift each of the 39.9-THz and the 19.2-THz line in P-doped fiber; and the final used a Ge-doped dispersion-shifted fiber with five shifts at the

**Table 7.3 A comparison of the efficiency of four different RFLs made with various input CPFL wavelengths and different types of Raman gain fibers. The P-doped CRRs use different combinations of Si and P gain peaks to cascade through to the desired 1.45-mm output wavelength. The results compare the CRR efficiency and the RFL efficiency (From Ref. 66)**

| Raman Fiber Type | Raman Shifts Used (THz) | Wavelengths of Yb-Doped CPF (nm) | Max. Power of Yb-Doped CPF (W) | Max. Output Power at 1.45 μm (W) | Conversion Efficiency CRR | Total Efficiency RFL |
|---|---|---|---|---|---|---|
| P-doped | 39+39 | 1049 | 4.8 | 1.6 | 0.33 | 0.16 |
| P-doped | 39+13.2+13.2 | 1089 | 6.8 | 1.1 | 0.16 | 0.11 |
| P-doped | 39+24 | 1108 | 6.5 | 0.65 | 0.1 | 0.065 |
| Ge-doped (DSF) | 4+13.2 | 1128 | 6.2 | 1.9 | 0.3 | 0.19 |

13.2-THz shift. This work concluded that for the 1450-nm region at least Ge-doped fibers had a higher overall efficiency.

## 7.2.3  Linewidth Considerations

The numerical model used in the previous section fails to predict the correct spectral output for a CRR. It produces a very narrow output linewidth, and as already seen in Figure 7.2d the spectral output of an RFL is fairly broad and contains spectral features, and its width increases as a function of output power. The incorrect predictions are made because the model does not account for the inhomogeneously broadened Raman gain. As the gain at any one of the longitudinal modes saturates, others can reach threshold [64, 65]. A model that can account for these features should provide more accurate quantitative predictions of the behavior. Such a model has been proposed, in which four-wave mixing (FWM) between the longitudinal modes of the laser provides a saturating effect [64].

Four-wave mixing describes a parametric process in which energy transfer takes place between four waves at two or more frequencies. It requires the conservation of energy, $\omega_1 + \omega_2 = \omega_3 + \omega_4$, and momentum (or phase matching), $k_1 + k_2 = k_3 + k_4$, between the participating waves, where $k_n = \omega_n/c = 2\pi/\lambda_n$ and $\lambda_n$ is the wavelength of the wave [52].

Even though the dispersion of the fiber used for RFL may be large (e.g., $D(1455 \text{ nm}) = -24$ ps/nm/km [68]) the close longitudinal mode spacing (approximately 200 kHz in a 500-m cavity length) still allows phase matching to occur. This is on a fundamental level equivalent to the broadening seen in optical pulses due to the nonlinear phenomena of self-phase modulation and was previously demonstrated by Hill et al. [67] in the context of light traveling down a fiber. A coherence length $L_{coh}$ over which efficient phase matching can occur can be defined as [52]

$$L_{coh} = \frac{4\pi^2 c}{|D|\lambda^2 \Omega_s^2},$$ (7.16)

where $\Omega_s^2$ is the cavity longitudinal spacing between modes. If the inequality $L \leq L_{coh}$ is satisfied, FWM can occur.

The 200-kHz longitudinal mode spacing should be compared to the frequency widths of the OC and HR. Figure 7.16 shows the spectral shape of an OC and a HR. Since the HR is much broader than the OC, differences in feedback to different longitudinal modes will be determined by the OC. For an OC with a full width at half the maximum bandwidth as narrow as

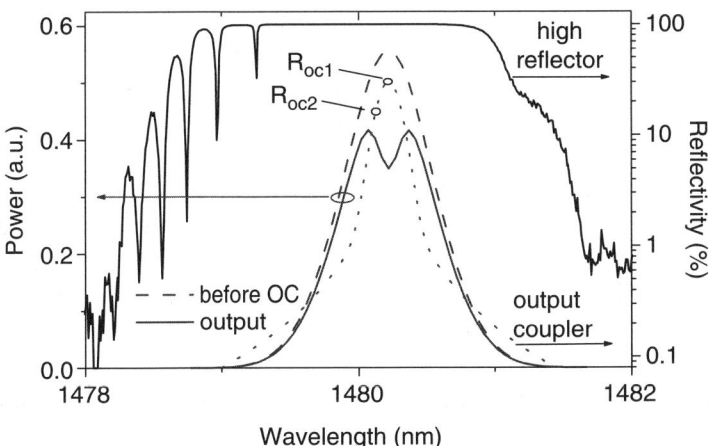

**Figure 7.16:** Experimental measurements of the reflectivity of an HR and OC, and a plot of simulation results showing the spectral output before and after the OC. (Figure courtesy of Ref. [65].)

0.1 nm, at 1455 nm, this represents a frequency width of 14 GHz; therefore, 70,000 longitudinal modes can fit within the spectral width of the OC. For the very broad Raman gain profile of doped silica fibers, it can be assumed that the gain of the fiber is uniform across the 1–2-nm widths typical of the output spectrum of these devices. Many of these modes will therefore have the same nominal gain and reflection coefficients. Considering only two modes with reflectivities given by $R_{OC1}$ (peak) and $R_{OC2}$ (off-peak) as shown in Figure 7.16, as the power in mode 1 grows it will transfer energy to mode 2 through FWM so that mode 2 will also begin to lase.

In Ref. [65] a numerical model was presented using this approach. In order to understand how Eq. (7.13) can be modified to account for FWM, four changes to Eq. (7.13) must be understood. First the many longitudinal modes at each of the Stokes orders are lumped together to form supermodes, each of width $\Delta\lambda$. This is to reduce the computing time in using this algorithm. Depending on the desired accuracy, these modes can have a width from 0.01 to 0.1 nm.

Next a different meaning is assigned to the subscript $j$ in Eq. (7.13). Whereas previously $j$ represented a given Stokes order, here it represents a supermode. The numbering of the modes begins at the pump light and is incremented continuously from the pump light to the output wavelength. So, for example, in the 1455-nm laser shown in Figure 7.2b, assuming each Stokes order contains five supermodes, the first longitudinal mode in the pump light is number 1, and the last mode in the 1455-nm light is number 30. This allows one to consider Raman interaction between modes within the same Stokes order; however, the FWM between different Stokes orders is negligible.

The third assumption is that the FWM is binary within some limit $\Delta\lambda_{limit}$. Therefore, if the wavelength difference between two modes is less than $\Delta\lambda_{limit}$ the two modes are fully matched; for larger wavelength differences, there is no FWM. The spacing limit should be set to include several modes but not to limit computing time. In Ref. [65], $\Delta\lambda_{limit}$ was set to 0.4 nm.

Finally, a simple phenomenological expression was found to define the power $P_j$ in the $j$th longitudinal mode and is given by [65]

$$\frac{dP_j}{dz} = \sum \frac{\delta\gamma K}{\Delta\lambda} \sqrt{P_j P_k P_l P_m} \left( \frac{1}{\sqrt{P_j}} + \frac{1}{\sqrt{P_k}} + \frac{1}{\sqrt{P_l}} + \frac{1}{\sqrt{P_m}} \right) P_T^2 \quad (7.17)$$

where the summation is over all possible mode combinations of $k$, $l$, and $m$, within $\Delta\lambda_{\text{limit}}$, that interact with the $j$th mode; $\delta$ is $+1(-1)$ if the power is being supplied (depleted) to (from) the $j$th mode; $\gamma$ is the nonlinear coefficient defined in Eq. 2.1.15 (from Chapter 2); and $P_T = P_j + P_k + P_l + P_m$; and $K$ is an experimentally determined constant.

Using this model, the performance of an RFL was modeled and matched to experimental results as shown in Figure 7.17a. Fairly accurate predictions about the shape of the RFL output spectra were made for different output pump power levels. The width of the OC used for these results was 0.2 nm. Figure 7.17a also shows that as the output spectrum eventually becomes broader than the OC, a dip appears in the center of the output spectrum of the RFL. This dip occurs at the wavelengths where the OC reflectivity is highest, since more power at these wavelengths is reflected back into the cavity.

The qualitative result shown in Figure 7.17a was quantified by defining an effective reflectivity $R_{\text{eff}}$ as follows:

$$R_{\text{eff}} \equiv \frac{P_R}{P_i} = \int S_{\text{out}}(\lambda)\frac{R_{\text{oc}}(\lambda)}{1 - R_{\text{oc}}(\lambda)}d\lambda, \tag{7.18}$$

where $P_i$ and $P_R$ are the powers incident and reflected at the output coupler, respectively, and $S_{\text{out}}(\lambda)$ is the linear spectral density in W/m measured on an optical spectrum analyzer. An experimental measurement of $R_{\text{eff}}$ can be obtained by measuring $R_{\text{oc}}(\lambda)$ and $S_{\text{out}}(\lambda)$ at various CRR output power levels. The $R_{\text{eff}}$ obtained in this way can be compared to that obtained in simulations. Such an evaluation was done in Ref. [65] for a 1428-nm CRR with an OC of 70% reflectivity and 0.1-nm bandwidth. The good agreement obtained is shown in Figure 7.17b. For low power levels, $R_{\text{eff}}$ is close to the peak reflectivity. As the power increases and light passes around the OC, $R_{\text{eff}}$ decreases, reaching almost 20%.

The model can now be used to predict the behavior of the spectral output of the CRR [68]. It has already been noted that as the output power increases, the spectrum broadens. This is quantified in Figure 7.18a for both experiments and simulations. This widening is due to the increase in the nonlinear transfer of energy between the modes due to the higher power in each mode.

Likewise, an increase in the linewidth of the CRR output occurs as the length of fiber is increased due to increased nonlinear interactions. This is

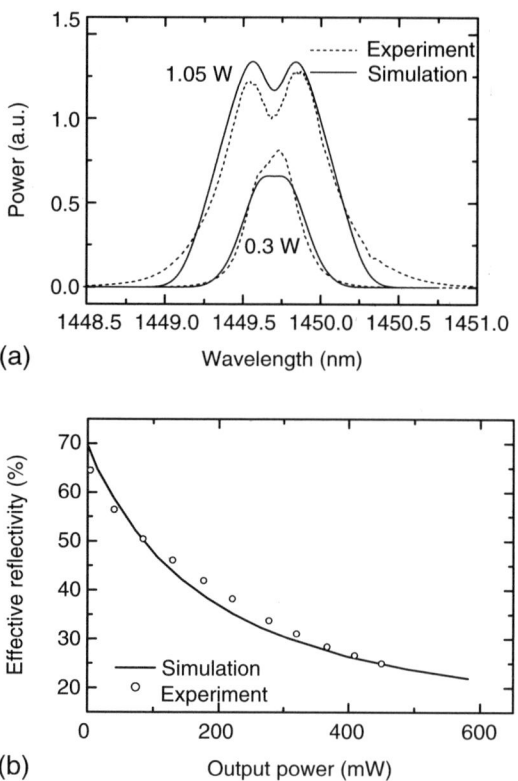

(a)

(b)

**Figure 7.17:** (a) Experimental and simulation plots of the output spectra of a CRR at two different power levels, and (b) effective reflectivity as a function of output power. (Figure courtesy of J.-C. Boutellier from Ref. [65].)

shown in Figure 7.18b under two conditions: constant output power and constant pump power. Under constant pump power, some of the changes in linewidth are due to increased output power due to a more optimum choice of fiber length.

It is also instructive to examine how the model predicts that the output spectral linewidth would be affected by the bandwidth of the OC. As seen in Figure 7.18c, which is a plot of the bandwidth of the output spectrum versus OC bandwidth of the RFL for constant output power levels, the model predicts that the output spectral width increases with increasing

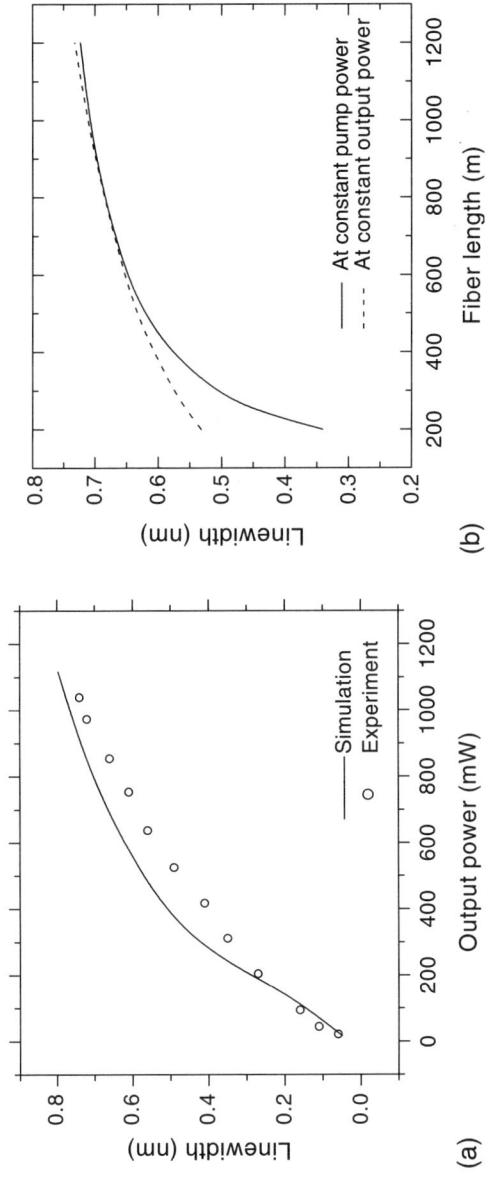

**Figure 7.18:** Simulation results showing the effect on linewidth of (a) output power, (b) fiber length, (c) OC bandwidth, and (d) $R_{OC}$. (Figure courtesy of J.-C. Boutellier from Ref. [68].)

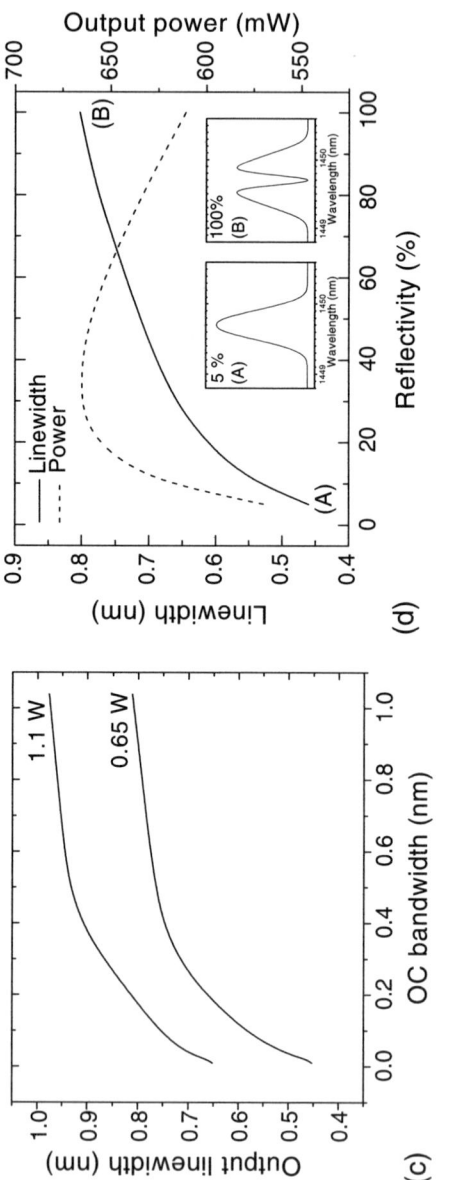

**Figure 7.18:** Continued.

OC bandwidth, but at a decreasing rate. This is shown when the output power level is held constant at 0.65 and 1 W. The increase in spectral width is intuitive as a large number of modes have enough feedback to reach threshold. The decreasing rate is expected since as the same amount of power is spread over a wider spectral bandwidth, the intermodal FWM will be reduced due to the lower peak powers.

Finally, the effect of reflectivity on the linewidth of the RFL is discussed. As was described in Section 7.2.1, the reflectivity will also affect the CRR output power. Figure 7.18d shows that as the OC reflectivity is increased, the laser bandwidth will increase but at a decreasing rate. This increase in linewidth continues even when the output power begins to decrease, though the rate of linewidth increase decreases. This behavior is explained by the increased intercavity power in the RFL, as more power is retained in the cavity as the reflectivity is increased. This will increase the FWM induced spectral broadening. It is important to note that the model predicts there will still be some output power even if the OC reflectivity is 100% as power will leak out of the sides of the OC, as shown in Figure 7.18d.

## 7.2.4   Noise Properties

Raman scattering is a very fast ($<1$ ps) process; thus, any pump fluctuations occurring on a time scale slower than 1 ps can cause fluctuations in the signal gain. This imposes stringent requirements on the noise of a pump laser [69,70], as was discussed in detail in Chapter 5. In a counterpumped configuration, because the pump and the signal travel in opposite directions, there is a strong averaging of pump power fluctuations that can occur for high-frequency noise oscillations. In the copumped configuration, the signal and pump propagate through the fiber together, and the only averaging effect is through the dispersive delay caused by walk-off between the pump and the signal. One measure of the noise in a pump is the relative intensity noise (RIN) of the pump. The RIN values at a frequency $\nu$ are equal to the mean square optical power fluctuations divided by the square of the mean signal power in a 1.0-Hz bandwidth. It has been shown that for a 100-km span of standard single mode fiber the required pump RIN is $<-119$ dB/Hz and $<-81$ dB/Hz for the co- and counterpumped cases, respectively, across a broad frequency range that extends up to 100s MHz [70]. While the exact

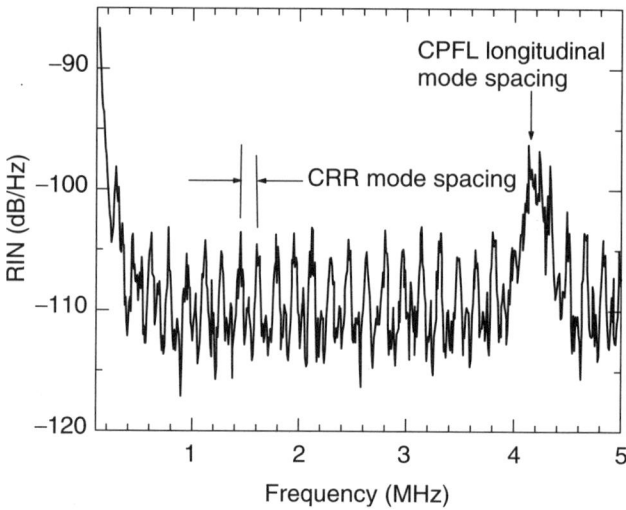

**Figure 7.19:** The RIN spectrum of a 1366-nm CRR. The low-frequency oscillations are the longitudinal modes of the CRR and the 4.2-MHz peak is the longitudinal mode spacing of the CPFL.

numbers will vary depending on fiber type and length these numbers are good estimates of the required pump RIN.

An example of the RIN spectrum for a 1366-nm CRR is shown in Figure 7.19. Evident in this spectrum are the peaks 170 kHz apart corresponding to the longitudinal modes of the CRR and the 4.2-MHz peak corresponding to the longitudinal modes of the CPFL. The RIN level for this device, and for CRR in general, is too high for copumped Raman amplification. Based on work with semiconductor diodes, as described in Chapter 6, it can be speculated that the higher RIN value is due to the larger number of modes in the fiber as compared to a semiconductor laser [71, 72]. This being the case, novel approaches are needed in order to significantly reduce the RIN of a CRR.

## 7.3  Multiple Wavelength Cascaded Raman Resonators

In previous chapters the use of multiple pump wavelengths to obtain a flat Raman gain profile was discussed. In this section, the adaptation of a single CRR to producing several wavelengths to fill this need is discussed [27, 73]. A CRR is capable of producing significantly more power than is necessary for Raman amplification in telecommunication, but it also has a relatively high threshold. By making a device that is capable of producing multiple output wavelengths from a single gain fiber, the relatively high output power is used advantageously, and because the total signal output is increased the fraction of the total pump power required to reach threshold is reduced. To date, two-, three-, and six-wavelength multiple-wavelength CRRs (MWCRRs) have been demonstrated [73–82].

Multiple-wavelength lasing in RFL was first demonstrated in a dual wavelength device [74]. Little work was done subsequent to this, but interest was renewed beginning with the publication of an MWCRR using a ring cavity [75]. Linear cavities showing two and three wavelengths were demonstrated shortly thereafter [76, 77]. For a three-wavelength CRR (3λCRR) device emitting 1427-, 1455-, and 1480-nm wavelengths in Ge-doped fibers, a threshold power of 400 mW and the total slope efficiency of 38% from 1100- to 1480-nm light have been reported [76, 77]. In a similar device using P-doped fibers, a threshold of 2.8 W and a slope efficiency of 50% were reported from the 1117-nm pump to the signal output [78].

A 3λCRR similar to that used in Ref. [79] is shown schematically in Figure 7.20. It is pumped by a Yb-doped CPFL at 1100 nm. It is similar to a 1λCRR in that it consists of a spool of enhanced Raman-gain single-mode fiber. Light is shifted from 1100 nm to 1347 nm in the Raman fiber with the efficiency aided by four nested pairs of HRs with fixed reflectivities of approximately 99%. The output grating set also contains an HR

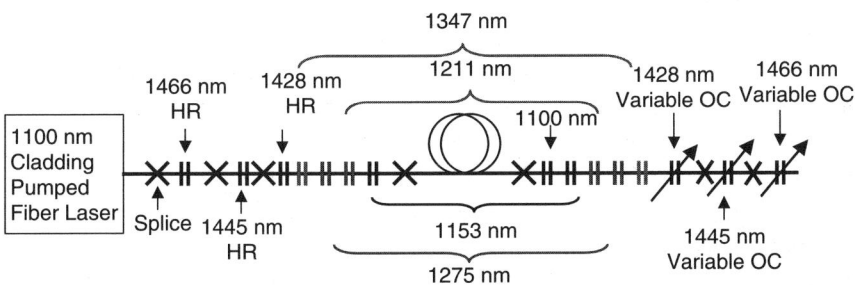

**Figure 7.20:** Schematic of a multiwavelength CRR with output wavelengths at 1428, 1445, and 1466 nm.

for the 1100-nm pump. The difference with a 1$\lambda$CRR is that resonant cavities at 1428, 1445, and 1466 nm ($\lambda_1,\lambda_2,\lambda_3$) are created by a set of OCs and a matching set of broadband HR gratings. Examples of the emission spectra obtained from this device are shown in Figure 7.21. The ability to reconfigure the power between the different wavelengths is evident.

The ability to control the distribution of power in a multiple-wavelength RFL (MWRFL) is critical for a practical device. Two techniques for accomplishing this have been proposed. In the first the HR and OC at a specific wavelength are deliberately misaligned, changing the efficiency of the cavity at that wavelength and hence its power. This is done by strain or temperature tuning the center wavelength of the grating [75, 80]. In the second approach the reflectivities of the OC are varied. This is accomplished by depositing a metal heater on the fiber with a tapered thickness. When current is applied to the metal, the resistance it sees is inversely proportional to the thickness of the metal film. This variation in resistance leads to a temperature gradient. The temperature gradient and hence the reflectivity can be varied by changing the applied current [77–79, 84]. Spectra for the OC at various reflectivities are shown in Figure 7.22a, with the corresponding output laser spectra at one of the wavelengths shown in Figure 7.22b. Note that in this approach there are only small changes in the center wavelength as the OC (power) at that wavelength is adjusted.

The redistribution of power between the different wavelengths as the OCs are adjusted is not straightforward. Figure 7.23 shows the normalized Raman gain profile for the 1347-, 1427-, 1445-, and 1466-nm wavelengths for the device shown in Figure 7.21. It is seen that the 1347-nm line can provide gain to all three output wavelengths, but the 1428-nm line can pump the 1445-nm and 1466-nm line, and the 1455-nm line can pump the 1466-nm line. Examples of the output power at each of the three wavelengths and the total output power as a function of the incident 1100-nm pump power are shown in Figure 7.24. In a nominal cavity configuration in which no voltages are applied to the OC, the total output power is nonlinear as seen in Figure 7.24a. This is because of the differences in quantum energy of the different wavelengths and the fact that the power is redistributed among the different wavelengths as the CPFL power is increased. Figures 7.24b and 7.24c show that if the power ratio between the wavelengths is kept the same, the input versus output power is linear for the individual wavelengths and the total power [79].

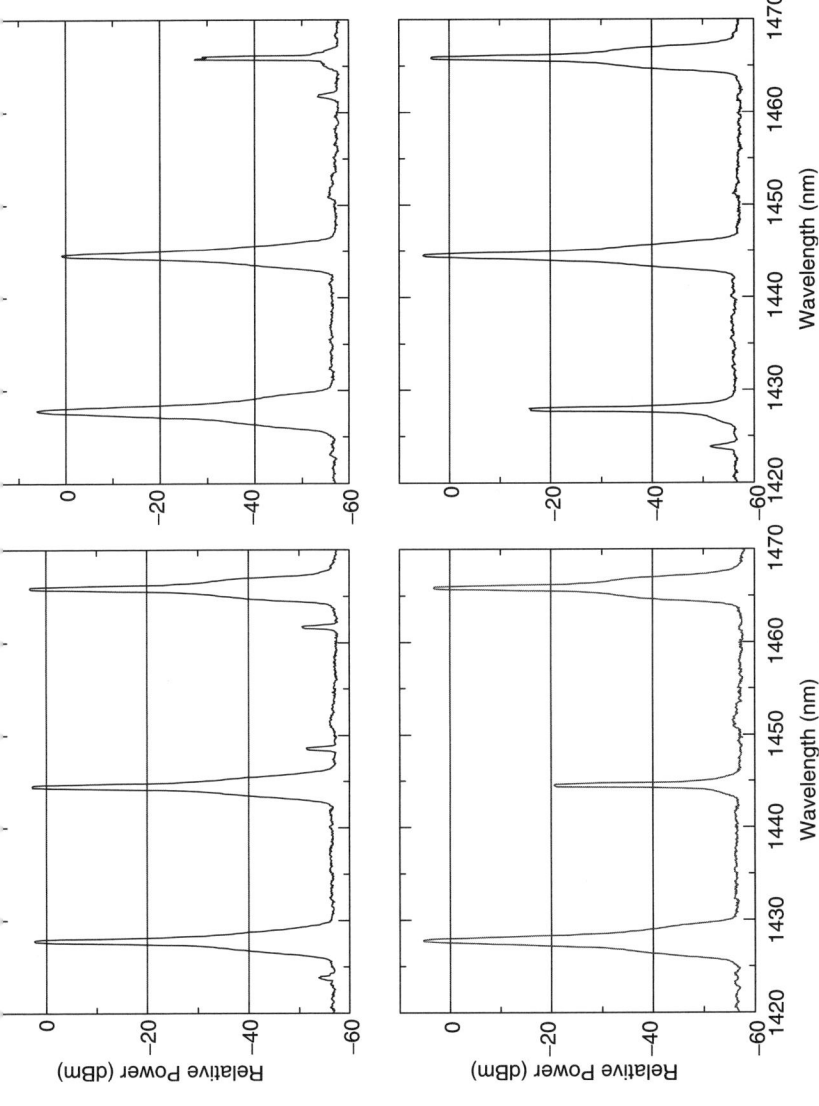

**Figure 7.21:** The output spectra of a $3\lambda$CRR for different combinations of OC reflectivity. The ability to reconfigure the distribution of power among the three wavelengths is evident.

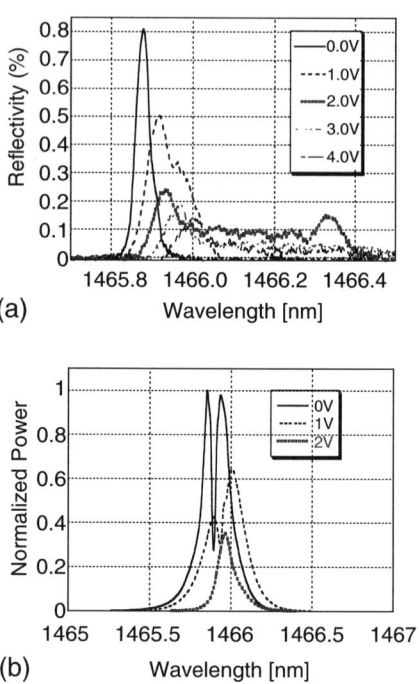

**Figure 7.22:** Spectral plots of (a) the OC reflectivity and (b) the output CRR power for various OC reflectivity settings. (Courtesy of M. Mermelstein OFS Fitel.)

With an understanding of how the different wavelengths behave as a function of pump power, attention is now turned to its behavior as the properties of the OC are varied. Regardless of the effect used to redistribute the power among the wavelengths, the qualitative behavior of the output power at each wavelength is similar to that shown in Figure 7.25, which shows the power at the three wavelengths as a function of the controlling voltage applied to one OC. In Figure 7.25a, as the voltage on the 1445-nm OC is increased, thereby decreasing the OC reflectivity, the power emitted at 1445 nm decreases. Simultaneously there is a rise in the power at 1428 nm since the depletion of its power by the 1445-nm light is decreasing. The power at 1466 nm remains fairly constant. Similar results are seen when the OC reflectivity of the 1466-nm line is reduced (Figure 7.25b). Now the power in emitted at 1428 and 1445 nm increases as they are no longer

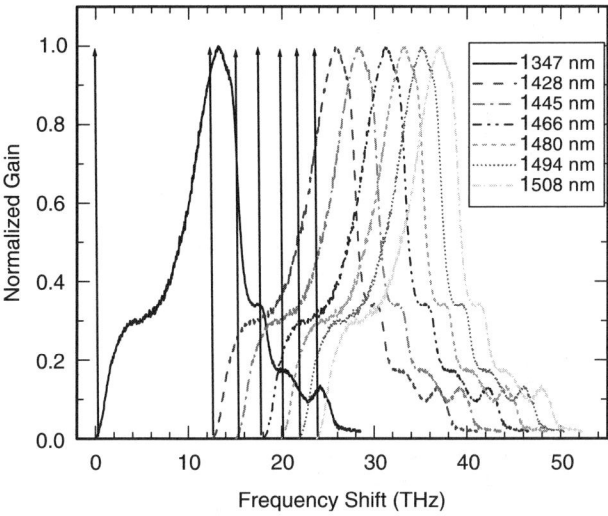

**Figure 7.23:** Schematic showing the normalized Raman gain curve at 1428 nm, 1445 nm, 1466 nm, 1480 nm, 1494 nm, and 1508 nm relative to a 1347-nm gain curve.

being depleted by the 1466-nm line. Note that in both cases the total power remains constant.

The interdependence of the optical powers at the different wavelengths raises two questions. Can a useful set of wavelength power distributions be reached, and how stable is the source at these operating points? The next two sections answer these questions using an MWCRR in which the power is divided between the wavelengths by varying the OC reflectivity. A $3\lambda$CRR was placed in the experimental setup shown in Figure 7.26. Eight C-band equally spaced signal lasers (1538.2–1560.6 nm) were launched down a 100-km span of OFS TrueWave® Fiber. Pump radiation from the $3\lambda$RFL was launched in the counterpropagating direction with the aid of a wavelength-division multiplexer (WDM). A 2% tap provides radiation for pump power monitoring by the optical spectrum analyzer (OSA-2). The amplified signal lasers passed through the WDM to OSA-1 for measurement of the gain and the gain ripple $\Delta G$. The OSAs and all voltage controls to the OCs were interfaced to a controlling computer for collection and storage of the gain-flattening and power-partitioning data. These

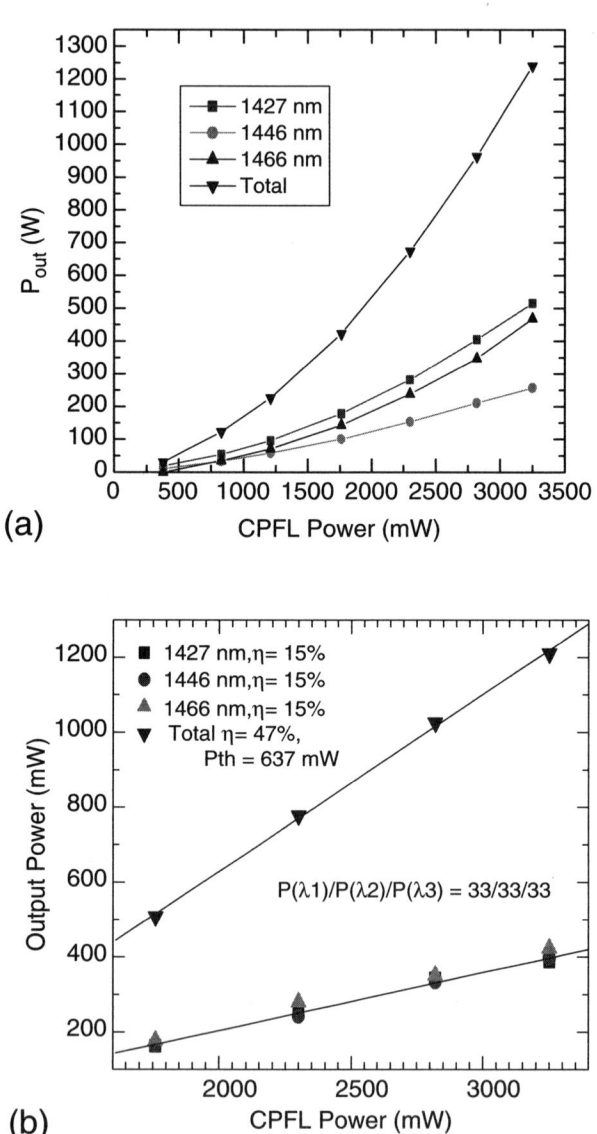

(a)

(b)

**Figure 7.24:** Output power versus input power of a CRR when the OC reflectivity is (a) with no voltages applied and with the OC reflectivity adjusted for a given input power to produce a (b) 33/33/33% and (c) 50/30/20% power ratio between the output powers of $\lambda_1/\lambda_2/\lambda_3$.

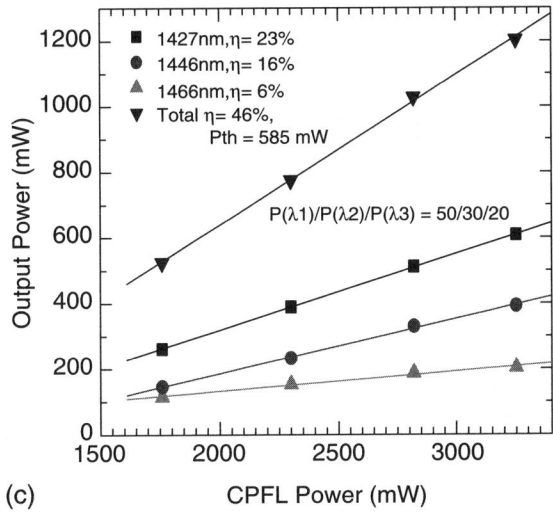

**Figure 7.24:** Continued.

results are not general in the sense that the stability of the source depends on the physical mechanism used to redistribute the wavelengths (e.g., varying reflectivity or misaligning gratings), as well as how that mechanism is exploited (e.g., strain or temperature tuning of gratings).

## 7.3.1 Obtainable Operating Points

Using the setup of Figure 7.26, the complete accessible power space was explored by varying the three OC voltages from 0 to 3 V in 0.2-V increments [79]. This was done for total output powers of 860 and 1330 mW. Figure 7.27 shows that the power points corresponding to each output power lie on a plane in a Cartesian optical power space confirming that the total power is constant.

The three-dimensional power state can be conveniently represented in two dimensions by the simplex diagram shown in Figure 7.28. An equilateral triangle is constructed with vertices $(x,y)$ equal to $(0,\sqrt{3}/2)$, $(-1/2,0)$, and $(0,1/2)$, representing the power states $(P_1,0,0)$, $(0,P_2,0)$, and $(0,0,P_3)$,

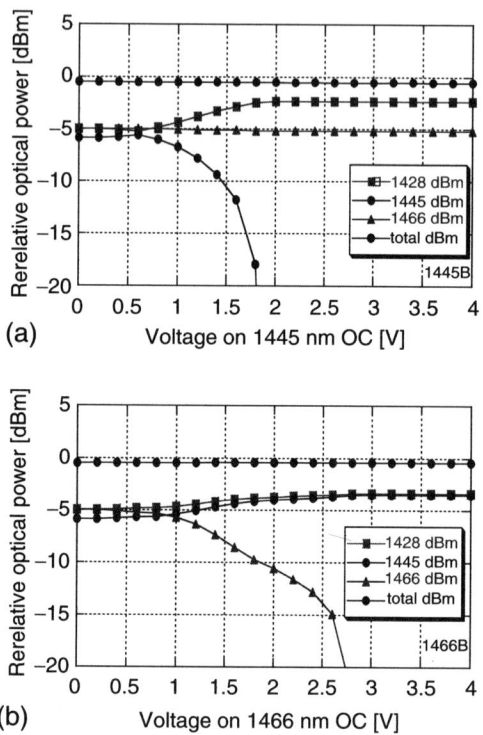

(a)

(b)

**Figure 7.25:** Changes in the output power of the three different wavelengths as the reflectivity of (a) the 1445-nm and (b) the 1466-nm OC is decreased by applying a voltage to them. (Courtesy of M. Mermelstein OFS Fitel.)

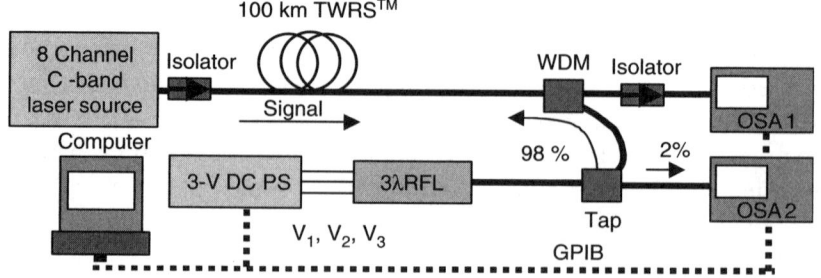

**Figure 7.26:** Experimental setup for evaluating the performance of an MWCRR. (Courtesy of M. Mermelstein from Ref. [78].)

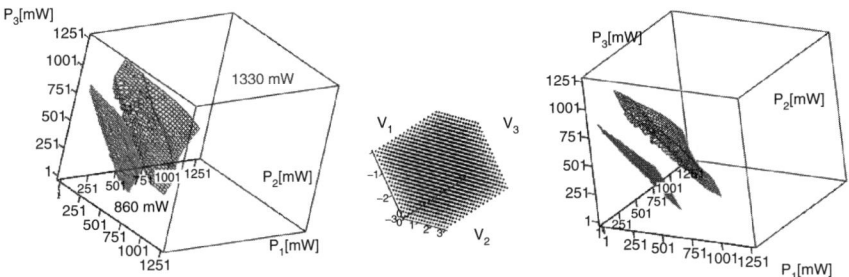

**Figure 7.27:** Two views of a three-dimensional plot showing the power distribution between the three wavelengths for two different output powers. The center plot shows the voltage distribution. (Courtesy of M. Mermelstein from Ref. [78].)

respectively. This construction can be generalized to an arbitrary power state $(P_1, P_2, P_3)$ according to the following equations:

$$x = -\frac{1}{2} \cdot \left(\frac{P_2}{P_0}\right) + \frac{1}{2} \cdot \left(\frac{P_3}{P_0}\right), \qquad (7.19a)$$

and

$$y = \frac{\sqrt{3}}{2} \cdot \left(\frac{P_1}{P_0}\right), \qquad (7.19b)$$

where $P_0 = P_1 + P_2 + P_3$. Therefore a point at the center of the triangle $(0, \sqrt{3}/6)$ represents an equal power distribution among all the wavelengths. A point at $(0,0)$ represents an even power distribution between $\lambda_2$ and $\lambda_3$. Likewise any other point represents some fractional distribution of the power among the wavelengths. The small dots in Figure 7.28 show the power states spanned by the 3λRFL at a total launch power of 620 mW. The simplex coverage is not a sensitive function of total power. Data were taken with a voltage resolution of (0.20, 0.20, 0.05) volts for $(V_1, V_2, V_3)$. The open symbols correspond to experimentally determined power distributions corresponding to minimum gain ripple, for 60-, 100-, and 140-km span lengths of OFS TWRS fiber. The adjacent solid symbols are simulation results for the same fiber and for standard single-mode fiber at similar lengths. The experimental data points and simulation data points

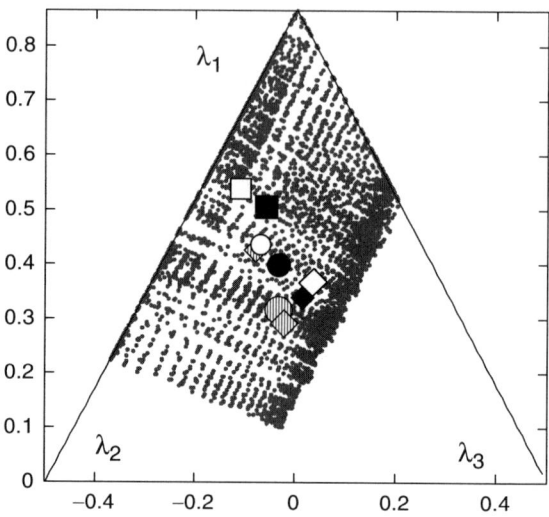

**Figure 7.28:** A Simplex plot showing the distribution of power between the three different wavelengths as the peak reflectivity of the OC is varied. The symbols have the following meaning.

TWRS, 60 km, Sim- ◆ $\Delta G = 0.7$ dB/Exp- ◇ $\Delta G = 0.7$ dB
TWRS, 100 km, Sim- • $\Delta G = 1.1$ dB/Exp- ○ $\Delta G = 1.1$ dB
TWRS, 140 km, Sim- ■ $\Delta G = 1.5$ dB/Exp- □ $\Delta G = 1.6$ dB
SMF-28, 60 km, Sim- ◆ $\Delta G = 0.7$ dB
SMF-28, 100 km, Sim- ◕ $\Delta G = 1.1$ dB
SMF-28, 140 km, Sim- ▦ $\Delta G = 1.5$ dB

fall within the laser's reach. This leads to the conclusion that the laser can access an adequate pump combination space.

The simplicity and ultimately the speed at which one is able to reconfigure the desired operating point are also interesting. One aspect of the problem that can be answered is the number of iterations in the OC reflectivity needed in tuning the cavity to a given wavelength distribution. Using the experimental setup shown in Figure 7.26, a given operating point was selected. Initially, the total power was determined by adjusting the injection current to the multimode laser diodes. The reflectivity of each OC was then adjusted in turn until the desired power at the wavelength corresponding to that OC was reached. The next OC was then adjusted until the power at its

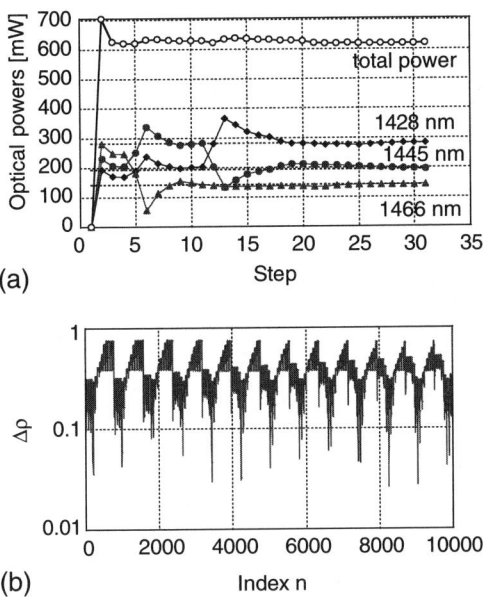

(a)

(b)

**Figure 7.29:** (a) A plot showing the number of iterations of the OC reflectivity required to reach a selected operating point. Each step represents one iteration of the voltage. (b) A plot of different combinations of OC reflectivities versus the deviation from the desired operating point. (Courtesy of M. Mermelstein OFS Fitel.)

wavelength was at the required power. Because changing the OC at each wavelength changes the power in the other wavelengths, this process was repeated until all the radiation at each wavelength was at the desired power level. Figure 7.29a is a plot of the number of steps required to reach the operating point. Each step corresponds to one adjustment of an OC. In this example as few as 20 steps were required to reach the desired operating point. The exact number of steps will depend on the prior and final operating points, and it may be possible to implement faster search algorithms; however, it is reasonable to assume the number of steps needed is on the order of tens. A more thorough answer to how fast the laser can be tuned depends on the electronics used, the physical mechanism for partitioning the powers, and the starting and finishing operating point.

It should also be noted that a given operating point is not a unique combination of reflectivities. This is shown by defining an operating point, $(P_1^0, P_2^0, P_3^0)$. Deviations from this operating point, $\Delta\rho$, were quantified by the equation

$$\Delta\rho = \frac{\sqrt{\left(P_1 - P_1^0\right)^2 + \left(P_2 - P_2^0\right)^2 + \left(P_3 - P_3^0\right)^2}}{P_0}, \qquad (7.20)$$

where $P_n$ is the actual power at wavelength $\lambda_n$. In the experimental setup of Figure 7.26, an operating point was defined, and the voltages applied to each OC were incremented in turn with each iteration of the voltage assigned a sequential number. The results of the error function versus the index number are plotted in Figure 7.29b. It shows that the configurations for a given operating point are cyclic, and various combinations of OC reflectivity lead to the same operating point.

## 7.3.2 Operating Point Stability

With some confidence that a 3λCRR can reach a multitude of operating points, the stability at these operating points to small changes in the voltage applied to the OC was considered in Ref. [79]. This was also studied using the experimental setup shown in Figure 7.26. The total optical power and power wavelength distribution were adjusted so that the difference between the signals with the highest and lowest gains, $\Delta G$, is a minimum, and the average gain supplied to the fiber is equal to its loss. A $\Delta G$ of 1.4 dB at transparency was achieved at an initial voltage operating point $(V_1^0, V_2^0, V_3^0)$ of (2.30, 1.30, 1.05) volts and a launch pump power operating point $(P_1^0, P_2^0, P_3^0)$ of (283, 194, 144) mW. The stability of this OP was interrogated by fixing one of the voltages at its operating point and varying the other two in a range of ±0.5 V in increments of 25 mV. Figure 7.30a is a contour plot of the gain ripple at constant $V_1$. The rectangle indicates a region of ±50 mV about the initial OP corresponding to 1.1 dB $< \Delta G <$ 1.8 dB, yielding a voltage sensitivity of ∼0.01 dB/mV. The gain ripple minimum was 1.13 dB. Contour plots for constant $V_2$ and constant $V_3$ show similar results with reduced voltage sensitivity.

The power space $(P_1, P_2, P_3)$ was explored in the vicinity of the power OP, with deviations in the power distribution measured as defined

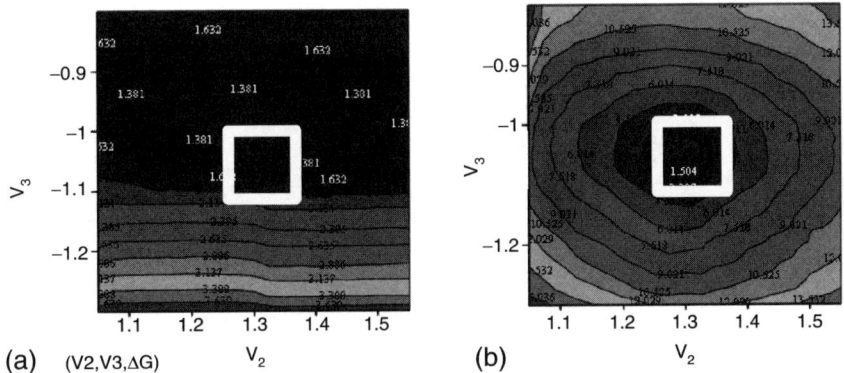

(a)  (V2,V3,ΔG)          $V_2$          (b)          $V_2$

**Figure 7.30:** Contour plots showing the stability of a 3λCRR as measured by (a) its gain ripple $\Delta G$ and (b) its power stability $\Delta\rho$. In each case the voltage $V_1$ is held constant as $V_2$ and $V_3$ are varied. The boxed interior indicates a voltage point of $\pm5$ mV. (Courtesy of M. Mermelstein from Ref. [78].)

in Eq. (7.20). Figure 7.30b shows that voltage excursions of $\pm50$ mV generate $<4\%$ variations in the 3λCRR spectral power distribution indicating a $\Delta\rho$ voltage sensitivity of $\sim0.04\%$/mV. Data at constant $V_2$ and constant $V_3$ show comparable results with lower voltage sensitivity.

The low sensitivity of $\Delta G$ to the applied OC voltage indicates that a 3λCRR can operate stably, and that either $\Delta G$ and/or $\Delta\rho$ can serve as an error signal in a suitable power configuring control algorithm.

## 7.3.3  Temporal Behavior of an MWRFL

The importance of the noise of a Raman pump source was discussed in Section 7.2.5 and in previous chapters. With an MWCRR there was a heightened concern that simultaneous lasing of different wavelengths could produce additional noise. The RIN of a 3λCRR was measured by Mermelstein et al. in Ref. [78], by separating the 1427-, 1455-, and 1480-nm wavelengths and measuring their individual RIN while they are all lasing. The results in a 100-kHz bandwidth are shown in Figure 7.31. This bandwidth was chosen to include the low and intermediate frequency regime where averaging of the pump power fluctuations does not completely

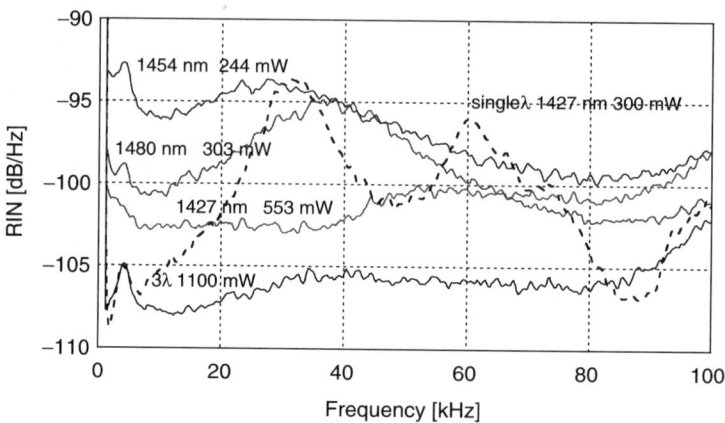

**Figure 7.31:** RIN spectra for the individual wavelengths and all three wavelengths of a 3λCRR. The RIN spectrum for a 1λCRR is shown for comparison. (Courtesy of M. Mermelstein from Ref. [77].)

remove this noise source in a counterpumped configuration [69, 70]. The solid lines in Figure 7.31 are the RIN spectra for each spectral component of the multiwavelength Raman fiber laser and for all three spectral components simultaneously. It is seen that the RIN values have a strong correlation with the output power, and all lie below −90 dB/Hz. The solid trace exhibiting the lowest noise level is that corresponding to the RIN of the three wavelengths measured simultaneously. A careful examination of the corresponding time records shows that the photovoltage standard deviation for all three spectral components is equal to the incoherent summation of the standard deviations for the individual spectral components indicating that the fluctuations in each spectral component are statistically independent. The dashed line corresponds to the RIN measured for a single wavelength 1427-nm Raman fiber laser identical to the multiwavelength laser but without the 1466-nm and 1480-nm OCs. This laser was operated at an optical power of 300 mW corresponding to the typical power in a single component of the multiwavelength fiber laser. The single-wavelength noise level is comparable to that measured for the three spectral components of the three-wavelength laser. Therefore, MWRFLs exhibit noise levels comparable to single-wavelength devices and are adequate for counterpumped Raman amplification.

### 7.3.4 Six-Wavelength Raman Fiber Lasers

The full potential of an MWCRR can be examined by looking at a 6λCRR, capable of amplifying both C- and L-band signals using a single device [80–82]. A 6λCRR is like the 3λCRR shown in Figure 7.20, except HRs in the input grating set and the variable OCs in the output grating set are added at 1480 nm, 1494 nm, and 1508 nm. The interactions of the output wavelengths within the cavity become even more complicated. With reference to Figure 7.23, the 1508-nm wavelength is removed in frequency from the Raman gain peak of the 1347-nm light, by nearly twice the gain curve linewidth, where the gain has fallen to approximately 10% of its peak value. Significant gain must be provided to the longer wavelength pumps from the shorter output wavelength pumps, suggesting that a controlled power distribution may have been difficult to achieve.

To test this control, the 6λCRR was used to counterpump a 100-km span of True Wave® RS fiber [82]. Figure 7.32 shows the six-wavelength pump power distribution required to achieve an optimal gain flatness of 1.7 dB peak–peak in the C+L bands. This power distribution was achieved

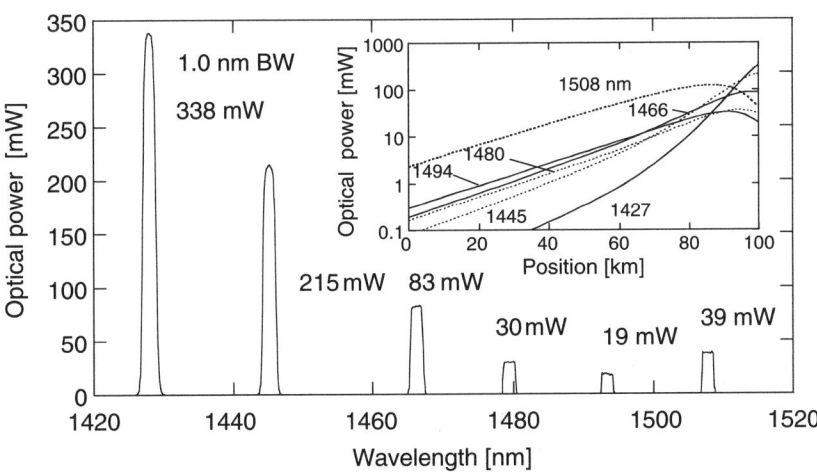

**Figure 7.32:** Pump power distribution of a 6λCRR needed to achieve optimal gain flatness in a 100-km span of True-Wave® RS fiber. The inset shows the pump power evolution in the fiber. (Courtesy of M. Mermelstein from Ref. [81].)

by judiciously varying the voltages to the six OCs. Note that the longer pump wavelengths have significantly less power than the shorter pump wavelengths. This is because the short wavelength pumps amplify the long wavelength pumps as the radiation propagates along the fiber length. This is illustrated by the inset in Figure 7.32, which shows a simulation of the pump power as a function of fiber length in a counterpumped configuration.

The RIN of this device was measured and is shown in Figure 7.33. Shown are the RIN spectra for the individual wavelengths and for a 1λCRR at 1427-nm wavelength. It should be noted that optical receivers typically have a low frequency cutoff of at least 10 kHz. Therefore the high RIN values at frequencies less than 10 kHz will not affect the system performance. Hence a 6λCRR can be used in a counterpumped Raman amplified optical system.

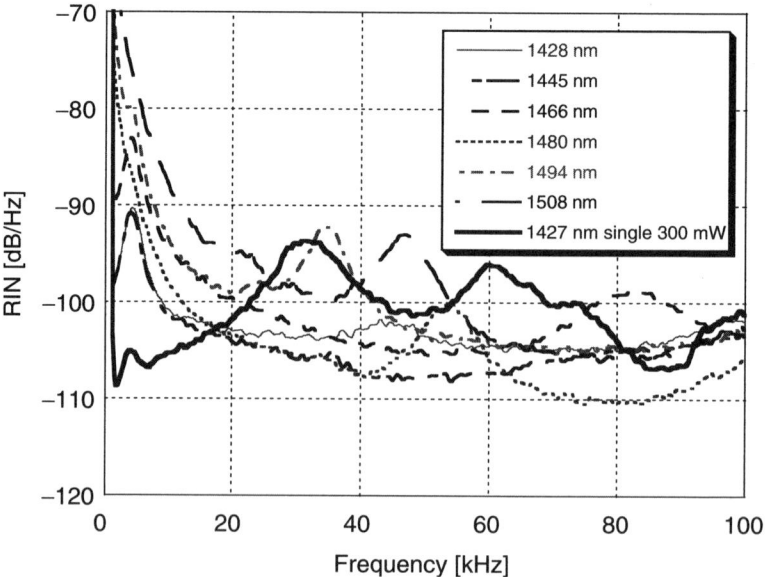

**Figure 7.33:** The individually measured RIN spectrum of each of the six wavelengths 6λCRR. Also shown is the RIN spectrum of a 1λCRR for comparison. (Courtesy of M. Mermelstein.)

## 7.3.5   Second-Order Raman Fiber Lasers

The use of higher order Raman pumping was described in Chapter 3. In this scheme at least two pumps, one of which is more than one Stokes shift away from the signal, are used. The shorter wavelength pumps amplify the longer wavelength pumps, which in turn amplify the signals. Higher order pumping reduces the noise figure (NF) of the amplifier by more evenly distributing the gain across the length of the fiber [28, 85–88]. When the pump is two Stokes shifts away from the signal, it is referred to as a second-order pump (SOP). Since the pump powers required for higher order pumping are significantly higher than with just first-order pumping, a CRR is a good candidate source. Even more advantageous is to be able to produce all of the pump wavelengths from a single cavity.

A dual-order device is conceptually similar to the MWCRR described in the previous section. One such device taken from Ref. [88] along with its output spectrum is shown in Figure 7.34. Light from an 1100-nm CPFL is converted to 1365 nm through a cascade of Raman shifts. The OC at 1365 nm allows a portion of the light at this wavelength to exit the cavity, but enough power is retained so that lasing also occurs at 1455 nm. Controlling the ratio of 1365-nm to 1455-nm power is accomplished by varying the properties of one or both of the OCs. For an SOP, the power needed at the first-order wavelength is low enough that it would be required to operate near threshold (e.g., 10 mW in Ref. [88]). Since the laser can be unstable at such low output levels, the device is actually run at higher power levels (70 mW in Ref. [88]) with the excess power ejected from the side of the cavity using a long period grating. For the device shown in Figure 7.34, only the OC at 1455 nm is adjustable, which requires a careful selection of the OC reflectivity of the first-order pump (FOP). The ability to manipulate the ratio between the two pumps is shown in Figure 7.35a, which is a plot of the power in the SOP versus FOP. As the total power is increased, the ratio between the powers in the two pumps is held constant.

Quantifying the performance of the dual-order source in order to compare its performance to that of a single or multiwavelength device is best done by comparing the total efficiency at the desired operating point. The slope efficiency is difficult since for a fixed FOP OC coupler reflectivity as the power is increased more power is emitted at the FOP wavelength. With the long-period grating in place, this would mean that

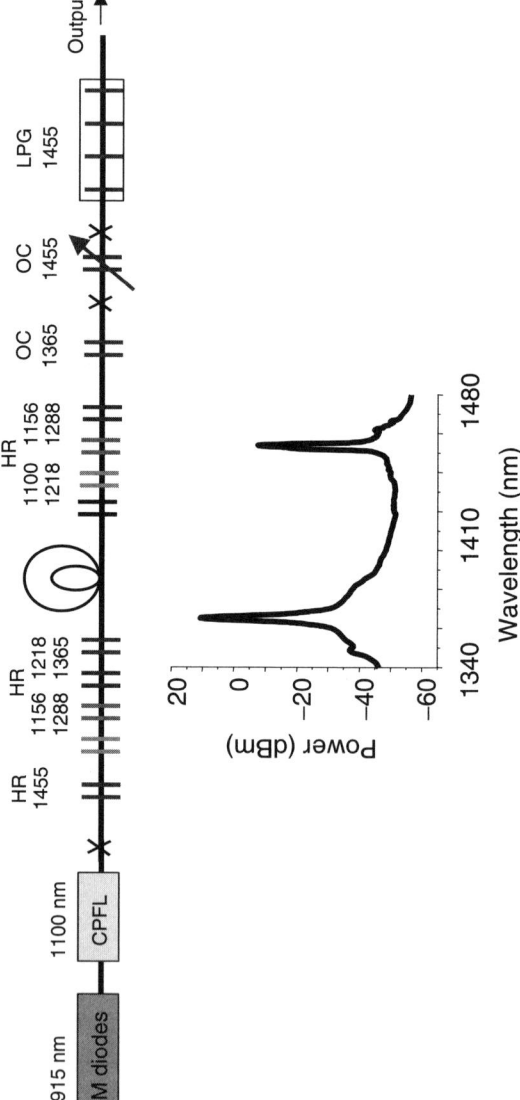

**Figure 7.34:** Schematic of a dual-order wavelength device along with its output spectrum.

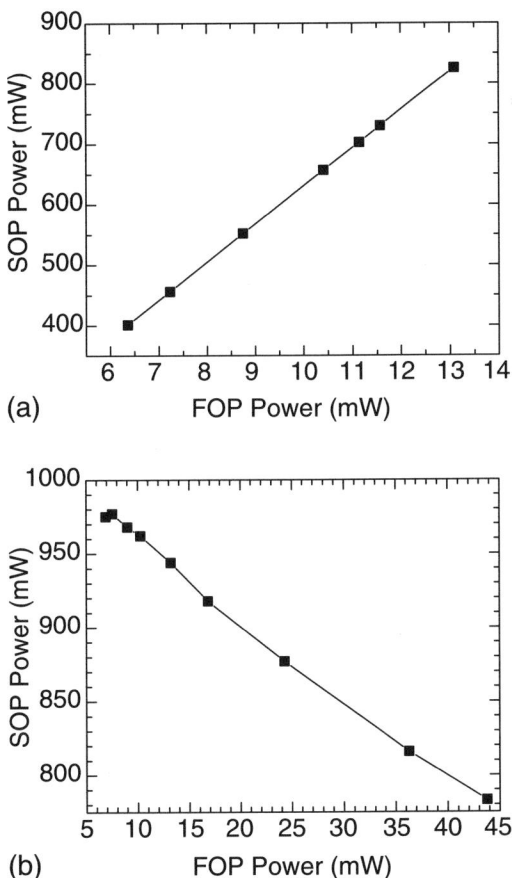

**Figure 7.35:** (a) A plot of FOP versus SOP when the ratio between the two pumps is kept constant as the pump power is increased. (b) A plot of the ratio of FOP to SOP as the gain of a system is held constant.

the efficiency of the device would decrease as a larger percentage of the power is being lost. For a Raman amplifier there are actually several different ratios of SOP-to-FOP powers that yield the same gain level. This is shown in Figure 7.35b, for an 80-km span standard single-mode fiber with 31 C-band channels pumped to transparency. In general the combinations that have the minimum amount of FOP will perform better, because (i) less power is being thrown away and (ii) lasing at 1365 nm is more efficient than at 1455 nm because it involves fewer orders.

The limitation of throwing away generated power was tackled in an experiment by Leplingard et al. [90] that used FWM in the CRR to help stabilize the low-power output of the FOP. In this scheme the zero dispersion wavelength of the Raman gain fiber was selected as 1384 nm, between the 1351-nm SOP ($\lambda_4$) and the 1428-nm FOP ($\lambda_5$). This is shown schematically in Figure 7.36. In this scheme degenerate FWM and nondegenerate FWM can occur at the 1428-nm wavelength as follows,

$$\frac{1}{\lambda_3} - \frac{1}{\lambda_4} = \frac{1}{\lambda_4} - \frac{1}{\lambda_5} \qquad (7.21a)$$

$$\frac{1}{\lambda_2} - \frac{1}{\lambda_3} = \frac{1}{\lambda_4} - \frac{1}{\lambda_5}, \qquad (7.21b)$$

where $\lambda_n$ are the various Stokes orders in the CRR shown in Figure 7.36a. The relative stability of this source at 1428 nm was compared to a

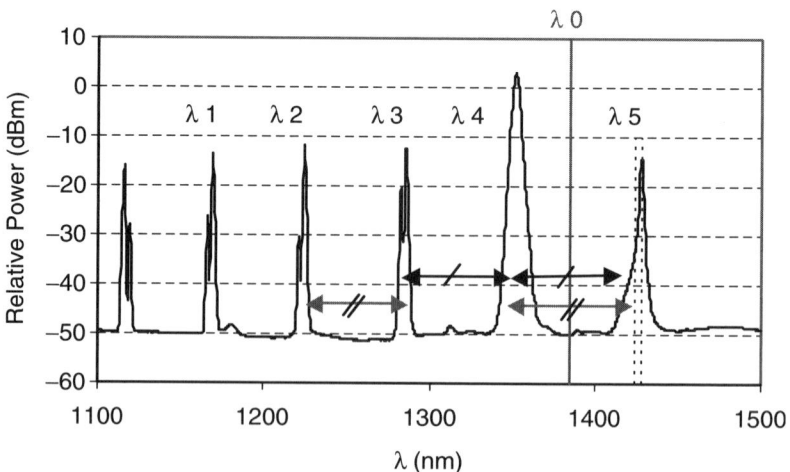

**Figure 7.36:** Plot of the spectral output of a second-order RFL. The location of the zero dispersion wavelength is indicated, and also shown are the degenerate (/) and nondegenerate (//) FWM components. (Courtesy of F. Leplingard from Ref. [90].)

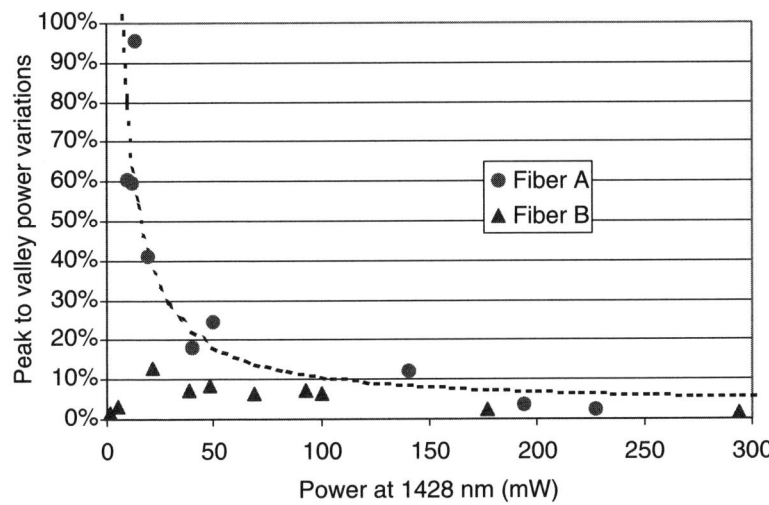

**Figure 7.37:** Stability measurements of the 1428-nm light as a function of its output power. In Fiber A there is no FWM, whereas Fiber B has FWM. The pump power and reflectivity at 1351 nm are fixed. (Courtesy of F. Leplingard from Ref. [90].)

conventional CRR, which typically has a large normal dispersion such that no FWM occurs between Stokes orders. The results are shown in Figure 7.37, which is a plot of the power at 1428 nm as the pump power is increased, versus peak-to-valley variations in the fiber. The source that uses FWM is indeed more stable at powers below 10 mW. This is due to the much lower threshold for FWM compared to SRS. Above 10 mW there is a rise in the fluctuations at the onset of Raman generated light, which eventually decreases as the 1428-nm light rises further above threshold. The degree of polarization of the source was measured to be less than 10%, a critical requirement for a Raman pump.

# References

[1]  E. P. Ippen, *Appl. Phys. Lett.* **16,** 303 (1970).

[2]  R. H. Stolen, E. P. Ippen, and A. R. Tynes, *Appl. Phys. Lett.* **20,** 62 (1972).

[3] K. O. Hill, B. S. Kawasaki, and D. C. Johnson, *Appl. Phys. Lett.* **29,** 181 (1976).

[4] C. Lin, R. H. Stolen, W. Pleibel, and P. Kaiser, *Appl. Phys. Lett.* **30,** 162 (1977).

[5] C. Lin, R. H. Stolen, W. G. French, and T. G. Malone, *Appl. Phys. Lett.* **31,** 97 (1977).

[6] C. Lin, L. G. Cohen, R. H. Stolen, G. W. Tasker, and W. G. French, *Opt. Commun.* **20,** 426 (1977).

[7] R. K. Jain, C. Lin, R. H. Stolen, and A. Ashkin, *Appl. Phys. Lett.* **31,** 89 (1977).

[8] K. O. Hill, Y. Fujii, D. C. Johnson, and B. S. Kawasaki, *Appl. Phys. Lett.* **32,** 647 (1978).

[9] R. Kashyap, *Fiber Bragg Gratings* (Academic Press, San Diego, CA, 1999).

[10] S. Grubb, T. Erdogan, V. Mizrahi, T. Strasser, W. Y. Cheung, W. A. Reed, P. J. Lemaire, A. E. Miller, S. G. Kosinski, G. Nykolak, and P. C. Becker, *Proc. Optical Amplifiers and Their Applications* (Daros, Switzerland, 1994), paper PD3-1, p. 187.

[11] S. G. Grubb, T. Strasser, W. Y. Cheung, W. A. Reed, V. Mizrachi, T. Erdogan, P. J. Lemaire, A. M. Vengsarkar, D. J. DiGiovanni, D. W. Peckham, and B. H. Rockney, *Proc. Optical Amplifiers and Their Applications* (1995), paper SA4, p. 197.

[12] H.-G. Treusch, A. Ovtchinnikov, X. He, M. Kankar, J. Mott, and S. Yang, *IEEE J. Selected Topics in Quantum Electronics*, **6,** 601 (2000).

[13] Syrbu, A. V., Yakovlev, V. P., Suruceanu, G. I., Mereutza, A. Z., Mawst, L. J., Bhattacharya, A., Nesnidal, M., Lopez, J., and Botez, D. *Proc. Conference on lasers and Electro-Optics* (1996), CTuC4.

[14] X. He, S. Srinivasan, S. Wilson, C. Mitchell and R. Patel, *Electron. Lett.* **34** 2126 (1998).

[15] E. Snitzer, H. Po, F. Hakimi, R. Tumminelli, and B. C. McCollum, *Proc. Optical Fiber Communication Conf.* (1988), paper PD5.

[16] H. Po, E. Snitzer, L. Tumminelli, F. Hakimi, N. M. Chu, and T. Haw, *Proc. Optical Fiber Communication Conf.* (1989), paper PD7.

[17] V. P. Gapontsev, P. I. Sadovsky, and I. E. Samartsev, *Proc. Conference on Lasers and Electro-Optics* (1990), paper CPDP-38.

[18] J. D. Minelly, E. R. Taylor, K. P. Jedrzejewski, J. Wang, and D. N. Payne, *Proc. Conference on Lasers and Electro-Optics* (1992), paper CWE-6.

[19] H. M. Pask, J. L. Archambault, D. C. Hanna, L. Reekie, P. St. J. Russell, J. E. Townsend, and A. C. Tropper, *Electron. Lett.* **30,** 863 (1994).

[20] D. Innis, D. J. DiGiovanni, T. A. Strasser, A. Hale, C. Headley, A. J. Stentz, R. Pedrazzani, D. Tipton, S. G. Kosinski, D. L. Brownlow, K. W. Quoi, K. S. Kranz, R. G. Huff, R. Espindola, J. D. Le Grange, and G. Jacobovitz-Veselka, *Proc. Conference on Lasers and Electro-Optics* (1997), CPD-31.

[21] E. M. Dianov, M. V. Grekov, I. A. Bufetov, S. A. Vasiliev, O. I. Medvedkov, V. G. Plotnichenko, V. V. Koltashev, A. V. Belov, M. M. Bubnov, S. L. Semjonov, and A. M. Prokhorov, *Electron. Lett.* **33**, 1542 (1997).

[22] V. I. Karpov, E. M. Dianov, A. S. Kurkov, V. M. Paramonov, V. N. Protopopov, M. P. Bachynski, and W. R. L. Clements, *Proc. Optical Fiber Communication Conf.* (1999), paper WM3, p. 202.

[23] E. M. Dianov, I. A. Bufetov, M. M. Bubnov, A. V. Shubin, S. A. Vasiliev, O. I. Medvedkov, S. I. Semjonov, M. V. Grekov, V. M. Paramonov, A. N. Gur'yanov, V. F. Khopin, D. Varelas, A. Iocco, D. Costantini, H. G. Limberger, and R.-P. Salathé, *Proc. Optical Fiber Communication Conf.* (1999), paper PD25.

[24] V. I. Karpov, E. M. Dianov, V. M. Paramonov, O. I. Medvedkov, M. M. Bubnov, S. L. Semyonov, S. A. Vasiliev, V. N. Protopopov, O. N. Egorova, V. F. Hopin, A. N. Guryanov, M. P. Bachynski, and W. R. L. Clements, *Opt. Lett.* **24**, 887 (1999).

[25] E. M. Dianov, I. A. Bufetov, M. M. Bubnov, M. V. Grekov, S. A. Vasiliev, and O. I. Medvedkov, *Opt. Lett.* **25**, 402 (2000).

[26] P. B. Hansen, L. Eskildsen, S. G. Grubb, A. J. Stentz, T. A. Strasser, J. Judkins, J. J. DeMarco, R. Pedrazzani, and D. J. DiGiovanni, *IEEE Photon. Technol. Lett.* **9**, 262 (1997).

[27] K. Rottwitt and H. D. Kidorf, *Proc. Optical Fiber Communication Conf.* (1998), paper PD6.

[28] K. Rottwitt, A. Stentz, T. Nielsen, P. Hansen, K. Feder, and K. Walker, *Proc. Eur. Conf. Optical Communication* (1999), p. II-144.

[29] See, for example, Alfalight, Boston Lasr Inc., IRE-Polus Group, JDSU, LaserTel, SLI, Unique-mode, Thales, Apollo Products List.

[30] S. Bedö, W. Lüthy, and H. P. Weber, *Opt. Commun.* **99**, 331 (1993).

[31] H. M. Pask, R. J. Carme, D. C. Hanna, A. C. Tropper, C. J. Mackechine, P. R. Barber, and J. M. Dawes, *IEEE J. Selected Topics Quantum Electron.* **1**, 2 (1995).

[32] D. J. DiGiovanni and A. J. Stentz, U.S. Patent 5 864 644 (Jan. 26, 1999).

[33] Michael Fishetyn, private communication.

[34] Th. Weber, W. Lüthy, H. P. Weber, V. Neuman, H. Berthou, and G. Kotrotsios, *Opt. Commun.* **115**, 99 (1995).

[35] Th. Weber, W. Lüthy, and H. P. Weber, *Appl. Phys. B* **63**, 131 (1996).

[36] L. Goldberg, B. Cole, and E. Snitzer, *Electron. Lett.* **33,** 2127 (1997).

[37] L. Goldberg and J. Koplow, *Electron. Lett.* **34,** 2027 (1998).

[38] L. Goldberg, J. P. Koplow, and D. A. V. Kliner, *Opt. Lett.* **24,** 673 (1999).

[39] J. P. Koplow, S. W. Moore, and D. A. Kliner, *IEEE J. Quantum Electron.* **39,** 529 (2003).

[40] C. Codermard, K. Yla-Jarkko, J. Singleton, P. W. Turner, I. Godfrey, S.-U. Alam, J. Nilsson, J. Sahu, and A. B. Grudinin, *Eur. Conf. Optical Commun.* (2002), PD1.6.

[41] F. L. Galeener, J. C. Mikkelsen, Jr., R. H. Geils, and W. J. Mosby, *Appl. Phys. Lett.* **32,** 34 (1978).

[42] N. Shibata, M. Horigudhi, and T. Edahiro, *J. Non-Crystalline Solids* **45,** 115 (1981).

[43] K. Rottwitt, J. Bromage, A. J. Stentz, L. Leng, M. E. Lines, and H. Smith, *J. Lightwave Technol.* **21,** 1652 (2003).

[44] D. Marcuse, *Light Transmission Optics* (Van Nostrand Reinhold, New York, 1982), Chap. 8.

[45] V. M. Mashinsky, E. M. Dianov, V. B. Neustruev, S. V. Lavrishchev, A. N. Guryanov, V. F. Khopin, N. N. Vechkanov, and O. D. Sazhin, in *Fiber Optic Materials and Components,* H. H. Yuce, D. K. Paul, R. A. Greenwell, Eds., *Proc. SPIE,* **2290,** 105 (1994).

[46] E. M. Dianov, V. M. Mashinsky, V. B. Neustruev, O. D. Sazhin, A. N. Guryanov, V. F. Khopin, N. N. Vechkanov, and S. V. Lavrishchev, *Optic. Fiber Technol.* **3,** 77 (1997).

[47] L. Grüner-Nielsen, High index fibers, thesis (Danish Academy of Technical Sciences), May 1998, EF 546/Ph.D. No. 94-0146-ATV.

[48] M. E. Lines, W. A. Reed, D. J. DiGiovanni, and J. R. Hamlins, *Electron. Lett.* 1009 **35,** (1999).

[49] Y. Qian, J. H. Povlsen, S. N. Knudsen, and L. Grüner-Nielsen, *Proc. Optical Amplifiers and Their Applications Conference* (2000), paper OMD18.

[50] H. Kogelnik and C. W. Shank, *J. Appl. Phys.* **43,** 2327 (1972).

[51] I. Bennion, J. A. R. Williams, L. Zhang, K. Sugden, and N. J. Doran, *Opt. Quantum Electron.* **28,** 93 (1996).

[52] G. P. Agrawal, *Nonlinear Fiber Optics* (Academic Press, San Diego, 1995), Chap. 10.

[53] Othonos and K. Kalli, *Fiber Bragg Gratings: Fundamentals and Applications in Telecommunications and Sensing* (Artech House, Norwood, MA, 1999).

[54] P. J. Lemaire, R. M. Atkins, V. Mizrahi, and W. A. Reed, *Electron. Lett.* **29,** 13 (1993).

[55] V. Mizrahi, P. J. Lemaire, T. Erdogan, W. A. Reed, D. J. DiGiovanni, and R. M. Atkins, *Appl. Phys. Lett.* **63,** 1727 (1993).

[56] G. Meltz, W. W. Morey, and W. H. Glenn, *Opt. Lett.* **14,** 823 (1989).

[57] K. O Hill, B. Malo, F. Bilodeau, D. C. Johnson, and J. Albert, *Appl. Phys. Lett.* **62,** 1035 (1993).

[58] D. Z. Anderson, V. Mizrahi, T. Erdogan, and A. E. White. *Electron. Lett.* **29,** 566 (1993).

[59] W. A. Reed, W. C. Coughran, and S. G. Grubb, *Opt. Fiber Communication Conf.* (1995), paper WD1, p. 107.

[60] M. Rini, I. Cristiani, and V. Degiorgio, *IEEE J. Quantum Elect.* **36,** 1117 (2000).

[61] A. Bertoni and G. C. Reali, *Appl. Phys. B* **67,** 5 (1998).

[62] G. Vareille, O. Audouin, and E. Desurvire, *Electron. Lett.* **34,** 675 (1998).

[63] S. D. Jackson and P. H. Muir, *J. Opt. Soc. Am. B* **18,** 1297 (2001).

[64] M. Krause, S. Cierullies, and H. Renner, *Opt. Commun.* **227,** 355 (2003).

[65] J.-C. Bouteiller, *IEEE Photon. Technol. Lett.* **15** No. 12, pp. 1698–1700 (2003).

[66] A. S. Kurkov, V. M. Paramonov, O. I. Medvedkov, S. A. Vasiliev, and E. M. Dianov, *Optical Amplifiers and Their Applications* (2000), paper OMB5, pp. 16–18.

[67] K. O. Hill, D. C. Johnson, B. S. Kawasaki, and R. I. MacDonald, *J. Appl. Phys.* **49,** 5098 (1978).

[68] J.-C. Bouteiller, *Electron. Lett.* **39,** pp. 1511–1512 (2003).

[69] C. R. S. Fludger, B. Handerek, and R. J. Mears, *J. Lightwave Technol.* **19,** 1140 (2001).

[70] M. D. Mermelstein, C. Headley, and J.-C. Bouteiller, *Electron. Lett.* **38,** 403 (2002).

[71] N. Tsukiji, N. Hayamizu, H. Shimizu, Y. Ohki, T. Kimura, S. Irino, J. Yoshida, T. Fukushima, and S. Namiki, *Proc. Optical Amplifiers and Their Applications* (2002), paper OMB4.

[72] L. L. Wang, R. E. Tench, L. M. Yang, and Z. Jiang, *Proc. Optical Amplifiers and Their Applications* (2002), paper OMB5.

[73] Y. Emori and S. Namiki, *Optical Fiber Conference Paper* (1999), paper PD19-1.

[74] D. I. Chang, D. S. Lim, M. Y. Jeon, H. K. Lee, K. H. Kim, and T. Park, *Electron. Lett.* **36,** 1365 (2001).

[75] S. B. Paperny, V. I. Karpov, and W. R. L. Clements, *Proc. Optical Fiber Communication Conf.* (2001), paper WDD15-1.

[76] M. D. Mermelstein, C. Headley, J.-C. Bouteiller, P. Steinvurzel, C. Horn, K. Fedder, and B. J. Eggleton, *Proc. Optical Fiber Communication Conf.* (2001), paper PD3-1.

[77] M. D. Mermelstein, C. Headley, J.-C. Bouteiller, P. Steinvurzel, K. Feder, and B. J. Eggleton, *IEEE Photon. Technol. Lett.* **13,** 1286 (2001).

[78] M. D. Mermelstein, C. Horn, Z. Huang, P. Steinvurzel, K. Feder, M. Luvalle, J.-C. Bouteiller, C. Headley, and B. J. Eggleton, *Proc. Optical Fiber Communication Conf.* (2002), paper TuJ2-1.

[79] X. Normandin, F. Leplingard, E. Bourova, C. Leclère, T. Lopez, J.-J. Guérin, D. Bayart, *Proc. Optical Fiber Communication Conf.* (2002), paper TuB2-1.

[80] M. D. Mermelstein, C. Horn, J.-C. Bouteiller, P. Steinvurzel, K. Feder, C. Headley, and B. J. Eggleton, *Proc. Conference on Lasers and Electro-Optics* (2002), paper CThJ1.

[81] M. D. Mermelstein, C. Horn, S. Radic, and C. Headley, *Electron. Lett.* **38,** 636 (2002).

[82] F. Leplingard, S. Borne, L. Lorcy, T. Lopez, J.-J. Guérin, C. Moreau, C. Martinelli, and D. Bayart, *Electron. Lett.* **38,** 886 (2002).

[83] B. J. Eggleton, J. A. Rogers, P. B. Westbrook, and T. A. Strasser, *IEEE Photon. Technol. Lett.* **11,** 854 (1999).

[84] V. Dominic, A. Mathur, and M. Ziari, *Proc. Optical Amplifiers and Their Applications* (2001), paper OMC6.

[85] L. Labrunie, F. Boubal, E. Brandon, L. Buet, N. Darbois, D. Dufournet, V. Havard, P. Le Roux, M. Mesic, L. Piriou, A. Tran, and J.-P. Blondel, *Proc. Optical Amplifiers and Their Applications* (2001), paper PD3-1.

[86] P. Le Roux, F. Boubal, E. Brandon, L. Buet, N. Darbois, V. Havard, L. Labrunie, L. Piriou, A. Tran, and J.-P. Blondel, *Proc. Eur. Conf. Optical Communication* (2001), paper PD.M. 1.5.

[87] J.-C. Bouteiller, K. Brar, S. Radic, J. Bromage, Z. Wang, and C. Headley, *Proc. Optical Fiber Communication Conf.* (2002), postdeadline paper FB3.

[88] S. B. Paperny, V. I. Karpov, and W. R. L. Clements, *Proc. Optical Fiber Communication Conf.* (2002), paper FB4-1.

[89] M. Prabhu, N. S. Kim, L. Jianren, and K. Ueda, *Opt. Commun.* **182,** 305 (2000).

[90] F. Leplingard, S. Borne, C. Martinelli, C. Leclère, T. Lopez, J. Guérin, and D. Bayart, *Proc. Optical Fiber Communication Conf.* (2003), paper ThB4.

# Index

active region, 270–273
  internal loss, 271
  kink, 273–274
  lattice-matched condition, 271
  lattice-strained layer, 271
  multi-quantum-well (MQW), 271
  optical confinement factor, 271
  quantum-size effect, 271
  quantum-well (QW), 271
  single-quantum-well (SQW), 271
  slope efficiency, 271
  strain, 271
amplified spontaneous emission, 10, 17, 53–59, 116, 173, 188, 189, 206, 207
anti-Stokes line, 34
asymmetric coating, 274, 275
auto-correlation, 263,
  properties of, 264

backward pumped (see counter-pumped)
bandwidth,
  optical filter, 53,
  Raman gain, 36–39, 49, 106
  receiver, 55, 236
  signal, 63
beat length, 70
BH-structure, 273–274
  current-blocking layer, 273
  leakage current, 273
  p-n-p-n thyristor, 273

birefringence, 67–72
Bragg gratings, 24, 47, 156, 287, 317,
  reflectivity, 319,
  bandwidth, 319–320
Bragg wavelength, 47, 318

cavity length, 270, 275–277
  activation energy, 276
  Arhenius relationship, 276, 272
  driving electrical power consumption, 276
  junction temperature, 276
  median lifetime, 276
  power consumption, 275
  small series resistance, 275
  thermal impedance, 275
  thermal saturation, 275
central limit theorem, 77
chain rule for partial differential equations, 224
chirp, 93, 96, 97
cladding-pumped fiber, 24, 105, 307
coherence length, 14, 257, 292, 333
composite gain, 201, 202, 205–206
corner frequency, 230, 234
correlation length, 70, 71
cross-phase modulation, 69, 87–93, 125
crosstalk, 59, 62
cutoff wavelength, 314